The Humongous Book of

Trigonometry Problems

Translated for People who don't Speak Math!!

by W. Michael Kelley

ALPHA BOOKS

Published by Penguin Group (USA) Inc.

Penguin Group (USA) Inc., 375 Hudson Street, New York, New York 10014, USA • Penguin Group (Canada), 90 Eglinton Avenue East, Suite 700, Toronto, Ontario M4P 2Y3, Canada (a division of Pearson Penguin Canada Inc.) • Penguin Books Ltd., 80 Strand, London WC2R 0RL, England • Penguin Ireland, 25 St. Stephen's Green, Dublin 2, Ireland (a division of Penguin Books Ltd.) • Penguin Group (Australia), 250 Camberwell Road, Camberwell, Victoria 3124, Australia (a division of Pearson Australia Group Pty. Ltd.) • Penguin Books India Pvt. Ltd., 11 Community Centre, Panchsheel Park, New Delhi—110 017, India • Penguin Group (NZ), 67 Apollo Drive, Rosedale, North Shore, Auckland 1311, New Zealand (a division of Pearson New Zealand Ltd.) • Penguin Books (South Africa) (Pty.) Ltd., 24 Sturdee Avenue, Rosebank, Johannesburg 2196, South Africa • Penguin Books Ltd., Registered Offices: 80 Strand, London WC2R 0RL, England

International Standard Book Number: 978-1-61564-182-6
Library of Congress Catalog Card Number: 2012933501

15   8 7 6 5 4

Interpretation of the printing code: The rightmost number of the first series of numbers is the year of the book's printing; the rightmost number of the second series of numbers is the number of the book's printing. For example, a printing code of 12-1 shows that the first printing occurred in 2012.

Printed in the United States of America

Contents

Introduction

Are you in a Trigonometry class? Are you trying to help a friend or loved one with a math class but discovering that you don't remember everything you used to know about math? Are you the curious sort who used to dislike math but now want to figure it out, once and for all?

If you answered YES to any of these questions, then you NEED this book. Here's why:

> **Fact #1: The best way to develop trigonometry skills is by working out trigonometry problems.**

There's no denying it. If you could figure this class out just by reading the textbook or taking good notes in class, everybody would pass with flying colors. Unfortunately, the harsh truth is that you have to buckle down and work problems out until your fingers are numb.

> **Fact #2: Most textbooks only tell you WHAT the answers to their practice problems are but not HOW to do them!**

Sure your textbook may have 175 problems for every topic, but most textbooks only give you the answers. That means if you don't get the answer right you're totally out of luck! Knowing you're wrong is no help at all if you don't know WHY you're wrong.

Math textbooks sit on a huge throne, like the Great and Powerful Oz and say, "Nope, try again," and you do. Over and over, usually getting the problem wrong. What a delightful way to learn! (Let's not even get into why they only tell you the answers to the odd problems. Does that mean the book's actual AUTHOR didn't even feel like working out the even ones?)

Fact #3: Even when math books try to show you the steps for a problem, they do a lousy job.

Math people love to skip steps. You'll be following along fine with an explanation and then all of a sudden BAM, you're lost. You'll think to yourself, "How did they do that?" or "Where the heck did that 42 come from? It wasn't there in the last step!"

Why do math textbooks insist that in order to work out a problem on page 200, you'd better know pages 1 through 199 like the back of your hand? You don't want to spend the rest of your life on homework! You just want to know when you're supposed to use the law of sines and when you're supposed to use the law of cosines, which you'll learn in Chapter 13.

Fact #4: Reading lists of facts is fun for a while, but then it gets old. Enough with this list—let's cut to the chase.

Just about every single kind of trigonometry problem you could possibly run into is in here. After all, this book is HUMONGOUS! If 750 problems aren't enough, then you've got some kind of crazy math hunger, my friend, and I'd seek professional help.

All of my notes are off to the side like this and point to the parts of the book I'm trying to explain.

This practice book was good at first, but to make it GREAT, I went through and worked out all the problems and took notes in the margins when I thought something was confusing or needed a little more explanation. I also drew little skulls next to the hardest problems, so you'd know not to freak out if they were too challenging. After all, if you're working on a problem and you've totally stumped, isn't it better to know that the problem is SUPPOSED to be hard? It's reassuring, at least for me.

The Humongous Book of Trigonometry Problems

I think you'll be pleasantly surprised by how detailed the answer explanations are, and I hope you'll find my little notes helpful along the way. Call me crazy, but I think that people who WANT to learn trigonometry and are willing to spend the time drilling their way through practice problems should actually be able to figure the problems out and learn as they go, but that's just my 2¢.

Good luck and make sure to come visit my website at www.calculus-help.com. If you feel so inclined, drop me an e-mail and give me your 2¢. (Not literally, though—real pennies clog up the internet pipes.)

—Mike Kelley

Acknowledgements

Special thanks to Alpha Books, who continue to support The Humongous Books, even as the publishing industry redefines itself in an era of electronic books. The first Humongous Book was released in 2006, and a mere six years later, the series is six books strong!

Thanks to Lisa Kelley and Dave Herzog, who have tirelessly worked for nearly a year to proofread, edit, and check all of these problems as I developed them.

Trademarks

Dedication

As always, this book is dedicated to the four most important people in my life, my family.

For my wife Lisa, who not only supported me emotionally during this book, she also supported me grammatically. Who knew that she had a secret proofreading/copy editing superpower? You are the only reason I make it through each and every day. Without you I would be lost and (now) full of spelling errors.

For my son Nicholas, who loves sports, video games, reading comic books with Dad, and sneaking up on birds. (That last one is new.) You're growing up to be more than just a well-behaved kid. You're becoming a young man I am incredibly proud of.

For my daughter Erin, the fastest girl in the continental United States and the most generous person I know. You'd willingly give up everything you have to make someone else happy, and it still warms my heart that you have to sit by Dad at every meal.

For my daughter Sara, who loves Curly from the Three Stooges, fat penguins, and writing profusely, including her new book "A New House for Old People." Since you were a baby, your smiles have been brilliant bursts of sunshine in my otherwise cloudy day.

Chapter 1
ANGLES AND ARCS

Pairs of rays and pieces of circle

Trigonometry is the study of triangles and the component pieces that form triangles: angles and sides. In this chapter, you will concern yourself with the former, exploring the construction of angles and the units of measurement that assign values to angles. Specific attention is given to angles in standard position, as the coordinate plane is an essential tool in the study of trigonometry. Finally, you will calculate the lengths of arcs based upon the angles that form them.

Welcome to trigonometry, an entire course dedicated to studying one shape: the triangle. If you thought you knew something about triangles before, well, you had no idea how much you didn't know about them. For the next 18 chapters, you and triangles are stuck on a desert island together, so you may as well start getting to know one another.

What is a triangle? It's a two-dimensional shape formed by three line segments. Specifically, each line segment intersects the other two segments only at their endpoints. A point of intersection is called a vertex (the plural is vertices), and two intersecting sides form an angle. In this chapter, you learn how to measure an angle using degrees or radians (or rotations, if you are feeling sassy), and how to convert between degrees and radians.

The big takeaway from this chapter is understanding what radians are, because you'll be using them a lot.

Standard Position

Angles, front and center

Note: Problems 1.1–1.3 refer to the diagram below, triangle ABC.

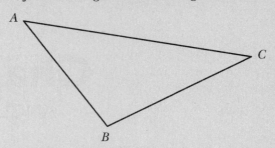

1.1   Identify the sides and vertices of the triangle.

> These little bars look like line segments, which is why they appear over the names of the segments. When you refer to the LENGTHS of line segments, you don't use the bars: AB, BC, and AC.

A triangle is a two-dimensional (or plane) figure that consists of three line segments intersecting only at their endpoints. Line segments are named according to the points at which they begin and end, so the sides of this triangle are $\overline{AB}$, $\overline{BC}$, and $\overline{AC}$. The vertices are the three points at which the segments intersect: A, B, and C.

Note: Problems 1.1–1.3 refer to the diagram illustrated in Problem 1.1, triangle ABC.

1.2   List three different names for the largest angle of the triangle.

The largest angle of the triangle appears at the bottom of the diagram and has vertex B. One valid name of the angle is simply B; angles may be named using their vertices as long as the angle you are describing is clear. For example, if multiple angles share the same vertex, then simply identifying the vertex is not enough information.

In these cases, you name an angle using three points in a specific order: the first point lies on one side of the angle, the second point is the vertex of the angle, and the third point lies on the other side of the angle. Therefore, ABC and CBA are also valid names for the largest angle of this triangle.

Note: Problems 1.1–1.3 refer to the diagram illustrated in Problem 1.1, triangle ABC.

> To draw angle ABC, draw a straight line connecting A to B, and then draw a straight line from B to C. The vertex has to be the middle letter in the name of the angle.

1.3   Are any of the angles in this triangle in standard position? Why or why not?

An angle in standard position satisfies the following conditions:

- One side of the angle overlaps the positive x-axis on the coordinate plane.
- The vertex of the angle overlaps the origin of the coordinate plane, the point at which the horizontal and vertical axes intersect.

Therefore, angles in standard position must be, by definition, drawn on the coordinate plane. Because this triangle is not plotted on a coordinate plane, none of its angles can be in standard position.

1.4   Draw an angle *RST* that is in standard position and explain your answer.

The diagram below represents one solution to this problem, though there are infinitely many correct answers. However, all answers must conform to two requirements. As Problem 1.3 explains, an angle in standard position has one side that lies along the positive *x*-axis. In other words, it overlaps the right half of the horizontal axis, like side $\overline{ST}$ in the diagram below. Note that the sides of angles in standard position are rays, which begin at a point (ray $\overrightarrow{ST}$ begins at point *S*), pass through a second specified point (in this case *T*), and continue in that direction.

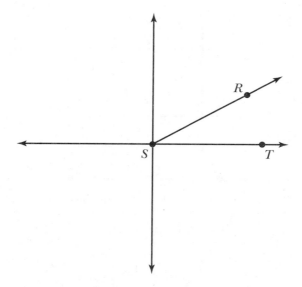

The other characteristic of an angle in standard position is a vertex that lies on the origin of the coordinate plane, the point with coordinates (0,0) at which the *x*- and *y*-axes intersect. Note that point *S*, the vertex of angle *RST*, lies on the origin.

Because the two conditions (regarding one side of the angle and its vertex) are met, you conclude that angle *RST* is in standard position.

1.5   Is angle *XYZ*, illustrated below, in standard position? Why or why not?

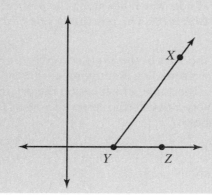

Angle *XYZ* meets one requirement of an angle in standard position, because side $\overrightarrow{YZ}$ lies along the positive *x*-axis. However, the second condition is not met; the vertex *Y* does not lie on the origin of the coordinate plane. Therefore, angle *XYZ* is not in standard position.

If you were to slide the angle to the left, keeping side $\overrightarrow{YZ}$ on the positive *x*-axis, so that point *Y* overlapped the origin, the result would be angle *XYZ* in the diagram below, which is in standard position. Any angle may be placed in standard position if it is rotated and/or shifted properly.

Or you could just draw it in standard position to start with and save yourself the trouble.

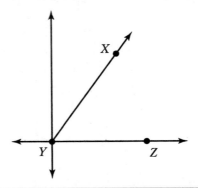

Note: Problems 1.6–1.8 refer to angle CDE in the diagram below. Note that angle CDE is in standard position.

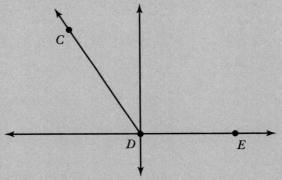

1.6 Identify the initial and terminal sides of angle *CDE*.

The initial and terminal sides of an angle are, respectively, the sides at which the angles begin (initiate) and end (terminate). All angles in standard position share the same initial side, which lies along the positive *x*-axis. Therefore, the initial side of angle *CDE* is $\overrightarrow{DE}$. The remaining side of the angle is the terminal side: $\overrightarrow{DC}$.

Make sure to name the sides of the angle correctly. Unlike line segments, whose endpoints may be reversed when identifying a segment, the first point in a ray must be the vertex of the angle, which lies on the origin when an angle is in standard position. Notice that *D* (the vertex of angle *CDE*) appears first in both of the rays named above.

Note: Problems 1.6–1.8 refer to the diagram in Problem 1.6, which illustrates angle CDE in standard position.

1.7 Identify two angles in the coordinate plane that share the initial and terminal sides identified in Problem 1.6.

According to Problem 1.6, the initial side of the angle is $\overrightarrow{DE}$ and the terminal side is $\overrightarrow{DC}$. Imagine that the angle is formed by a single ray, anchored at the origin, that begins along the positive *x*-axis and then rotates until it reaches its terminal position. Because you may rotate clockwise or counterclockwise from the initial side, you produce one of the two angles illustrated below.

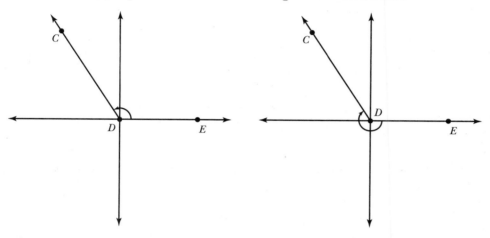

In trig, only the terminal side matters—not how you got to that terminal side. Both of the angles in this problem will have the same trigonometric values (which you'll learn about in Chapters 2 and 3), even though they're not the same size.

Rotating counterclockwise produces a smaller angle (represented by the arrowed arc in the left portion of the diagram above) than rotating clockwise. Both of these angles share the same name: *CDE*. Furthermore, both of the angles share the same terminal side, so they are described as "coterminal."

Note: Problems 1.6–1.8 refer to the diagram in Problem 1.6, which illustrates angle CDE in standard position.

1.8 Draw an angle in standard position whose measure is the opposite of angle *CDE*.

One of the fundamental assumptions of Euclidean geometry is that all angles have a positive angle measurement. However, angles assigned to the coordinate plane and placed in standard position allow you to measure in a more precise manner. In Problem 1.7, you identify two different angles with the same initial and terminal sides; the only difference between the two angles is the direction you travel, either clockwise or counterclockwise.

If you travel in a counterclockwise direction from the initial side of an angle to its terminal side, the angle formed has a positive measurement. Alternately, if you travel in a clockwise direction, the angle has a negative measurement. Assume, for the moment, that angle *CDE* is formed by traveling counterclockwise, beginning at the positive *x*-axis, rotating through the first quadrant, and coming to rest in the second quadrant.

The x- and y-axes divide the plane into four sections called quadrants, which are numbered like this:

$$
\begin{array}{c|c}
2 & 1 \\
\hline
3 & 4
\end{array}
$$

To draw an angle in standard position that has a measure opposite of angle *CDE*, you simply rotate in the opposite direction. In the figure below, angle *FDE* is in standard position, so its initial side is the positive *x*-axis. The angle is formed by rotating clockwise through the fourth quadrant and coming to rest in the third quadrant at its terminal side, $\overline{DF}$. This angle has the same measure as angle *CDE*, but it is negative rather than positive. Therefore, the angles have opposite measures.

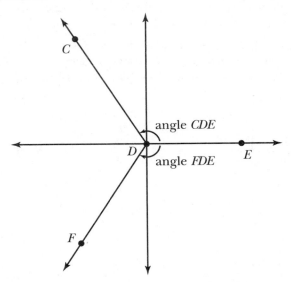

Measuring Angles

Degrees, radians, rotations

Note: Problems 1.9–1.11 refer to angle ABC in the diagram below. The angle's initial side and terminal side both lie on the positive x-axis.

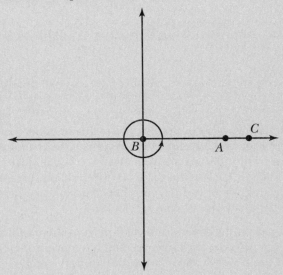

1.9 Express the measure of angle *ABC* in revolutions.

According to the arrowed arc at the center of this diagram, angle *ABC* begins at the positive *x*-axis, rotates exactly once around the coordinate plane, and terminates where it began, at the positive *x*-axis. To visualize this angle, stand up and point at a specific spot on the wall. Then, while standing in the same location, slowly turn in the counterclockwise direction, keeping your arm straight in front of you as you turn. Once you have turned all the way back around to your original starting point, your finger will once again point at that initial location you chose.

If you start at a fixed ray, rotate counterclockwise, and then end at the precise moment you re-encounter the initial ray, you have revolved exactly one time. Angle *ABC* represents exactly one full revolution about the origin in the coordinate plane. Therefore, *ABC* measures exactly 1 revolution.

Note: Problems 1.9–1.11 refer to the diagram in Problem 1.9, in which both the initial and terminal sides of angle ABC lie on the positive x-axis.

1.10   Express the measure of angle *ABC* in degrees.

Problem 1.9 explains that angle *ABC* measures exactly one full revolution about the origin in the coordinate plane. Therefore, angle *ABC* measures 360°, which is read "360 degrees." ←

> If someone changes their mind about something and does the exact opposite, you may call that "pulling a 180." Half of one revolution is 360° ÷ 2 = 180°, so "pulling a 180" means facing the opposite of your original direction. (A 180° angle in standard position has a terminal side on the negative x-axis.)

Note: Problems 1.9–1.11 refer to the diagram in Problem 1.9, in which both the initial and terminal sides of angle ABC lie on the positive x-axis.

1.11   Express the measure of angle *ABC* in radians.

Degrees are used more commonly than revolutions to measure angles, and many students intuitively understand the concept of degrees. Another unit of angle measurement, called the radian, is not as innately familiar to new trigonometry students, but it is nonetheless an important concept. In fact, most contemporary textbooks use radian measurements significantly more than degree measurements.

In Problem 1.12, you are introduced to the radian in greater detail. For this problem, however, you need only understand the radian equivalent for 360°. Angle *ABC* represents one full revolution on the coordinate plane, so it measures 2π radians. ←

> 1 revolution = 360° = 2π radians

1.12   Explain what a radian is, and include an illustration to supplement your definition.

Before you can measure an angle in radians, you must first understand what a radian is. Consider the following diagram, which depicts an angle *QCP* in standard position and a circle centered at *C*, the origin of the coordinate plane. Note that $\overline{CQ}$ and $\overline{CP}$ are radii of the circle, so both have length *r*. Because they are radii, those segments share one common endpoint (the center *C*) and each have one other endpoint that lies on the circle itself. Those points, *Q* and *P*, are boundaries of a small arc lying entirely in the first quadrant: $\overset{\frown}{QP}$.

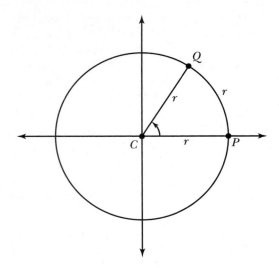

Because the vertex of angle *QCP* is also the center of the circle, *QCP* is a "central angle" of the circle; when the sides of a central angle intersect a circle to define an arc (like $\overset{\frown}{QP}$ in this diagram), the angle is said to "subtend" the arc. In other words, an angle subtends an arc when its sides slice into a circle, identifying a specific piece of the circle that is bounded by the points at which the angle's sides intersect that circle.

In this illustration, angle *QCP* intersects circle *C* in a very specific way. The length of the subtended arc $\overset{\frown}{QP}$ is exactly equal to the radius of the circle. Therefore, angle *QCP* measures exactly 1 radian. Roughly speaking, a radian measures approximately 57.29578°, so there are far fewer radians in 1 rotation than there are degrees. One rotation is equal to more than 6, but less than 7 radians:

$$1 \text{ rotation} = 2\pi \text{ radians}$$
$$\approx 2(3.14159265) \text{ radians}$$
$$\approx 6.28319 \text{ radians}$$

In the diagram below, each of the shaded central angles has a measure of 1 radian. Notice that 6 full radians fit within 1 rotation about the coordinate plane, with an additional angle of approximately 0.28319 radians remaining.

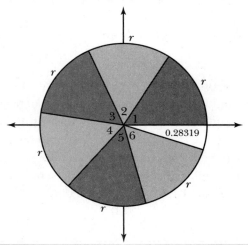

1.13 Apply the circumference formula for a circle ($C = \pi d$), in which d is the diameter of a circle with circumference C, to justify your answer to Problem 1.11.

According to Problem 1.11, 1 rotation is equal to 2π radians. Whereas Problem 1.12 visually demonstrates that 1 revolution is equal to slightly more than 6 radians, it does not offer convincing proof that the number of radians is exactly equal to 2π. To verify this value, you need to apply the circumference formula for a circle, $C = \pi d$, where d is the diameter. Recall from geometry that the diameter of a circle is twice the length of a radius: $d = 2r$. Apply the substitution property to replace d with $2r$ in the circumference formula.

$$C = \pi \cdot d$$
$$C = \pi \cdot (2r)$$
$$C = 2\pi r$$

As Problem 1.12 explains, a central angle measures 1 radian when it subtends an arc that is the same length as the radius of a circle. Therefore, dividing the circumference of the circle C by the radius r calculates the exact number of arcs (with length r) that the circle can be divided into.

$$\frac{C}{r} = \frac{2\pi r}{r}$$
$$\frac{C}{r} = \frac{2\pi \cancel{r}}{\cancel{r}}$$
$$\frac{C}{r} = \frac{2\pi}{1}$$
$$\frac{C}{r} = 2\pi$$
$$\frac{C}{r} \approx 6.28319$$

This is like dividing 50 pieces of candy by 4 to see how many pieces of candy 4 children would get if you attempted to divide the 50 pieces evenly among them.

Every circle can be divided into approximately 6.28319 arcs that are the same length as the radius of that circle. Each of those arcs corresponds to a central angle that measures 1 radian. Therefore, 1 rotation is equal to $2\pi \approx 6.28319$ radians. Note that it is common to leave a radian measure in terms of π rather than expressing it as a decimal (such as 6.28319); radians left in terms of π are more accurate than answers expressed a finite number of decimal places.

1.14 Angle *XYZ* is in standard position in the diagram below. Express its measure in rotations, degrees, and radians.

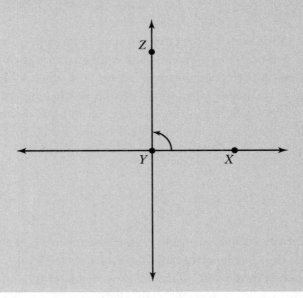

Angle *XYZ* has an initial side lying on the positive *x*-axis and a terminal side that lies on the positive *y*-axis. In other words, it begins and ends at the horizontal and vertical lines that define the first quadrant. In Problems 1.9–1.11, you measured angle *ABC*, which extends through all four quadrants of the coordinate plane. Because *XYZ* in this diagram extends through exactly one quadrant, it is exactly one-fourth the size of angle *ABC*.

To calculate the measure of the angle in revolutions, degrees, and radians, divide the solutions to Problems 1.9–1.11 by 4:

$$m\angle ABC = (1 \text{ revolution}) \div 4 = \frac{1}{4} = 0.25 \text{ revolutions}$$

$$m\angle ABC = (360°) \div 4 = 90°$$

$$m\angle ABC = (2\pi) \div 4 = \frac{2\pi}{4} = \frac{\pi}{2} \text{ radians}$$

> *This notation is read "The measure of angle ABC is equal to."*

Angles with terminal sides that lie on the *x*- or *y*-axes (like angle *XYZ* in this example) are called "quadrantals." Their measures are multiples of 90° (or $\pi/2$ radians).

1.15 If an angle measuring 0.45 rotations is in standard position on the coordinate plane, in what quadrant does its terminal side lie?

According to Problem 1.14, an angle terminating on the positive *y*-axis has a measure of 0.25 revolutions. By extension, the negative *x*-axis is the terminal side of an angle measuring 0.5 revolutions, and the negative *y*-axis marks an angle of 0.75 revolutions. In the following diagram, 1 revolution is divided into 8 equal parts to help you visualize where the terminal sides of angles measured in revolutions lie in the coordinate plane.

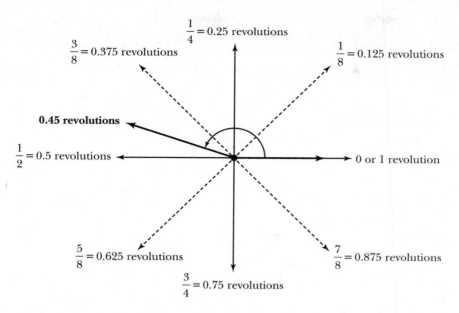

An angle measuring 0.45 rotations extends beyond the first quadrant (which ends at 90° = 0.25 rotations) but is not large enough to extend beyond the second quadrant (which ends at 180° = 0.5 rotations). Therefore, the terminal side of the angle lies in the second quadrant.

1.16 If an angle measuring 225° is in standard position on the coordinate plane, in what quadrant does its terminal side lie?

According to Problem 1.14, the first quadrant spans 90°. Therefore, as you continue to rotate counterclockwise, each time you pass through another quadrant, you rotate another 90°. An angle with a terminal side on the negative *x*-axis (which marks the end of the second quadrant) has a measure of 90° + 90° = 180°. Similarly, an angle with a terminal side that lies on the negative *y*-axis measures 180° + 90° = 270°.

Because 225° is greater than 180° but less than 270°, the terminal side of a 225° angle must lie within the third quadrant. Consider the diagram below as further verification of the answer. It presents the angle in standard position and illustrates a number of other terminal sides and their respective angle measurements in all four quadrants of the coordinate plane.

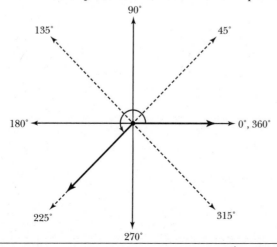

1.17 If angles measuring $\dfrac{5\pi}{3}$ and $-\dfrac{5\pi}{3}$ radians are in standard position on the coordinate plane, in what quadrants do their terminal sides lie?

According to Problem 1.11, 1 complete revolution in the coordinate plane measures 2π radians. Reference the diagram below, which illustrates common angle measurements expressed in radians, and note that $5\pi/3$ lies in the fourth quadrant.

> Frankly, degrees are easier to deal with than radians in this kind of problem—you just need to know which degree range an angle falls in: 0–90 (quadrant I), 90–180 (quadrant II), 180–270 (quadrant III), or 270–360 (quadrant IV). In Problems 1.24–1.30, you learn how to convert degrees to radians, and because $\dfrac{5\pi}{3} = 300°$, the angle terminates in the fourth quadrant.

$\dfrac{2\pi}{3}$ radians $\dfrac{\pi}{2}$ radians $\dfrac{\pi}{3}$ radians

$\dfrac{3\pi}{4}$ radians $\dfrac{\pi}{4}$ radians

$\dfrac{5\pi}{6}$ radians $\dfrac{\pi}{6}$ radians

π radians $0, 2\pi$ radians

$\dfrac{7\pi}{6}$ radians $\dfrac{11\pi}{6}$ radians

$\dfrac{5\pi}{4}$ radians $\dfrac{7\pi}{4}$ radians

$\dfrac{4\pi}{3}$ radians $\dfrac{3\pi}{2}$ radians $\dfrac{5\pi}{3}$ radians

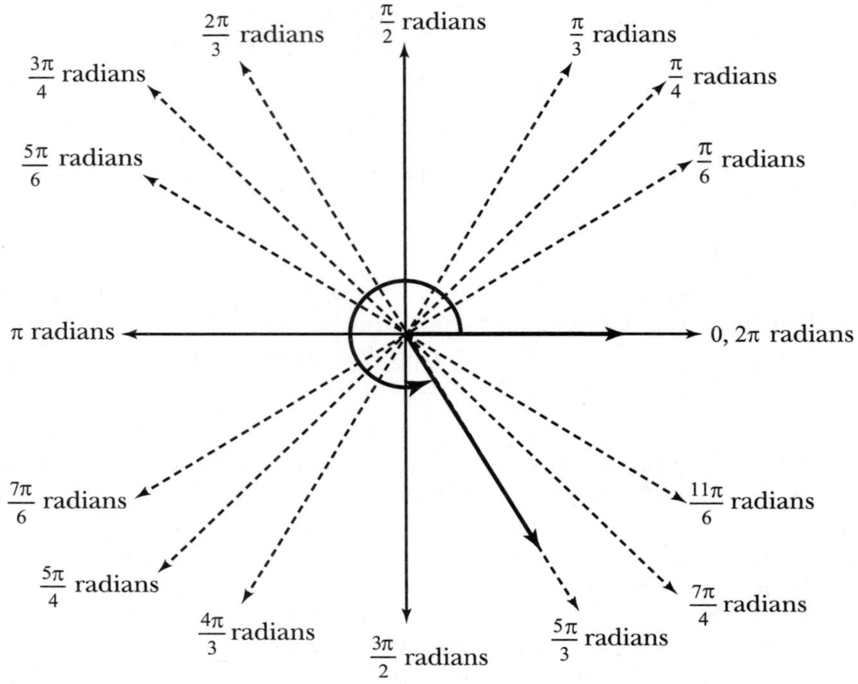

An angle in standard position measuring $-5\pi/3$ rotates the same distance in the opposite direction, terminating in the first quadrant, as illustrated in the following diagram.

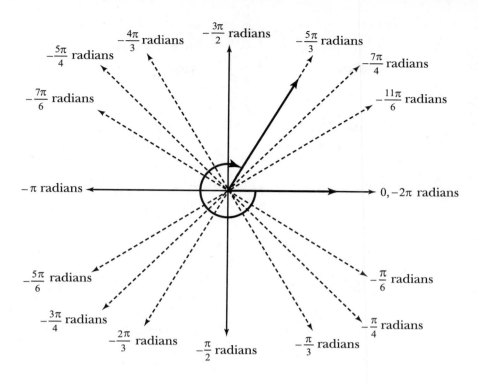

1.18 Convert the angle measurement into decimal degrees: 43° 18′.

Degrees, like hours, are divided into minutes and seconds. In fact, just as one hour contains 60 minutes and each of those minutes contains 60 seconds, one degree contains 60 minutes and each of those minutes contains 60 seconds. The measurement 43° 18′ is read "43 degrees 18 minutes."

To convert the angle measurement $d°\ m'\ s''$ (where d = degrees, m = minutes, and s = seconds), substitute the values into the following formula:

$$d°\ m'\ s'' = d + \frac{m}{60} + \frac{s}{3,600}$$

By dividing the minutes by 60 and the seconds by 3,600, you are converting each into decimal degrees. The angle in this problem does not contain any seconds, so you need only substitute $d = 43$ and $m = 18$ into the formula.

$$43°\ 18' = 43 + \frac{18}{60}$$

Reduce the fraction to lowest terms, dividing 18 and 60 by 6, their greatest common factor.

$$= 43 + \frac{18 \div 6}{60 \div 6}$$
$$= 43 + \frac{3}{10}$$

A quarter of a degree, like a quarter of an hour, is 15 minutes. In other words, 15 minutes = 0.25 hours or degrees. The angle in this problem has 18 minutes, so the decimal part of the answer will be larger than 0.25.

Note that the fraction 3/10 is read "three tenths," which is equivalent to 0.3.

Or you could just type 3 ÷ 10 into your calculator and you get 0.3.

$$= 43 + 0.3$$
$$= 43.3$$

Therefore, $43° 18' = 43.3°$.

1.19 Convert the angle measurement into a decimal accurate to the thousandths place: $165° 52' 39''$.

Substitute $d = 165$, $m = 52$, and $s = 39$ into the conversion formula $d° \, m' \, s'' = d + \dfrac{m}{60} + \dfrac{s}{3,600}$ introduced in Problem 1.18.

$$165° \, 52' \, 39'' = 165 + \frac{52}{60} + \frac{39}{3,600}$$

If you're one of those "Calculators are evil!" sorts, you might as well start getting over that now. If you spend all your time getting bogged down with arithmetic, the concepts of trig get lost in the process. Embrace technology!

Because these fractions are more complex than the fractions in Problem 1.18, it is appropriate to use a calculator to compute the decimal equivalents. Include as many decimals as your calculator provides in each step—rounding any of them before the final answer may produce an inaccurate answer.

$$= 165 + \frac{52}{60} + \frac{39}{3,600}$$
$$= 165 + 0.866666667 + 0.010833333$$
$$\approx 165.8775$$

The problem directs you to round to the thousandths place, which is three digits right of the decimal point: $165.8775 \approx 165.878$.

Converting Between Angle Measurements
Especially degrees and radians

1.20 Express the angle measurement in degrees: 3 revolutions.

In order to convert from one unit of angle measurement to another, you multiply it by a fraction of the following form:

$$\frac{\text{an angle measured in the new units}}{\text{the same angle measured in the old units}}$$

Select a simple angle when you create the fraction. For example, Problems 1.9–1.10 state that 1 revolution = 360°. This problem instructs you to convert from revolutions to degrees, so "an angle measured in the new units" would be 360°, and "the same angle measured in the old units" would be 1 revolution. Note that the units themselves are omitted from the fraction:

$$\frac{\text{an angle measured in the new units}}{\text{the same angle measured in the old units}} = \frac{360}{1}$$

Thus, to convert 3 revolutions into degrees, you multiply by 360/1, or more simply, 360.

$$3 \cdot \frac{360}{1} = 3 \cdot 360 = 1,080$$

You conclude that an angle measurement of 3 revolutions is equivalent to an angle measurement of 1,080°.

1.21 Express the angle measurement in revolutions: 210°.

According to Problem 1.20, to convert from one unit of measurement to another, you multiply by the fraction whose numerator is an angle expressed in the new units *to which* you are converting and whose denominator is the same angle expressed in the old units *from which* you are converting. In this case, you are converting from degrees to revolutions, so the new units are revolutions and the old units are degrees.

Once again, it is appropriate to select a simple angle for the equal measurements. Apply the equivalency relationship 1 revolution = 360°, like you did in Problem 1.20.

$$\frac{\text{an angle measured in the new units}}{\text{the same angle measured in the old units}} = \frac{1}{360}$$

Multiply 210° by the conversion fraction identified above.

$$210 \cdot \frac{1}{360} = \frac{210}{1} \cdot \frac{1}{360}$$
$$= \frac{210}{360}$$

Reduce the fraction to lowest terms by dividing its numerator and denominator by the greatest common factor of 210 and 360, which is 30.

$$= \frac{210 \div 30}{360 \div 30}$$
$$= \frac{7}{12}$$

Can't remember how to reduce fractions? Check out Problems 7.27–7.34 in The Humongous Book of Basic Math and Pre-Algebra Problems.

Therefore, 210° is equal to $\frac{7}{12}$, or $0.58\bar{3}$, revolutions.

1.22 Express the angle measurement in radians: 0.5 revolutions.

As Problems 1.20–1.21 demonstrate, to convert from one unit of angle measurement to another, you need to identify a fraction with values representing the same angle in the old and new units. In this problem, you are converting from revolutions to radians, so the new units are radians and the old units are revolutions. Because radians are typically expressed as fractions, rather than decimals, you should rewrite 0.5 as a fraction:

$$0.5 = \frac{1}{2}$$

That little bar over the 3 means that the digit repeats infinitely: 0.58333333333333....

Like Problems 1.20–1.21, you can use the measure of 1 full revolution about the coordinate plane to create the necessary conversion fraction: 1 revolution = 2π radians.

$$\frac{\text{an angle measured in the new units}}{\text{the same angle measured in the old units}} = \frac{2\pi}{1} = 2\pi$$

Therefore, to convert 1/2 revolutions into radians, multiply by 2π.

$$\frac{1}{2} \cdot 2\pi = \frac{1}{2} \cdot \frac{2\pi}{1}$$
$$= \frac{2\pi}{2}$$
$$= \frac{2\pi}{2}$$
$$= \pi$$

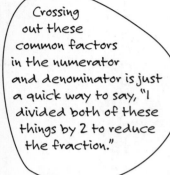

Crossing out these common factors in the numerator and denominator is just a quick way to say, "I divided both of these things by 2 to reduce the fraction."

1.23 Express the angle measurement in revolutions: $\dfrac{7\pi}{4}$ radians.

Apply the same angle measurements that you used in Problem 1.22 (1 revolution = 2π radians). In this problem, however, the new units are revolutions and the old units are radians.

$$\frac{\text{an angle measured in the new units}}{\text{the same angle measured in the old units}} = \frac{1}{2\pi}$$

Multiply $7\pi/4$ by the conversion fraction above.

$$\frac{7\pi}{4} \cdot \frac{1}{2\pi} = \frac{7\pi}{8\pi}$$

Reduce the product to lowest terms, dividing the numerator and the denominator by the greatest common factor (π).

$$= \frac{7\pi}{8\pi}$$
$$= \frac{7}{8}$$

You conclude that $7\pi/4$ radians = 7/8 revolutions.

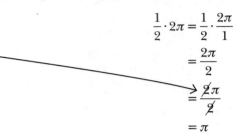

You're not going to see many more references to revolutions from now on, because you will work in degrees and radians for the remainder of the book.

1.24 Express the angle measurement in radians: 120°.

Like the preceding problems in this section, you must create a conversion fraction whose numerator and denominator are measures of the same angle in the new and old units, respectively. In this problem, the new units are radians and the old units are degrees. According to Problems 1.10–1.11, 1 full revolution measures 2π radians and 360°.

$$\frac{\text{an angle measured in the new units}}{\text{the same angle measured in the old units}} = \frac{2\pi}{360}$$

This fraction should be reduced to lowest terms, because the numerator and denominator are both divisible by 2.

$$\frac{2\pi}{360} = \frac{(2\pi) \div 2}{360 \div 2}$$
$$= \frac{\pi}{180}$$

Therefore, you should multiply an angle measurement by $\pi/180$ to convert it from degrees to radians.

$$120 \cdot \frac{\pi}{180} = \frac{120}{1} \cdot \frac{\pi}{180}$$
$$= \frac{120\pi}{180}$$

Reduce the fraction to lowest terms.

$$= \frac{120\pi \div 60}{180 \div 60}$$
$$= \frac{2\pi}{3}$$

Therefore, $120° = 2\pi/3$ radians.

> And, as you'll see in Problems 1.27–1.30, you multiply by $\frac{180}{\pi}$ to convert from radians to degrees.

1.25 Express the angle measurement in radians: 270°.

As Problem 1.24 explains, to convert an angle measurement from degrees to radians, multiply it by $\pi/180$ and (when possible) reduce the resulting fraction to lowest terms.

$$270 \cdot \frac{\pi}{180} = \frac{270\pi}{180}$$
$$= \frac{(270\pi) \div 90}{180 \div 90}$$
$$= \frac{3\pi}{2}$$

You conclude that $270° = 3\pi/2$ radians.

1.26 Express the angle measurement in radians: −330°.

The sign of an angle measurement does not affect the technique you apply to convert between different units. Therefore, you should apply the same technique modeled in Problems 1.24–1.25: Multiply the angle measurement by $\pi/180$. Note that the radian angle measurement must have the same sign as the degree angle measurement, so your final answer will be negative.

According to this, the numerator is negative. If a fraction has a negative numerator or denominator, the negative sign is usually written in front of the fraction, like this.

$$-330 \cdot \frac{\pi}{180} = \frac{-330\pi}{180}$$

$$= \frac{(-330\pi) \div 30}{180 \div 30}$$

$$= \frac{-11\pi}{6}$$

You conclude that $-330° = -\dfrac{11\pi}{6}$ radians.

1.27 Express the angle measurement in degrees: $\dfrac{\pi}{4}$ radians.

To convert from radians to degrees, multiply the radian angle measurement by $180/\pi$ and reduce the resulting fraction to lowest terms.

$$\frac{\pi}{4} \cdot \frac{180}{\pi} = \frac{180\pi}{4\pi}$$

$$= \frac{180\,\not\pi}{4\,\not\pi}$$

$$= \frac{180 \div 4}{4 \div 4}$$

$$= \frac{45}{1}$$

$$= 45$$

So you don't have to write "radians" next to radian measurements to specify the units.

Therefore, $\pi/4$ radians = 45°. Because most radian measurements contain π, they are easily differentiated from degree measurements. (Degree measurements always include the degree symbol, as well.) Therefore, the word "radians" is usually omitted when reporting an angle measurement. In trigonometry, if no units are written next to an angle, you should assume that the angle is measured in radians. Hence, it is more common to write the answer to this problem as $\pi/4 = 45°$.

1.28 Express the angle measurement in degrees: $-\dfrac{5\pi}{6}$.

Apply the same procedure demonstrated in Problem 1.27: Multiply the radian angle measurement by $180/\pi$. Note that an angle with a negative radian measure will also have a negative degree measure.

$$-\frac{5\pi}{6} \cdot \frac{180}{\pi} = -\frac{5\pi \cdot 180}{6\pi}$$

Before you multiply 5 and 180 in the numerator of the product, reduce the numerator value 180 and the denominator value 6, which share a greatest common factor of 6.

$$= -\frac{5\pi \cdot (180 \div 6)}{(6 \div 6)\pi}$$

$$= -\frac{5\pi(30)}{(1)\pi}$$

$$= -\frac{5\cancel{\pi}(30)}{1\cancel{\pi}}$$

$$= -\frac{150}{1}$$

$$= -150$$

You conclude that $-5\pi/6 = -150°$.

1.29 Express the angle measurement in degrees: $\dfrac{15\pi}{4}$.

One revolution is equal to 2π. Calculate the value equivalent to 2π that has a denominator of 4, by multiplying the numerator and denominator of $2\pi/1$ by 4.

$$\frac{2\pi}{1} \cdot \frac{4}{4} = \frac{8\pi}{4}$$

Because $15\pi/4 > 8\pi/4$, you know that $15\pi/4$ is larger than 1 revolution. This does not affect the technique you use to convert the angle measurement—you still multiply by $180/\pi$, but you should expect a degree measurement larger than $360°$.

$$\frac{15\pi}{4} \cdot \frac{180}{\pi} = \frac{(15\pi) \cdot 180}{4\pi}$$

Note that 180 and 4 share a greatest common factor of 4. Eliminate it to reduce the fraction.

$$= \frac{(15\pi) \cdot (180 \div 4)}{(4 \div 4) \cdot \pi}$$

$$= \frac{15\pi \cdot 45}{1 \cdot \pi}$$

$$= \frac{15\cancel{\pi} \cdot 45}{1 \cdot \cancel{\pi}}$$

$$= \frac{15 \cdot 45}{1}$$

$$= 675°$$

All of the other radian measurements so far have had π in them somewhere, which makes an angle of 2 radians feel weird. Not to worry—you still multiply by $\frac{180}{\pi}$ to convert it into degrees.

1.30 Express the angle measurement in degrees: 2.

As Problem 1.27 explains, any angle measurement written without units is assumed to be expressed in radians. Therefore, this problem instructs you to convert 2 radians into degrees.

$$2 \cdot \frac{180}{\pi} = \frac{2}{1} \cdot \frac{180}{\pi}$$
$$= \frac{360}{\pi}$$

You conclude that 2 radians = $(360/\pi)°$, which is approximately equal to 114.592°.

Complementary and Supplementary Angles
Sums of 90° and 180°

Note: Problems 1.31–1.33 refer to the diagram below, in which ∠1 and ∠2 are complementary.

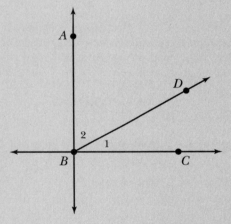

1.31 Create an equation that describes the relationship between the numbered angles in the diagram.

If two angles are complementary, then the sum of their measures is 90°. In the given diagram, angles CBD (∠1) and ABD (∠2) together form right angle ABC. Recall from geometry that perpendicular lines (like the x- and y-axes) form right angles, which measure 90°. Therefore, the sum of the measures of the angles is 90°:

$$m\angle 1 + m\angle 2 = 90°$$

In the equation above, $m\angle 1$ and $m\angle 2$ represent the measures of ∠1 and ∠2, respectively.

Note: Problems 1.31–1.33 refer to the diagram in Problem 1.31, in which ∠1 and ∠2 are complementary.

1.32 Given $m\angle 1 = 31°$, calculate $m\angle 2$.

The problem states that $m\angle 1$ and $m\angle 2$ are complementary, so the sum of their measures is 90°.

$$m\angle 1 + m\angle 2 = 90°$$

Substitute $m\angle 1 = 31°$ into the equation and solve for $m\angle 2$.

$$31° + m\angle 2 = 90°$$
$$m\angle 2 = 90° - 31°$$
$$m\angle 2 = 59°$$

Note: Problems 1.31–1.33 refer to the diagram in Problem 1.31, in which ∠1 and ∠2 are complementary.

1.33 Given $m\angle 2 = \dfrac{\pi}{11}$, calculate $m\angle 1$.

> This is just like Problem 1.32, but this time you're working in radians instead of degrees.

Complementary angles have measures that sum to 90°. This problem expresses the degree measure in radians, so you should begin by converting 90° to radians as well.

$$90 \cdot \frac{\pi}{180} = \frac{90\pi}{180}$$
$$= \frac{(90 \div 90)\pi}{(180 \div 90)}$$
$$= \frac{1\pi}{2}$$
$$= \frac{\pi}{2}$$

> Problem 1.24 explains this step—how to convert from degrees to radians.

Therefore, complementary angles measured in radians have a sum of $\pi/2$.

$$m\angle 1 + m\angle 2 = \frac{\pi}{2}$$

Substitute $m\angle 2 = \pi/11$ into the equation

$$m\angle 1 + \frac{\pi}{11} = \frac{\pi}{2}$$

Multiply both sides of the equation by 22, the least common multiple of 2 and 11, to eliminate the fractions.

$$22\left(m\angle 1 + \frac{\pi}{11}\right) = 22\left(\frac{\pi}{2}\right)$$
$$22 \cdot m\angle 1 + \frac{22\pi}{11} = \frac{22\pi}{2}$$

Reduce the fractions to lowest terms and solve for $m\angle 1$.

$$22m\angle 1 + 2\pi = 11\pi$$
$$22m\angle 1 = 11\pi - 2\pi$$
$$22m\angle 1 = 9\pi$$
$$m\angle 1 = \frac{9\pi}{22}$$

The complement of $\pi/11$ is $9\pi/22$.

1.34 In the diagram below, right triangle XYZ has acute angle Z that measures $\dfrac{\pi}{6}$. Calculate the measure of the remaining acute angle.

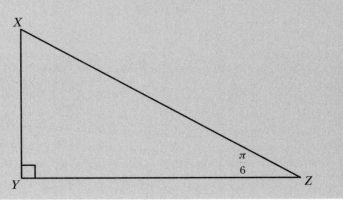

The small square located at Y indicates that $m\angle Y = 90° = \pi/2$, so $\angle Y$ is a right angle. According to a geometric theorem, the acute angles of a right triangle are complementary. Recall that acute angles measure less than 90° (or $\pi/2$ radians), so in this triangle, $\angle X$ and $\angle Z$ are acute.

$$m\angle X + m\angle Z = \frac{\pi}{2}$$

According to the diagram, $m\angle Z = \pi/6$. Substitute this value into the equation above and solve for $m\angle X$.

$$m\angle X + \frac{\pi}{6} = \frac{\pi}{2}$$
$$m\angle X = \frac{\pi}{2} - \frac{\pi}{6}$$

Express the fractions in terms of the least common denominator, 6.

$$m\angle X = \frac{\pi}{2} \cdot \frac{3}{3} - \frac{\pi}{6}$$
$$= \frac{3\pi}{6} - \frac{\pi}{6}$$
$$= \frac{2\pi}{6}$$

The measures of the angles in a triangle add up to 180°. If you subtract the 90° right angle from this total, the two remaining acute angles have to add up to the remaining 90°, so they are complementary. For more details, see Problem 7.28 in The Humongous Book of Geometry Problems.

Reduce the fraction to lowest terms.

$$= \frac{(2 \div 2)\pi}{(6 \div 2)}$$

$$= \frac{\pi}{3}$$

You conclude that $m\angle X = \pi/3$.

Note: Problems 1.35–1.37 refer to the diagram below, in which $\angle 3$ and $\angle 4$ are supplementary.

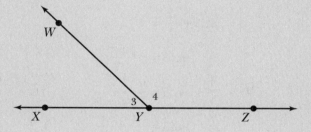

1.35 Create an equation that describes the relationship between the numbered angles in the diagram.

A pair of angles is supplementary if their measures have a sum of 180°. In the given diagram, angles 3 and 4 combine to create a straight angle—the line $\overline{XZ}$. Lines are also described as "straight angles" and have a measure of 180°. Therefore, angles 3 and 4 are supplementary, and their measures have a sum of 180°.

$$m\angle 3 + m\angle 4 = 180°$$

Note: Problems 1.35–1.37 refer to the diagram in Problem 1.35, in which $\angle 3$ and $\angle 4$ are supplementary.

1.36 Given $m\angle 3 = 75°$, calculate $m\angle 4$.

According to Problem 1.35, angles 3 and 4 are supplementary. Therefore, their measures have a sum of 180°.

$$m\angle 3 + m\angle 4 = 180°$$

Substitute $m\angle 3 = 75°$ into the equation and solve for $m\angle 4$.

$$75° + m\angle 4 = 180°$$
$$m\angle 4 = 180° - 75°$$
$$m\angle 4 = 105°$$

Note: Problems 1.35–1.37 refer to the diagram in Problem 1.35, in which ∠3 and ∠4 are supplementary.

1.37   Given $m\angle 4 = \dfrac{5\pi}{7}$, calculate $m\angle 3$.

$\boxed{180° = \pi \text{ radians}}$

The measures of supplementary angles have a sum of π.

$$m\angle 3 + m\angle 4 = \pi$$

Substitute $m\angle 4 = 5\pi/7$ into the equation and solve for $m\angle 3$.

$$m\angle 3 + \frac{5\pi}{7} = \pi$$

$$m\angle 3 = \pi - \frac{5\pi}{7}$$

$$m\angle 3 = \frac{\pi}{1} \cdot \frac{7}{7} - \frac{5\pi}{7}$$

$$m\angle 3 = \frac{7\pi}{7} - \frac{5\pi}{7}$$

$$m\angle 3 = \frac{2\pi}{7}$$

1.38   In the diagram below, parallel lines l and m are intersected by line p, which produces same-side interior angles 1 and 2. Given $m\angle 1 = \dfrac{\pi}{5}$, calculate $m\angle 2$.

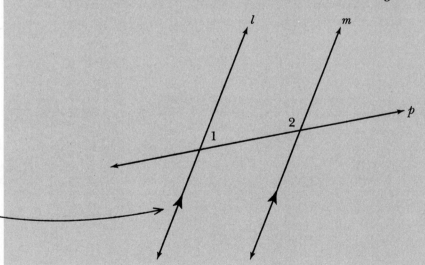

Arrowheads in the middle of lines indicate that the lines are parallel to each other.

According to a geometric theorem, when two parallel lines are intersected by a third, non-parallel line (called the transversal), same-side interior angles are supplementary. Same-side interior angles are defined as two angles on the same side of the transversal located between the parallel lines.

In other words, angles 1 and 2 are same-side interior angles because they are both on the same side of p (both are above the line) and both are located between parallel lines l and m. Therefore, the angles are supplementary. Note that the problem presents an angle measured in radians, so rather than set the sum of the supplementary angles equal to 180°, you should set it equal to π.

$$m\angle 1 + m\angle 2 = \pi$$

Substitute $m\angle 1 = \pi/5$ into the equation and solve for $m\angle 2$. Note that you will need to use common denominators to simplify the result.

$$\frac{\pi}{5} + m\angle 2 = \pi$$

$$m\angle 2 = \pi - \frac{\pi}{5}$$

$$m\angle 2 = \frac{5\pi}{5} - \frac{\pi}{5}$$

$$m\angle 2 = \frac{4\pi}{5}$$

1.39 Calculate x in the diagram below. Express the answer in terms of radians.

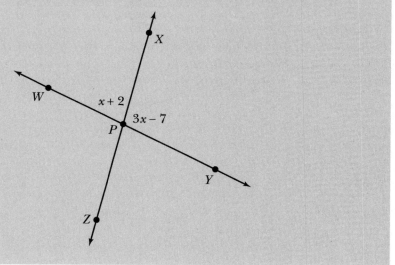

Because $\angle WPX$ and $\angle XPY$ combine to form a straight line—which measures π radians—you can conclude that the angles are supplementary. Thus, the sum of their measures is equal to π.

$$m\angle WPX + m\angle XPY = \pi$$
$$(x+2) + (3x-7) = \pi$$
$$x + 2 + 3x - 7 = \pi$$
$$(x+3x) + (2-7) = \pi$$
$$4x - 5 = \pi$$

Solve for *x*.

$$4x = \pi + 5$$

$$x = \frac{\pi + 5}{4}$$

> The answer $\frac{1}{4}(\pi + 5)$ is also correct.

Arc Length

To measure arcs, you first measure angles

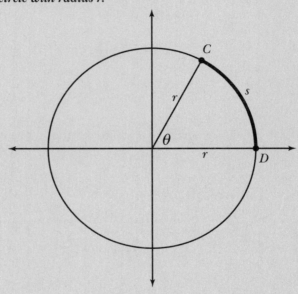

Note: Problems 1.40–1.41 refer to the diagram below, in which θ is a central angle that subtends $\widehat{CD}$ on a circle with radius r.

1.40 Assuming θ is measured in degrees, express $m\widehat{CD}$ in degrees.

> This is the Greek letter theta, pronounced "THAY-tuh." In trigonometry, angles are often represented by lowercase Greek letters. Other common letters: α (alpha) and β (beta).

Recall, from Problem 1.12, that a central angle subtends an arc when its sides intersect a circle. In this diagram, central angle θ subtends $\widehat{CD}$ because the sides of the angle intersect the circle at points *C* and *D*.

This problem asks you to calculate the *measure* of an arc, which is an entirely different task than calculating the *length* of the arc. An arc's measure is exactly equal to the measure of the central angle that subtends it. Therefore, because the central angle subtending $\widehat{CD}$ measures θ°, you conclude that the arc has the same measure: $m\widehat{CD} = \theta°$.

Note: Problems 1.40–1.41 refer to the diagram in Problem 1.40, in which θ is a central angle that subtends $\overset{\frown}{CD}$ on a circle with radius r.

1.41   Assuming θ is measured in radians, calculate s, the length of $\overset{\frown}{CD}$.

As Problem 1.40 explains, calculating the *measure* of an arc and the *length* of an arc produce entirely different results. To calculate the length of an arc, multiply the length of the radius of the circle by the measure of the central angle (in radians) that subtends the arc. In other words, apply the formula below, in which s represents arc length, r represents the radius, and θ represents the radian measure of the central angle.

$$s = r\theta$$

You conclude that the length of $\overset{\frown}{CD}$ in the given diagram is equal to $r\theta$. Because no values are given for either of the variables, you cannot simplify or expand upon that answer any farther.

1.42   The circumference C of a circle is equal to πd, where d is the diameter of the circle. Verify this value using the arc length formula.

According to Problem 1.41, you apply the formula $s = r\theta$ to calculate the length, s, of an arc given the radius r and the measure of its central angle θ in radians. The circumference of a circle is simply an "arc" with a central angle of 2π radians—a full revolution about the coordinate plane—so substitute this value into the arc length formula.

$$s = r\theta$$
$$s = r(2\pi)$$

The radius of a circle is exactly half the length of its diameter: $r = d/2$. Substitute this value into the formula and simplify.

$$s = \frac{d}{2}(2\pi)$$
$$= \frac{\cancel{2}\pi d}{\cancel{2}}$$
$$= \pi d$$

According to the arc length formula, the circumference of a circle, the "arc" with a central angle of 2π and a radius of $d/2$, is πd. This exactly matches the circumference formula for a circle, and therefore verifies that the circumference formula is true.

1.43   Calculate the length of $\overset{\frown}{AB}$ in the diagram below.

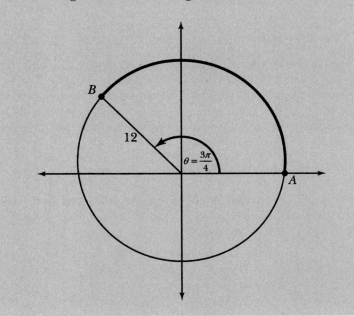

According to the diagram, the radius of $\overset{\frown}{AB}$ is $r = 12$ and its central angle is $\theta = 3\pi/4$. Apply the arc length formula.

$$s = r\theta$$
$$= 12\left(\frac{3\pi}{4}\right)$$
$$= \frac{12(3\pi)}{4}$$
$$= \frac{(12 \div 4)(3\pi)}{(4 \div 4)}$$
$$= \frac{3 \cdot 3\pi}{1}$$
$$= 9\pi$$

1.44 A central angle θ on a circle with diameter 10 subtends an arc of length $\dfrac{35\pi}{6}$. Calculate the measure of θ in degrees.

In this problem, you are given $s = 35\pi/6$ and $d = 10$. Recall that the radius of a circle is half the length of the diameter, so $r = d/2 = 10/2 = 5$. Substitute these values into the arc length formula to calculate θ.

$$s = r\theta$$

$$\frac{35\pi}{6} = 5\theta$$

Multiply both sides of the equation by 1/5 to solve for θ.

$$\left(\frac{1}{5}\right)\left(\frac{35\pi}{6}\right) = \left(\frac{1}{5}\right)\left(\frac{5\theta}{1}\right)$$

$$\frac{35\pi}{5 \cdot 6} = \frac{5\theta}{5}$$

$$\frac{(35 \div 5)\pi}{(5 \div 5) \cdot 6} = \frac{(5 \div 5)\theta}{(5 \div 5)}$$

$$\frac{(7)\pi}{(1)6} = \frac{(1)\theta}{(1)}$$

$$\frac{7\pi}{6} = \theta$$

Therefore, the central angle is $7\pi/6$. However, the problem requests the angle measurement in degrees. Multiply θ by $180/\pi$ to convert it into degrees.

Check the problem when you're done to make sure you answer the question it actually asks.

$$\frac{7\pi}{6} \cdot \frac{180}{\pi} = \frac{7\cancel{\pi}(180)}{6\cancel{\pi}}$$

$$= \frac{7(180)}{6}$$

$$= \frac{7(180 \div 6)}{(6 \div 6)}$$

$$= \frac{7(30)}{1}$$

$$= 210°$$

1.45 Calculate the length of $\overset{\frown}{XYZ}$ in the diagram below, given the radius of the circle is $r = 7$.

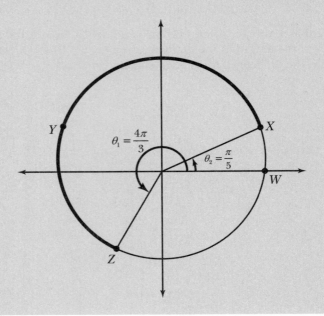

Note that the arc you are asked to measure is named with three letters instead of two letters, which were used to name the other arcs in this section. This is because the arc measures more than 180° (π radians) and is therefore classified as a "major arc."

The outer points in the name (X and Z in this case) still represent the endpoints of the arc. The middle point indicates which direction around the circle you travel to get from one endpoint to the other. Therefore, in the diagram above, $\overset{\frown}{XZ}$ refers to the smaller "minor arc" bounded by X and Z that passes through W.

To calculate the length of $\overset{\frown}{XYZ}$, the darkened arc on the circle, you must first calculate the length of $\overset{\frown}{WYZ}$, which is subtended by central angle θ_1, and subtract the length of $\overset{\frown}{WX}$, which is subtended by central angle θ_2.

$$\text{arc length of } \overset{\frown}{XYZ} = \left(\text{arc length of } \overset{\frown}{WYZ}\right) - \left(\text{arc length of } \overset{\frown}{WX}\right)$$
$$= \left(r \cdot \theta_1\right) - \left(r \cdot \theta_2\right)$$
$$= \left(7 \cdot \frac{4\pi}{3}\right) - \left(7 \cdot \frac{\pi}{5}\right)$$
$$= \frac{28\pi}{3} - \frac{7\pi}{5}$$

Express the fractions in terms of the least common denominator (15) and simplify.

$$= \left(\frac{28\pi}{3} \cdot \frac{5}{5}\right) - \left(\frac{7\pi}{5} \cdot \frac{3}{3}\right)$$

$$= \frac{140\pi}{15} - \frac{21\pi}{15}$$

$$= \frac{119\pi}{15}$$

Chapter 2
RIGHT TRIANGLE TRIGONOMETRY

Including all six trig functions

Once you can measure angles using degrees and radians, as explored in Chapter 1, you are sufficiently equipped to extend that knowledge into the primary geometric figure of interest in trigonometry: the triangle. In this chapter, you focus exclusively on right triangles, and in so doing, you explore how all of the basic trigonometric functions are defined (both with and without the use of technology to aid you).

When you think "trigonometry," you probably think "sine and cosine." (And you're right.) You may not know that once you calculate sine and cosine values of an angle, you can calculate four other function values as well: tangent, cotangent, secant, and cosecant. In this chapter, you'll learn how to evaluate all of those functions with and without a calculator, but first it's time to review the Pythagorean theorem.

Pythagorean Theorem

$a^2 + b^2 = c^2$

Note: Problems 2.1–2.3 refer to right triangle XYZ in the diagram below.

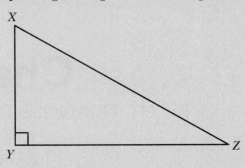

2.1 Identify the hypotenuse of the triangle and justify your answer. Then, classify the remaining sides of the triangle.

The hypotenuse of a right triangle is its longest side, and it is always located opposite the right angle. A small square drawn at a vertex of a figure indicates a right angle, so in this diagram, $\angle Y$ is a right angle. Therefore, the side opposite angle Y is the hypotenuse: $\overline{XZ}$. Note that $\overline{XZ}$ is visibly longer than either of the two remaining sides of the triangle, $\overline{XY}$ and $\overline{YZ}$, which are called the "legs" of the right triangle.

Note: Problems 2.1–2.3 refer to the diagram in Problem 2.1.

2.2 State the Pythagorean theorem in terms of right triangle XYZ.

According to the Pythagorean theorem, if you square the lengths of the legs and then add those squared values together, it is equal to the square of the hypotenuse's length. In other words, if a and b represent the lengths of the legs of a right triangle and c represents the length of the hypotenuse, then $a^2 + b^2 = c^2$.

> You use a bar to name a segment ($\overline{XY}$), but not when you describe the LENGTH of that segment: XY. In other words, XY is the length of $\overline{XY}$.

The lengths of the legs in triangle XYZ are XY and YZ; the length of hypotenuse is XZ. Therefore, according to the Pythagorean theorem, $(XY)^2 + (YZ)^2 = (XZ)^2$.

Note: Problems 2.1–2.3 refer to the diagram in Problem 2.1.

2.3 Calculate YZ in terms of XY and XZ.

This problem directs you to calculate the length of one leg (YZ) in terms of the lengths of the other leg (XY) and the hypotenuse (XZ). Problem 2.2 states the relationship between these lengths, as dictated by the Pythagorean theorem.

$$(XY)^2 + (YZ)^2 = (XZ)^2$$

Solve this equation for *YZ*. Begin by subtracting $(XY)^2$ from both sides of the equation in order to isolate the term containing *YZ*.

$$(YZ)^2 = (XZ)^2 - (XY)^2$$

The equation is now solved for $(YZ)^2$, but you need to solve for *YZ*. To eliminate the exponent, take the square root of both sides of the equation.

$$\sqrt{(YZ)^2} = \sqrt{(XZ)^2 - (XY)^2}$$
$$YZ = \pm\sqrt{(XZ)^2 - (XY)^2}$$

Because length must be a positive value, you omit the possible negative solutions (and, therefore, the "±" symbol on the right side of the equation).

$$YZ = \sqrt{(XZ)^2 - (XY)^2}$$

> You have to include the "plus or minus" sign whenever you take the square root (or any even root) of both sides of an equation. In this problem, however, you drop it in the next step.

2.4 Calculate the length of the hypotenuse of a right triangle if the lengths of the legs are 5 and 12.

In the formula for the Pythagorean theorem, $a^2 + b^2 = c^2$, the lengths of the legs are represented by *a* and *b*. Therefore, if you substitute $a = 5$ and $b = 12$ into the formula, you can calculate the length of the hypotenuse *c*. (Note that substituting $a = 12$ and $b = 5$ produces the same result—it does not matter which leg lengths are substituted into *a* and *b*.)

$$a^2 + b^2 = c^2$$
$$5^2 + 12^2 = c^2$$
$$25 + 144 = c^2$$
$$169 = c^2$$

To solve the equation, take the square root of both sides. Because length is a positive value, you should omit possible negative solutions.

$$\sqrt{169} = \sqrt{c^2}$$
$$13 = c$$

> In other words, don't worry about the "plus or minus" symbol until further notice.

If a right triangle has legs of length 5 and 12, the length of the hypotenuse is 13.

2.5 Calculate the lengths of the sides in the triangle below.

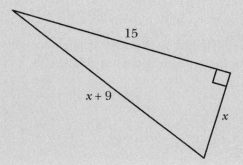

This right triangle has legs of length 15 and x; the length of the hypotenuse is $x + 9$. In order to calculate the lengths of the sides, you will first need to know the value of x. Substitute these values into the Pythagorean theorem.

$$a^2 + b^2 = c^2$$
$$15^2 + x^2 = (x + 9)^2$$
$$225 + x^2 = (x + 9)^2$$

$(x + 9)^2$
$= (x + 9)(x + 9)$
$= (x)(x) + (x)(9) +$
    $(9)(x) + (9)(9)$
$= x^2 + 9x + 9x + 81$
$= x^2 + 18x + 81$

Apply the FOIL method to square the quantity $(x + 9)$.

$$225 + x^2 = x^2 + 18x + 81$$

Subtract x^2 from both sides of the equation and solve for x.

$$225 = 18x + 81$$
$$225 - 81 = 18x$$
$$144 = 18x$$
$$\frac{144}{18} = x$$
$$8 = x$$

If $x = 8$, the lengths of the legs are 15 and $x = 8$; the length of the hypotenuse is $x + 9 = 8 + 9 = 17$.

2.6 Given a right triangle with hypotenuse length 7 and one leg with a length of 4, calculate the length of the remaining leg.

You are given the length of one leg and the hypotenuse, so substitute $a = 4$ and $c = 7$ into the Pythagorean theorem and solve for b, the length of the remaining leg. (Alternatively, you could substitute $b = 4$ and $c = 7$ into the Pythagorean theorem and solve for a, but in either case, c must be equal to 7, as it is the length of the hypotenuse.)

$$a^2 + b^2 = c^2$$
$$4^2 + b^2 = 7^2$$
$$16 + b^2 = 49$$
$$b^2 = 49 - 16$$
$$b^2 = 33$$
$$\sqrt{b^2} = \sqrt{33}$$
$$b = \sqrt{33}$$

If one leg and the hypotenuse of a right triangle have lengths 4 and 7, respectively, then the remaining leg has length $\sqrt{33}$.

2.7 Calculate the length of the legs in the isosceles right triangle below.

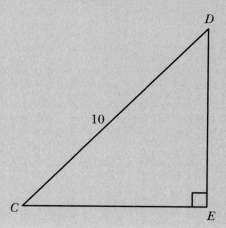

An isosceles triangle has at least two sides that are the same length. It is not possible for a right triangle to have three sides of the same length, so in an isosceles right triangle, *exactly* two sides are the same length—the legs. Therefore, in this triangle, $CE = DE$.

Let x represent the length of each leg in the triangle; in other words, let $x = CE = DE$. Substitute $CE = x$, $DE = x$, and $c = 10$ into the Pythagorean theorem.

$$(CE)^2 + (DE)^2 = (CD)^2$$
$$x^2 + x^2 = 10^2$$
$$2x^2 = 100$$
$$x^2 = \frac{100}{2}$$
$$x^2 = 50$$

Solve for x and simplify the radical.

$$\sqrt{x^2} = \sqrt{50}$$
$$x = \sqrt{25 \cdot 2}$$
$$x = \sqrt{25} \cdot \sqrt{2}$$
$$x = 5\sqrt{2}$$

> A right triangle HAS to have one side that is longer than the others, the hypotenuse. If all three sides were the same length, you wouldn't have a hypotenuse. (There is a more formal way to prove this—in math, there almost always is—but that's not the point of the problem, so don't worry about it.)

> If you need help simplifying radicals, check out Problems 13.1–13.20 in The Humongous Book of Algebra Problems. In fact, Problem 13.5 shows you how to simplify this exact square root.

Note: Problems 2.8–2.9 refer to right triangle PQR in the diagram below.

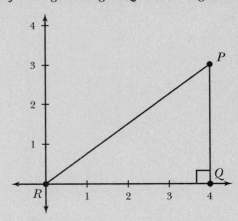

2.8 Identify the coordinates of points P, Q, and R, and use them to calculate PQ and RQ.

Points P and Q lie on the same vertical line, which is 4 units right of the origin; thus, they both have an x-coordinate of 4. While point Q is located on the x-axis (and therefore has a y-coordinate of 0), point P is 3 units above the x-axis (and therefore has a y-coordinate of 3). Point R is located at the origin, so its x- and y-coordinates are both 0.

$$P = (4,3) \qquad Q = (4,0) \qquad R = (0,0)$$

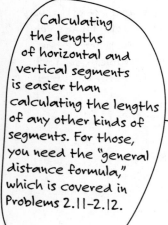

Calculating the lengths of horizontal and vertical segments is easier than calculating the lengths of any other kinds of segments. For those, you need the "general distance formula," which is covered in Problems 2.11–2.12.

Note that $\overline{PQ}$ is a vertical segment, a portion of the vertical line $x = 4$. To calculate the length of the segment, simply subtract the y-coordinate of Q from the y-coordinate of P.

$$PQ = 3 - 0 = 3$$

Similarly, $\overline{QR}$ is a horizontal segment, a portion of the horizontal x-axis. Therefore, subtracting the x-coordinate of R from the x-coordinate of Q calculates the length of the segment, QR.

$$QR = 4 - 0 = 4$$

Note: Problems 2.8–2.9 refer to the diagram in Problem 2.8.

2.9 Apply the Pythagorean theorem to calculate PR.

Note that triangle PQR has two perpendicular sides, $\overline{PQ}$ and $\overline{QR}$. Perpendicular lines form right angles, so PQR is a right triangle. In Problem 2.8, you calculate the lengths of the legs: $PQ = 3$ and $QR = 4$.

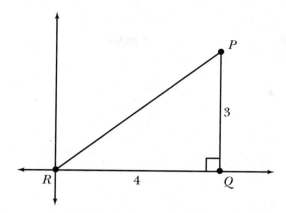

Substitute the lengths of the legs into the Pythagorean theorem and solve for PR, the length of the hypotenuse.

$$(PQ)^2 + (QR)^2 = (PR)^2$$
$$3^2 + 4^2 = (PR)^2$$
$$9 + 16 = (PR)^2$$
$$25 = (PR)^2$$
$$\sqrt{25} = \sqrt{(PR)^2}$$
$$5 = PR$$

2.10 Apply the technique demonstrated in Problems 2.8–2.9 to calculate the distance between the origin and a point with coordinates (7,24).

The line segment connecting the origin to point (7,24) is the hypotenuse of a right triangle whose horizontal leg has a length of 7 and whose vertical leg has a length of 24, as illustrated in the following diagram.

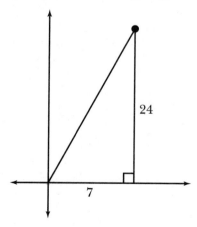

To calculate the length of the hypotenuse, apply the Pythagorean theorem $a^2 + b^2 = c^2$, in which $a = 7$, $b = 24$, and c is the length of the hypotenuse—the value for which you are solving the equation.

$$a^2 + b^2 = c^2$$
$$7^2 + 24^2 = c^2$$
$$49 + 576 = c^2$$
$$625 = c^2$$
$$\sqrt{625} = \sqrt{c^2}$$
$$25 = c$$

You conclude that the distance between the origin and a point with coordinates (7,24) is 25.

2.11 Apply the general distance formula to calculate the distance between points $A = (2,3)$ and $B = (-1,5)$.

The general distance formula is, as its name suggests, a more general implementation of the process applied in Problems 2.8–2.10. It does not require that one of the endpoints of a segment is the origin, nor does it require that you construct a right triangle in the coordinate plane. Essentially, both of those steps are completed automatically.

According to the general distance formula, the distance between points $A = (x_1, y_1)$ and $B = (x_2, y_2)$ is $D = \sqrt{(x_2 - x_1)^2 + (y_2 - y_1)^2}$. Note that subscripts are used to differentiate the coordinates in the formula. In other words, the x- and y-coordinates of point A both have the subscript "1" and point B coordinates have the subscript "2."

The subtraction parts of the formula, $x_2 - x_1$ and $y_2 - y_1$, calculate the horizontal and vertical distances between the points, which are the lengths of the legs of the right triangle whose hypotenuse is the distance between the points.

In this problem, point $A = (2,3)$, so $x_1 = 2$ and $y_1 = 3$. Point $B = (-1,5)$, so $x_2 = -1$ and $y_2 = 5$. Substitute x_1, y_1, x_2, and y_2 into the distance formula and calculate D.

$$D = \sqrt{(x_2 - x_1)^2 + (y_2 - y_1)^2}$$
$$= \sqrt{(-1-2)^2 + (5-3)^2}$$
$$= \sqrt{(-3)^2 + (2)^2}$$
$$= \sqrt{9+4}$$
$$= \sqrt{13}$$

The distance between points A and B is $\sqrt{13}$.

2.12 Apply the general distance formula to calculate the distance between points $C = (-6, -9)$ and $D = (4, -7)$.

Substitute $x_1 = -6$, $y_1 = -9$, $x_2 = 4$, and $y_2 = -7$ into the general distance formula and calculate D.

$$
\begin{aligned}
D &= \sqrt{(x_2 - x_1)^2 + (y_2 - y_1)^2} \\
&= \sqrt{(4 - (-6))^2 + (-7 - (-9))^2} \\
&= \sqrt{(4 + 6)^2 + (-7 + 9)^2} \\
&= \sqrt{10^2 + 2^2} \\
&= \sqrt{100 + 4} \\
&= \sqrt{104}
\end{aligned}
$$

Reduce the radical expression.

$$
\begin{aligned}
&= \sqrt{4 \cdot 26} \\
&= \sqrt{4}\sqrt{26} \\
&= 2\sqrt{26}
\end{aligned}
$$

2.13 Calculate the height of an isosceles triangle with a base of length 6 and legs of length 8.

The height of an isosceles triangle is the length h of the segment that is perpendicular to the base of the triangle and extends to the vertex of the opposite angle. As the diagram below illustrates, the segment representing height divides the isosceles triangle in half—including the base. Therefore, the two halves of the base in the diagram below both have length 3.

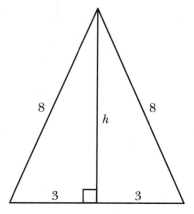

Notice that the height of the isosceles triangle divides the figure into two right triangles, each with a hypotenuse of 8 and legs of length h and 3. Apply the Pythagorean theorem to calculate h.

$$a^2 + b^2 = c^2$$
$$h^2 + 3^2 = 8^2$$
$$h^2 + 9 = 64$$
$$h^2 = 64 - 9$$
$$h^2 = 55$$
$$\sqrt{h^2} = \sqrt{55}$$
$$h = \sqrt{55}$$

The height of the isosceles triangle is $\sqrt{55}$.

2.14   Calculate the lengths of the sides of triangle ABC in the diagram below.

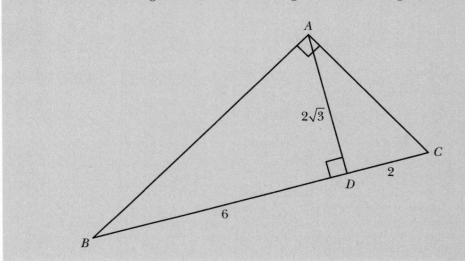

You are asked to calculate the lengths of the sides of right triangle ABC; in other words, you are to calculate AB, AC, and BC. The length of the hypotenuse is the easiest of the three lengths to compute, as it is equal to the sum of the lengths of segments $\overline{BD}$ and $\overline{DC}$.

$$BC = BD + DC$$
$$= 6 + 2$$
$$= 8$$

Notice that $\overline{AD}$ splits right triangle ABC into two smaller right triangles: ABD and ACD. In each of these triangles, two sides have known lengths and the Pythagorean theorem can be applied to calculate the remaining side. For example, right triangle ABD has legs $AD = 2\sqrt{3}$ and $BD = 6$. Substitute these values into the Pythagorean theorem to calculate AB.

$$(AD)^2 + (BD)^2 = (AB)^2$$
$$\left(2\sqrt{3}\right)^2 + 6^2 = (AB)^2$$
$$(2)^2 \cdot \left(\sqrt{3}\right)^2 + 6^2 = (AB)^2$$
$$4(3) + 36 = (AB)^2$$
$$12 + 36 = (AB)^2$$
$$48 = (AB)^2$$
$$\sqrt{48} = \sqrt{(AB)^2}$$
$$\sqrt{48} = AB$$

Simplify the radical expression.

$$\sqrt{16 \cdot 3} = AB$$
$$\sqrt{16} \cdot \sqrt{3} = AB$$
$$4\sqrt{3} = AB$$

Now that you know the lengths of two sides of right triangle ABC, you can once again apply the Pythagorean theorem to calculate AC.

$$(AB)^2 + (AC)^2 = (BC)^2$$
$$\left(4\sqrt{3}\right)^2 + (AC)^2 = (8)^2$$
$$(4)^2 \cdot \left(\sqrt{3}\right)^2 + (AC)^2 = (8)^2$$
$$16 \cdot 3 + (AC)^2 = 64$$
$$48 + (AC)^2 = 64$$
$$(AC)^2 = 64 - 48$$
$$(AC)^2 = 16$$
$$\sqrt{(AC)^2} = \sqrt{16}$$
$$AC = 4$$

Or you can apply the Pythagorean theorem using the smaller right triangle ADC. You know that AD = $2\sqrt{3}$ and DC = 2, so substitute them into the equation $(AD)^2 + (DC)^2 = (AC)^2$ and solve for AC. You'll get the same answer.

The lengths of the sides of right triangle ABC are $AB = 4\sqrt{3}$, $AC = 4$, and $BC = 8$.

Trigonometric Functions

sin, cos, tan, cot, sec, and csc

Note: Problems 2.15–2.19 refer to the right triangle in the diagram below.

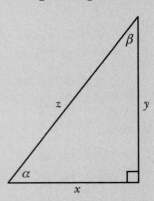

2.15 Identify one leg that is adjacent to, and one leg that is opposite, angle α.

Trigonometric ratios (like sine, cosine, and tangent) are defined in terms of the relationship between the acute angles of a right triangle and the lengths of the sides of that right triangle. Therefore, you need to know how to describe the legs of a right triangle as either "opposite" or "adjacent to" an acute angle.

> *Acute angles measure less than 90°, so the acute angles of a right triangle are the two angles that aren't right angles.*

In this triangle, side y is opposite angle α. Both of the other sides in the triangle (x and z) are also the sides of angle α; only the side opposite an angle is not part of the angle itself. Consider the sides that do form the angle, x and z. One of those sides is the hypotenuse of the triangle (z); however, the problem specifically states that you are to identify the *leg* that is adjacent to α. Therefore, you conclude that x is the leg adjacent to angle α.

> *The "hypotenuse" is always the longest side. Only the legs are classified as "adjacent" or "opposite."*

Note: Problems 2.15–2.19 refer to the diagram in Problem 2.15.

2.16 Identify the leg that is adjacent to, and the leg that is opposite, angle β.

Angle β is formed by sides y and z. Note that z is the hypotenuse—and not a leg—of the right triangle. Because y is a leg of the triangle and also a side of angle β, you conclude that y is the leg adjacent to β. Leg x is opposite β.

Note: Problems 2.15–2.19 refer to the diagram in Problem 2.15.

2.17 Calculate $\sin \alpha$ and $\sin \beta$.

The sine function (abbreviated "sin") assigns a value to acute angles within a right triangle. Specifically, the sine of an angle is equal to the length of the leg opposite that angle divided by the length of the hypotenuse.

$$\sin \alpha = \frac{\text{length of leg opposite } \alpha}{\text{length of hypotenuse}}$$

According to Problem 2.15, the length of the side opposite α is y and the length of the hypotenuse is z. Substitute these values into the equation above.

$$\sin\alpha = \frac{y}{z}$$

To calculate $\sin\beta$, you apply the same technique.

$$\sin\beta = \frac{\text{length of leg opposite }\beta}{\text{length of hypotenuse}}$$

The hypotenuse is still z, but the length of the side opposite β (according to Problem 2.16) is x.

$$\sin\beta = \frac{x}{z}$$

Note: Problems 2.15–2.19 refer to the diagram in Problem 2.15.

2.18 Calculate $\cos\alpha$ and $\cos\beta$.

The cosine function (abbreviated "cos") is very similar to the sine function. Its values are also assigned to acute angles of a right triangle, and those values are also derived from a ratio. However, the cosine of an angle is equal to the length of the *adjacent* side divided by the hypotenuse.

$$\cos\alpha = \frac{\text{length of leg adjacent to }\alpha}{\text{length of hypotenuse}}$$

According to Problem 2.15, the side adjacent to α has length x and the length of the hypotenuse is z.

$$\cos\alpha = \frac{x}{z}$$

To compute $\cos\beta$, divide the length of its adjacent side (y) by the length of the hypotenuse (z).

$$\cos\beta = \frac{y}{z}$$

Note: Problems 2.15–2.19 refer to the diagram in Problem 2.15.

2.19 Calculate $\tan\alpha$ and $\tan\beta$.

Like the sine and cosine functions described in Problems 2.17–2.18, the tangent function (abbreviated "tan") is defined based upon the lengths of the sides of a right triangle. Unlike the sine and cosine functions, however, it only uses the lengths of the legs, not the hypotenuse:

$$\tan\alpha = \frac{\text{length of the leg opposite }\alpha}{\text{length of the leg adjacent to }\alpha}$$

$$= \frac{y}{x}$$

Similarly, calculate the tangent of angle β. Note that the opposite and adjacent legs of a right triangle reverse when you are referring to different acute angles within the triangle.

You can remember which trig function goes with which ratio using the four-syllable "word" SOH-CAH-TOA, pronounced "SO-cah-TOE-ah." (Catchy, right?) It stands for "<u>S</u>ine equals <u>O</u>pposite over <u>H</u>ypotenuse, <u>C</u>osine equals <u>A</u>djacent over <u>H</u>ypotenuse, and <u>T</u>angent equals <u>O</u>pposite over <u>A</u>djacent."

$$\tan \beta = \frac{\text{length of the leg opposite } \beta}{\text{length of the leg adjacent to } \beta}$$

$$= \frac{x}{y}$$

Therefore, the tangent values of the acute angles in a right triangle are reciprocals of each other: x/y and y/x. (Recall that the reciprocal of a fraction reverses its numerator and denominator.)

Note: Problems 2.20–2.22 refer to right triangle ABC in the diagram below.

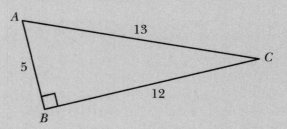

2.20   Calculate cos C and sin C.

> "Ratio" means fraction, so "the ratio of the adjacent length and the hypotenuse" just means "the adjacent length divided by the hypotenuse."

The length of the leg opposite acute angle C has length $AB = 5$, and the length of the adjacent leg is $BC = 12$. The hypotenuse of the triangle has length $AC = 13$. According to Problem 2.18, cos C is equal to the ratio of the lengths of the adjacent leg and hypotenuse.

$$\cos C = \frac{\text{length of leg adjacent to angle } C}{\text{length of the hypotenuse}}$$

$$= \frac{BC}{AC}$$

$$= \frac{12}{13}$$

The sine of acute angle C is equal to the length of the leg opposite C divided by the length of the hypotenuse.

$$\sin C = \frac{\text{length of leg opposite } C}{\text{length of hypotenuse}}$$

$$= \frac{AB}{AC}$$

$$= \frac{5}{13}$$

Note: Problems 2.20–2.22 refer to the diagram in Problem 2.20.

2.21 Calculate tan C and cot C.

According to Problem 2.19, the tangent of an acute angle is equal to the quotient of the opposite and adjacent leg lengths.

$$\tan C = \frac{\text{length of leg opposite } C}{\text{length of leg adjacent to } C}$$
$$= \frac{AB}{BC}$$
$$= \frac{5}{12}$$

The cotangent function (abbreviated "cot") is one of three basic trigonometric functions that are defined as reciprocals of other trigonometric functions: Cotangent is equal to the reciprocal of the tangent function. Therefore, its value is the quotient of the adjacent and opposite leg lengths.

$$\cot C = \frac{\text{length of leg adjacent to } C}{\text{length of leg opposite } C}$$
$$= \frac{BC}{AB}$$
$$= \frac{12}{5}$$

Note: Problems 2.20–2.22 refer to the diagram in Problem 2.20.

2.22 Calculate sec A and csc A.

> Don't get confused: The last two problems dealt with angle C. This problem involves the OTHER acute angle in the triangle.

The secant and cosecant functions (abbreviated "sec" and "csc," respectively) are—like cotangent—defined as the reciprocals of other trig functions. Specifically, secant is the reciprocal of cosine and cosecant is the reciprocal of sine.

$$\sec A = \frac{\text{length of the hypotenuse}}{\text{length of leg adjacent to } A} \qquad \csc A = \frac{\text{length of the hypotenuse}}{\text{length of leg opposite } A}$$

In the diagram, acute angle A has an opposite leg of length $BC = 12$, an adjacent leg of length $AB = 5$, and a hypotenuse of length $AC = 13$.

$$\sec A = \frac{AC}{AB} \qquad \csc A = \frac{AC}{BC}$$
$$= \frac{13}{5} \qquad\qquad = \frac{13}{12}$$

> The first letters of reciprocal functions DON'T MATCH. In other words Secant is NOT the reciprocal of Sine—it's the reciprocal of cosine. Cosecant is NOT the reciprocal of Cosine—it's the reciprocal of sine.

Note: Problems 2.23–2.24 refer to right triangle LMN in the diagram below.

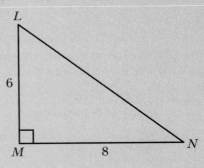

2.23 Calculate tan L.

The tangent of acute angle L is equal to the length of the opposite leg divided by the length of the adjacent leg.

$$\tan L = \frac{MN}{LM}$$
$$= \frac{8}{6}$$

Reduce the fraction to lowest terms.

$$= \frac{8 \div 2}{6 \div 2}$$
$$= \frac{4}{3}$$

Note: Problems 2.23–2.24 refer to the diagram in Problem 2.23.

2.24 Calculate csc N.

The cosecant function is defined as the reciprocal of the sine function. Therefore, csc N is equal to the length of the hypotenuse divided by the length of the opposite leg.

> Sine equals opposite over hypotenuse, and cosecant flips that fraction upside down: hypotenuse over opposite.

$$\csc N = \frac{LN}{LM}$$

According to the diagram, $LM = 6$. However, LN is not given. You need to apply the Pythagorean theorem to calculate the length of the hypotenuse.

$$(LM)^2 + (MN)^2 = (LN)^2$$
$$6^2 + 8^2 = (LN)^2$$
$$36 + 64 = (LN)^2$$
$$100 = (LN)^2$$
$$\sqrt{100} = \sqrt{(LN)^2}$$
$$10 = LN$$

Substitute $LN = 10$ into the cosecant equation.

$$\csc N = \frac{LN}{LM}$$
$$= \frac{10}{6}$$
$$= \frac{10 \div 2}{6 \div 2}$$
$$= \frac{5}{3}$$

Note: Problems 2.25–2.26 refer to isosceles right triangle RST in the diagram below.

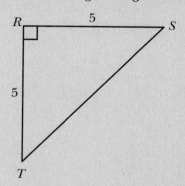

2.25 Calculate cot S.

The cotangent of an acute angle is equal to the length of the adjacent leg divided by the length of the opposite leg.

$$\cot S = \frac{RS}{RT}$$
$$= \frac{5}{5}$$
$$= 1$$

Note: Problems 2.25–2.26 refer to the diagram in Problem 2.25.

2.26 Calculate sec T.

The cosine of an acute angle is equal to the length of the adjacent leg divided by the length of the hypotenuse. Because the secant is defined as the reciprocal of cosine, secant is equal to the length of the hypotenuse divided by the length of the adjacent leg.

$$\sec T = \frac{ST}{RT}$$

According to the diagram, $RT = 5$; apply the Pythagorean theorem to calculate ST.

$$(RS)^2 + (RT)^2 = (ST)^2$$
$$5^2 + 5^2 = (ST)^2$$
$$25 + 25 = (ST)^2$$
$$50 = (ST)^2$$
$$\sqrt{50} = \sqrt{(ST)^2}$$
$$\sqrt{25 \cdot 2} = ST$$
$$\sqrt{25} \cdot \sqrt{2} = ST$$
$$5\sqrt{2} = ST$$

Substitute the values of RT and ST into the secant equation above.

$$\sec T = \frac{ST}{RT}$$
$$= \frac{5\sqrt{2}}{5}$$
$$= \frac{\cancel{5}\sqrt{2}}{\cancel{5}}$$
$$= \frac{\sqrt{2}}{1}$$
$$= \sqrt{2}$$

Note: Problems 2.27–2.28 refer to right triangle XYZ in the diagram below.

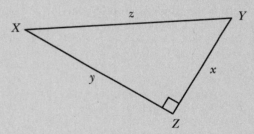

2.27 Demonstrate that the sine of one acute angle in a right triangle is equal to the cosine of the other acute angle. Explain why this is true.

Note that the angles and the sides of this triangle are named using the same letters, but the names are assigned in a very specific way. Each angle and its opposite side share the same letter; the angles are named with capital letters, and the sides are named using lowercase letters. For example, the shortest side x is opposite angle X, the smallest angle in the triangle.

The two acute angles in the right triangle are X and Y. Identify the sine of one angle and the cosine of the other. In the solution below, $\sin X$ and $\cos Y$ are compared, although a solution comparing $\cos X$ and $\sin Y$ would also be correct.

$$\sin X = \frac{YZ}{XY} \qquad \cos Y = \frac{YZ}{XY}$$
$$= \frac{x}{z} \qquad\qquad = \frac{x}{z}$$

Notice that the side opposite one acute angle is always the side adjacent to the other acute angle—side $\overline{YZ}$ in this example—and the length of the hypotenuse (XY) does not change. Therefore, the sine of one acute angle is always equal to the cosine of the other acute angle.

Note: Problems 2.27–2.28 refer to the diagram in Problem 2.27.

2.28 Demonstrate that the tangent of an acute angle in a right triangle is equal to the quotient of the sine and cosine of that angle.

Although tangent may be defined as the length of the opposite side divided by the length of the adjacent side, it is equally correct to state that tangent is equal to the quotient of sine and cosine.

$$\tan \theta = \frac{\sin \theta}{\cos \theta}$$

Cotangent is the reciprocal of tangent: $\cot \theta = \dfrac{\cos \theta}{\sin \theta}.$

Select one of the acute angles in the right triangle—the solution below investigates acute angle X—and calculate its sine, cosine, and tangent values.

$$\sin X = \frac{YZ}{XY} \qquad \cos X = \frac{XZ}{XY} \qquad \tan X = \frac{YZ}{XZ}$$

$$= \frac{x}{z} \qquad\qquad = \frac{y}{z} \qquad\qquad = \frac{x}{y}$$

According to the problem, tan X is equal to the quotient of sin X and cos X.

$$\tan X = \frac{\sin X}{\cos X}$$

Substitute the values of sin X, cos X, and tan X into the equation above.

$$\frac{x}{y} = \frac{x / z}{y / z}$$

This equation contains a complex fraction—a fraction that, itself, contains frations. To simplify the right side of the equation, multiply the numerator and denominator by z.

$$\frac{x}{y} = \frac{\left(\dfrac{x}{z}\right)\left(\dfrac{z}{1}\right)}{\left(\dfrac{y}{z}\right)\left(\dfrac{z}{1}\right)}$$

$$= \frac{\left(\dfrac{xz}{z}\right)}{\left(\dfrac{yz}{z}\right)}$$

The fractions in the numerator and denominator have a greatest common factor of z. Reduce both fractions.

$$\frac{x}{y} = \frac{\left(\dfrac{x\cancel{z}}{\cancel{z}}\right)}{\left(\dfrac{y\cancel{z}}{\cancel{z}}\right)}$$

$$= \frac{\left(\dfrac{x}{1}\right)}{\left(\dfrac{y}{1}\right)}$$

$$= \frac{x}{y}$$

Therefore, $\tan X = \dfrac{\sin X}{\cos X} = \dfrac{x}{y}$.

Trigonometric Tables

When "close enough" is good enough

2.29 Use the trigonometric table in Appendix A to calculate sin 88°, accurate to five decimal places.

Before scientific and graphing calculators were readily available to scientists, engineers, and mathematics students, other tools were necessary to calculate trigonometric values. Greek mathematicians constructed the first trigonometric tables, and they have been used for over two thousand years to assist in the approximation and estimation of trigonometric functions. Although you will use a calculator for the vast majority of this book, this section of problems demonstrates that you can still derive surprisingly accurate estimations using historical methods.

Here's the real reason trig tables are included in the book: Because some teachers still teach them, even though technology gives you faster, more accurate answers.

In Appendix A, angles between 0° (0 radians) and 90° ($\pi/2$ radians) are listed in quarter degree increments. The first column represents an angle in degrees; the second column represents the angle in radians; and then the next six columns, in order, list the cosine, sine, tangent, secant, cosecant, and cotangent values of that angle. Locate the row of the table that begins with 88°.

Degrees	Radians	Cosine	Sine	Tangent	Secant	Cosecant	Cotangent
88.00	1.53589	0.0349	0.99939	28.63625	28.65371	1.00061	0.03492

According to this row of the table, sin 88° ≈ 0.99939. All values in the table are accurate to five decimal places.

2.30 Use the trigonometric table in Appendix A to approximate the measure of θ in radians, given tan θ = 0.68728.

In order to approximate the angle θ that has a tangent value of 0.68728, consult the tangent column in the table and look for the value 0.68728. Notice that the tangent values increase as the angle increases, so the larger the angle, the larger the tangent value.

The row corresponding to 34.50° (0.60214 radians) contains the precise tangent value for which you are searching: 0.68728.

Degrees	Radians	Cosine	Sine	Tangent	Secant	Cosecant	Cotangent
34.50	0.60214	0.82413	0.56641	0.68728	1.21341	1.76552	1.45501

Therefore, you conclude that $\theta \approx 0.60214$ radians if tan θ = 0.68728. Remember that this trigonometric table is accurate to five decimal places, but nearly all of the values in the table are rounded. Thus, you cannot conclude that θ is *exactly* equal to 0.60214 radians.

2.31 Determine whether each of the trigonometric functions listed in Appendix A increase or decrease as the angles increase from 0° to 90°.

Different trigonometric functions exhibit different behavior on different intervals, so this problem specifically identifies the interval described in Appendix A, 0° to 90°. In Problems 2.32–2.36, you use the table in Appendix A to approximate unlisted trigonometric values; when you do, it is important to know whether successive rows in the table increase or decrease for the specific trigonometric function you are investigating.

Three of the trigonometric functions increase between 0° and 90°: sine, tangent, and secant. The remaining three functions, termed the "cofunctions" because their names are derived by adding the prefix "co-" to the three functions already listed, decrease: cosine, cotangent, and cosecant.

Here's one way to remember which functions decrease: "If it's a CO-, down it goes!"

Note: Problems 2.32–2.33 explain how to approximate tan 45.11° using the trigonometric table in Appendix A.

2.32 Identify two consecutive rows in the table that bound the given angle, 45.11°. List the angles that bound 45.11° in the table and their corresponding tangent values.

You are given an angle measured in degrees, and each row of the table represents a degree measurement exactly 0.25° larger than the preceding row. Therefore, the rows that bound 45.11° are 45.00° and 45.25°; they are duplicated below.

Degrees	Radians	Cosine	Sine	Tangent	Secant	Cosecant	Cotangent
45.00	0.78540	0.70711	0.70711	1	1.41421	1.41421	1
45.25	0.78976	0.70401	0.71019	1.00876	1.42042	1.40808	0.99131

In other words, because 45.11 is between 45 and 45.25, its tangent value is between the tangent values of 45 and 45.25.

As Problem 2.31 explains, the tangent function increases as the angles increase between 0° and 90°. According to the shaded values in the previous table, tan 45° = 1 and tan 45.25° ≈ 1.00876. Thus, 1 < tan 45.11° < 1.00876.

Note: Problems 2.32–2.33 explain how to approximate tan 45.11° using the trigonometric table in Appendix A.

2.33 Construct a proportion to interpolate tan 45.11°, based on the bounding values you identified in Problem 2.32. Round the estimate to the nearest hundred thousandth.

According to Problem 2.32, 45.11° falls between the 45.00° and 45.25° rows in the trigonometric table, so its value, x = tan 45.11°, falls between 1 and 1.00876. It may be helpful to imagine a new row of the table that represents 45.11°, as illustrated by the shaded row in the table below.

Degrees	Radians	Cosine	Sine	Tangent	Secant	Cosecant	Cotangent
45.00	0.78540	0.70711	0.70711	1	1.41421	1.41421	1
45.11	—	—	—	x	—	—	—
45.25	0.78976	0.70401	0.71019	1.00876	1.42042	1.40808	0.99131

Estimating trigonometric values based upon values that bound them is called "interpolating." You assume, for the sake of approximation, that the angles and their corresponding trigonometric values have a linear relationship on very small intervals. While this is not true, it does provide a reasonably rough estimate.

Trig graphs are not straight lines, but any graph sort of looks like a straight line if you look at a very tiny piece of it. This is a concept called "local linearity."

Consider the three equations below, which summarize the important information from the trigonometric table. The left sides of the equations contain the three angles in the problem (45°, 45.11°, and 45.25°), and the right sides contain the corresponding tangent values for those angles (1, the unknown value x, and 1.00876).

$$\tan 45.00° \quad = \quad 1.00000$$
$$\mathbf{tan\ 45.11°} \quad = \quad x$$
$$\tan 45.25° \quad = \quad 1.00876$$

In the proportion, you will compare differences: How different are two angles when compared to their tangent values? Begin by comparing the values given in the trig table. When an angle increases from 45.00° to 45.25° (for a total increase of 45.25° − 45.00° = 0.25°), the tangent value increases from 1.0 to 1.00876 (for a total increase of 1.00876 − 1 = 0.00876).

$$0.25 \left\{ \begin{array}{lcl} \tan 45.00° & = & 1.00000 \\ \mathbf{tan\ 45.11°} & = & x \\ \tan 45.25° & = & 1.00876 \end{array} \right\} 0.00876$$

Now conduct the same comparison for the new angle, 45.11°, and the bounding angle below it, 45.00°. In other words, note that an angle increase of $45.11° - 45.00° = 0.11°$ corresponds to a tangent value increase of $x - 1$.

$$0.25 \left\{ 0.11 \left\{ \begin{array}{lcl} \tan 45.00° & = & 1.00000 \\ \mathbf{\tan 45.11°} & = & x \\ \tan 45.25° & = & 1.00876 \end{array} \right\} x - 1 \right\} 0.00876$$

You are left with four different values, two that compare angle differences and two that compare tangent value differences. Substitute them into the following proportion:

$$\frac{\text{small angle difference}}{\text{large angle difference}} = \frac{\text{small trigonometric value difference}}{\text{large trigonometric value difference}}$$

$$\frac{0.11}{0.25} = \frac{x - 1}{0.00876}$$

Cross-multiply and solve for x.

$$(0.11)(0.00876) = (0.25)(x - 1)$$
$$0.0009636 = (0.25)(x - 1)$$
$$\frac{0.0009636}{0.25} = x - 1$$
$$0.0038544 = x - 1$$
$$0.0038544 + 1 = x$$
$$1.0038544 = x$$

> Multiply the top of the left fraction by the bottom of the right fraction and vice versa. Set those two products equal.

The problem instructs you to round this answer to the nearest hundred thousandth, so you should round to the fifth decimal place: $x \approx 1.00385$. Therefore, $\tan 45.11° \approx 1.00385$.

The actual value of $\tan 45.11°$ is 1.0038471150116471310256758144812..., so the estimate is very accurate for the first four decimal places and less so after that.

Note: Problems 2.34–2.35 explain how to approximate $\cos \dfrac{\pi}{11}$ using the trigonometric table in Appendix A.

2.34 List the angles that bound $\cos \dfrac{\pi}{11}$ in the table, as well as their corresponding cosine values.

Although the angles in the table are listed in increments of 0.25°, the radian equivalents of those angles are listed in the second column. Use a calculator to compute the decimal equivalent of the angle.

$$\frac{\pi}{11} \approx 0.285599$$

Now identify the two rows within the trigonometric table that bound 0.285599 radians. These rows are copied on the next page.

Degrees	Radians	Cosine	Sine	Tangent	Secant	Cosecant	Cotangent
16.25	0.28362	0.96005	0.27983	0.29147	1.04161	3.57361	3.43084
16.50	0.28798	0.95882	0.28402	0.29621	1.04295	3.52094	3.37594

Note that the angle 0.285599 falls between the rows containing radian values of 0.28362 and 0.28798. Therefore, the cos 0.285599 will fall between 0.95882 and 0.96005.

Degrees	Radians	Cosine	Sine	Tangent	Secant	Cosecant	Cotangent
16.25	0.28362	0.96005	0.27983	0.29147	1.04161	3.57361	3.43084
—	0.285599	x	—	—	—	—	—
16.50	0.28798	0.95882	0.28402	0.29621	1.04295	3.52094	3.37594

Note: Problems 2.34–2.35 explain how to approximate $\cos \dfrac{\pi}{11}$ *using the trigonometric table in Appendix A.*

2.35 Construct a proportion to interpolate $\cos \dfrac{\pi}{11}$ based on the bounding values you identified in Problem 2.34. Round the estimate to the nearest thousandth.

According to Problem 2.34, $\cos \pi/11$ lies somewhere within the interval bounded above by 0.96005 and below by 0.95882.

$$\cos 0.28362 = 0.96005$$
$$\mathbf{\cos 0.285599} = x$$
$$\cos 0.28798 = 0.95882$$

Apply the technique described in Problem 2.33, comparing angle differences to trigonometric value differences, as illustrated below.

$$0.00436 \left\{ 0.001979 \left\{ \begin{array}{l} \cos 0.28362 = 0.96005 \\ \mathbf{\cos 0.285599} = x \\ \cos 0.28798 = 0.95882 \end{array} \right. \left. \begin{array}{l} \\ \end{array} \right\} x - 0.96055 \right\} -0.00123$$

0.285599 − 0.28362

0.28798 − 0.28362

0.95882 − 0.96005

Make sure to subtract in the same order each time. In the computations above, values from the upper rows are always subtracted from values in rows below them. This leads to the negative value −0.00123 among the results, because the larger the angle in the table, the smaller its cosine value.

Substitute the four differences into the interpolation proportion.

$$\frac{\text{small angle difference}}{\text{large angle difference}} = \frac{\text{small trigonometric value difference}}{\text{large trigonometric value difference}}$$

$$\frac{0.001979}{0.00436} = \frac{x - 0.96055}{-0.00123}$$

Remember, "co" functions decrease in the table, so the "large trigonometric value difference" in the proportion is negative.

Cross-multiply and solve for x.

$$(0.001979)(-0.00123) = (0.00436)(x - 0.96055)$$

$$-0.00000243417 = (0.00436)(x - 0.96055)$$

$$-\frac{0.00000243417}{0.00436} = x - 0.96055$$

$$-0.0005582959 = x - 0.96055$$

$$-0.0005582959 + 0.96055 = x$$

$$0.95999 \approx x$$

You conclude that $\cos \pi/11 \approx 0.960$. The actual value is 0.9594929736, so the estimate is reasonably accurate.

2.36 Given $\sin \theta = 0.89835$, construct a proportion to approximate θ using the trigonometric table in Appendix A. Report the answer in degrees, rounded to the nearest thousandth.

In this problem, you are given a trigonometric value and are asked to identify the corresponding angle. Begin by identifying the two consecutive rows in the table with sine values that bound 0.89835 above and below.

Degrees	Radians	Cosine	Sine	Tangent	Secant	Cosecant	Cotangent
63.75	1.11265	0.44229	0.89687	2.02780	2.26097	1.11499	0.49315
θ	—	—	0.89835	—	—	—	—
64.00	1.11701	0.43837	0.89879	2.05030	2.28117	1.1126	0.48773

Apply the procedure modeled in Problems 2.33 and 2.35, calculating the proportional differences between angles and their trigonometric values.

$$0.25 \left\{ \theta - 63.75 \left\{ \begin{array}{lcl} \sin 63.75 & = & 0.89687 \\ \mathbf{\sin \theta} & = & \mathbf{0.89835} \\ \sin 64.00 & = & 0.89879 \end{array} \right\} 0.00148 \right\} 0.00192$$

$$64.00 - 63.75 \qquad\qquad\qquad 0.89879 - 0.89687$$

Construct a proportion that relates the differences calculated above.

$$\frac{\text{small angle difference}}{\text{large angle difference}} = \frac{\text{small trigonometric value difference}}{\text{large trigonometric value difference}}$$

$$\frac{\theta - 63.75}{0.25} = \frac{0.00148}{0.00192}$$

Solve the proportion for θ.

$$(\theta - 63.75)(0.00192) = (0.25)(0.00148)$$
$$(\theta - 63.75)(0.00192) = 0.00037$$
$$\theta - 63.75 = \frac{0.00037}{0.00192}$$
$$\theta - 63.75 = 0.1927083333$$
$$\theta = 0.1927083333 + 63.75$$
$$\theta = 63.94270833$$

You conclude that $\theta \approx 63.943°$.

The actual angle (accurate to seven decimal places) is 63.9420226°.

Calculator-Generated Trigonometric Values
So long, trig tables

2.37 Use a scientific or graphing calculator to compute the following value, accurate to eight decimal places: sin 294.59°.

Different calculators require different rules of syntax. On a basic scientific calculator, you type the angle first and then press the sine button. Graphing calculators usually require you to type the expression in order as you would read it, using parentheses to surround the angle: "sin(294.59)." Make sure that your calculator is in degrees mode; if you do not know how to do this, consult your calculator owner's manual. If you lost your manual, it is most likely available to download on the manufacturer's website.

$$\sin 294.59° \approx -0.90930875$$

2.38 Use a scientific or graphing calculator to compute the following value, accurate to eight decimal places: $\tan \frac{14\pi}{9}$.

Ensure that your calculator is in radians mode and use one of the following sequences of keystrokes, depending upon the type of calculator you have:

- Scientific calculator: Type "$14 \times \pi$," divide by 9, and then press the tangent button
- Graphing calculator: Type "tan(14π / 9)"

The calculator should report a tangent value of −5.67128182.

2.39 Use a scientific or graphing calculator to compute the following value in radians, accurate to eight decimal places: $\csc \dfrac{10\pi}{13}$.

Recall that cosecant is the reciprocal of sine.

$$\csc \frac{10\pi}{13} = \left(\sin \frac{10\pi}{13} \right)^{-1}$$

> You can't just reverse the numerator and denominator of the angle. In other words, $\csc \dfrac{10\pi}{13} \neq \sin \dfrac{13}{10\pi}$.

Use one of the following keystroke sequences to compute this value on your calculator, and make sure that your calculator is in radians mode:

- Scientific calculator: Type "10 * π", divide by 13, press the sine button, and then press the reciprocal button, which is usually labeled "$1/x$"

- Scientific calculator (Alternate technique): Type "10 * π", divide by 13, press the sine button, press the exponent button (usually labeled "x^y" or "^"), type 1, and then make 1 negative by pressing the "+/–" button

- Graphing calculator: Type "(sin(10π/13))^–1"

> Raising a value to the –1 power produces its reciprocal.

The value of csc 10π/13, accurate to eight decimal places, is 1.50801664.

2.40 Given $\sin \theta = 0.3304651$, calculate θ in radians, accurate to eight decimal places.

This problem lists a trigonometric value, and you are asked to identify the corresponding angle. Whenever the desired result is an angle, you need to apply the *inverse* trigonometric function on your calculator. The inverse sine function is written "arcsin" or "sin^{-1}." Many calculators list inverse functions on the same button as the functions themselves, and to access them you must press a "shift," "alt," or "2nd" button—followed by the function button—to reach them.

Use one of the following keystroke sequences to calculate θ, ensuring that your calculator is in radians mode:

- Scientific calculator: Type "0.3304651," and press the "sin^{-1}" button

- Graphing calculator: Type "sin^{-1}(0.3304651)"

Given $\sin \theta = 0.3304651$, you conclude that $\theta \approx 0.33679632$.

> Small angles have sine and tangent values that are almost equal to the angles, themselves, in radians. Check out the margin note on the first page of Appendix A for more information.

2.41 Given $\sec \theta = -1.81390613$, calculate θ in degrees, accurate to the thousandths place.

Recall that secant and cosine are reciprocal functions. Therefore, if $\sec \theta = -1.81390613$, you can conclude $\cos \theta = -1/1.81390613$. Ensure that your calculator is in degrees mode and complete one of the following keystroke sequences.

- Scientific calculator: Type "1.81390613," make the value negative by pressing the "+/−" button, press the reciprocal button (usually labeled "1/x"), and then press the "cos⁻¹" button

- Scientific calculator (Alternate technique): Press "1", make the value negative by pressing the "+/−" button, divide by 1.81390613, and then press the "cos⁻¹" button

- Graphing calculator: Type "cos⁻¹(−1/1.81390613)"

Given $\sec\theta = -1.81390613$, you conclude that $\theta \approx 123.456$.

Note: Problems 2.42–2.43 refer to the diagram below.

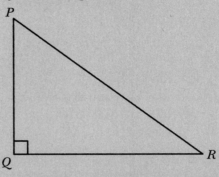

2.42 Given $m\angle P = 71°$ and $QR = 9$, calculate PR and round the answer to the nearest thousandth.

You are given the measure of one acute angle of the right triangle (P) and the length of the opposite leg (QR). You are asked to calculate the length of the hypotenuse (PR). To solve this problem, you must first identify the trigonometric ratio that relates an angle measurement, the opposite leg, and the hypotenuse. Recall that sine is equal to the ratio of the lengths of the opposite leg and the hypotenuse.

$$\sin\theta = \frac{\text{length of leg opposite } \angle\theta}{\text{length of hypotenuse}}$$

$$\sin P = \frac{QR}{PR}$$

Substitute the given information into the equation.

$$\sin 71° = \frac{9}{PR}$$

Multiply both sides of the equation by PR to eliminate the fraction. Then, solve for PR.

$$(PR)(\sin 71°) = \left(\frac{\cancel{PR}}{1}\right)\left(\frac{9}{\cancel{PR}}\right)$$

$$(PR)(\sin 71°) = 9$$

$$PR = \frac{9}{\sin 71°}$$

Use a calculator to divide 9 by sin 71°. Remember to set the calculator to degrees mode.

$$PR \approx 9.519$$

Note: Problems 2.42–2.43 refer to the diagram in Problem 2.42.

2.43 Given $PR = 13.28$ and $PQ = 4.6$, calculate $m\angle P$ in radians and round the answer to the nearest thousandth.

In this problem, you are asked to calculate the measure of an angle given the lengths of its adjacent leg and the hypotenuse. Recall that cosine ratio relates these sides in a right triangle.

$$\cos P = \frac{\text{length of leg adjacent to } \angle P}{\text{length of hypotenuse}}$$

$$\cos P = \frac{PQ}{PR}$$

$$\cos P = \frac{4.6}{13.28}$$

$$\cos P = 0.3463855422 \leftarrow$$

Write out all the decimal places your calculator can display. To get the most accurate answer possible, don't round to the thousandths place until the very end.

To identify the angle that has a cosine of 0.3463855422, use the inverse cosine function on your calculator. Make sure the calculator is set to radians mode.

$$P = \cos^{-1} 0.3463855422$$

$$P \approx 1.217$$

You conclude that $m\angle P \approx 1.217$ radians.

2.44 A golfer standing at point A (in the figure below), wishing to estimate the distance to the flag at point P, walks a distance of 15 yards due north to point R and determines that $m\angle R = 79°$. Apply a trigonometric ratio to compute AP, accurate to the nearest yard.

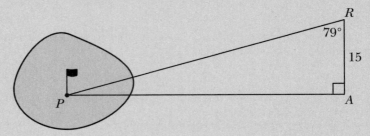

The golfer is attempting to calculate the length of the leg opposite acute angle R given the measure of R and the length of its adjacent leg. Apply the tangent function.

$$\tan R = \frac{\text{length of leg opposite } R}{\text{length of leg adjacent to } R}$$

$$\tan R = \frac{AP}{AR}$$

$$\tan 79° = \frac{AP}{15}$$

Multiply both sides of the equation by 15 (to eliminate the fraction) and solve for AP. Ensure that your calculator is in degrees mode, because angle R is measured in degrees.

$$15(\tan 79°) = \left(\frac{\cancel{15}}{1}\right)\left(\frac{AP}{\cancel{15}}\right)$$

$$15(\tan 79°) = AP$$

$$77.16831024 \approx AP$$

You conclude that the distance between the golfer's original position (A) and the flag (P) is approximately 77 yards.

2.45 A high dive platform that towers 100 feet above the ground is secured by 135-foot-long guy-wires, as illustrated below. Calculate the measure of θ, the angle formed by a guy-wire and the ground, in radians, and report the answer accurate to three decimal places.

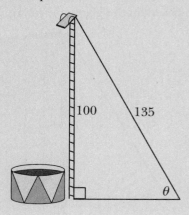

In order to calculate the measure of an acute angle based on the lengths of the opposite leg and the hypotenuse, you should apply the sine function. Ensure that your calculator is in radians mode, as directed by the problem.

$$\sin \theta = \frac{100}{135}$$
$$\sin \theta = 0.7407407407$$
$$\theta = \sin^{-1}(0.7407407407)$$
$$\theta \approx 0.8341723248$$

You conclude that the guy-wire makes an angle with the ground that measures $\theta \approx 0.834$ radians.

2.46 The angle of elevation to the top of a building from a point on the ground 50 feet away is 81°. Calculate the height of the building and round your answer to the nearest foot.

Construct a diagram based on the given information.

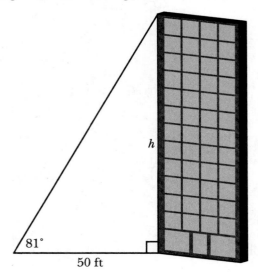

Apply the tangent ratio and solve for h.

$$\tan 81° = \frac{h}{50}$$

$$\frac{\tan 81°}{1} = \frac{h}{50}$$

$$50 \cdot \tan 81° = h$$

$$315.6875757 \approx h$$

$$316 \approx h$$

The height of the building is approximately 316 feet.

Chapter 3
THE UNIT CIRCLE

A little circle, a lot of memorizing

In Chapter 2, you explore trigonometric ratios (via tables and calculator technology), but no rationale for the values of the functions is provided. Why do sine and cosine values range between –1 and 1, whereas the tangent and cotangent functions may grow infinitely large, in both the positive and negative directions? These are not questions you are yet equipped to answer, but this chapter provides a significant step in the right direction.

The unit circle is simply a circle, centered at the origin of the coordinate plane, with a radius equal to 1 unit in length. By graphing angles in standard position and examining the points at which their terminal sides intersect the unit circle, you generate the cosine and sine values of those angles (as well as the secant, cosecant, tangent, and cotangent values, which are based upon sine and cosine).

Now that you have worked through Chapter 2, you have some experience with sine and cosine functions. However, tables and calculators don't tell the whole story. In this chapter, you work with unit circles, which are circles that have a center of $(x,y) = (0,0)$ and a radius of $r = 1$. Believe it or not, these simple circles are the birthplace of sine and cosine—trig functions get their values from the coordinates of points on the circle. This chapter leads you step-by-step through the process, so take your time and soak it in.

By the end of the chapter, you will know the sine and cosine values of the 16 most commonly used angles in trigonometry. Forget tables—you'll know these values off the top of your head with nary a calculation. Sure, there's some memorization involved, but it's worth it in the same way memorizing multiplication tables was worth it back in grade school.

45°–45°–90° Triangles
(H = √2 · L)

> *Note: Problems 3.1–3.3 refer to the diagram below, featuring right triangle XYZ, in which XY = YZ.*

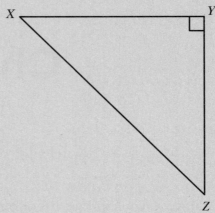

3.1 Given $m\angle X = m\angle Z$ in the triangle, explain why both acute angles measure 45°.

All triangles contain three angles whose measures have a sum of $180° = \pi$ radians. Apply this geometric theorem to the given triangle by adding the measures of angles X, Y, and Z, and setting the sum equal to 180°.

$$m\angle X + m\angle Y + m\angle Z = 180°$$

In this diagram, angle Y is a right angle, so $m\angle Y = 90°$. Furthermore, you are given $m\angle X = m\angle Z$. Therefore, you can substitute 90° for $m\angle Y$ and $m\angle Z$ for $m\angle X$ in the angle sum equation above.

$$m\angle X + m\angle Y + m\angle Z = 180°$$
$$m\angle Z + 90° + m\angle Z = 180°$$

Note that $m\angle Z + m\angle Z = 2(m\angle Z)$.

$$2m\angle Z + 90° = 180°$$

Solve the equation for $m\angle Z$.

$$2m\angle Z = 180° - 90°$$
$$2m\angle Z = 90°$$
$$m\angle Z = \frac{90°}{2}$$
$$m\angle Z = 45°$$

You conclude that $m\angle X = m\angle Z = 45°$. Therefore, triangle XYZ is a 45°–45°–90° triangle, a geometric classification based upon the measures of its angles.

> So once you subtract the right angle from 180°, you are left with 90°. Because the two remaining angles have the same measure, you have to divide the leftover 90° by 2. That gives you two angles that each measure 45°.

Note: Problems 3.1–3.3 refer to the diagram in Problem 3.1, in which XY = YZ.

3.2 Let l represent the length of a leg of triangle XYZ. Apply the Pythagorean theorem to calculate XZ in terms of l.

According to the Pythagorean theorem, given a right triangle with legs of length a and b and a hypotenuse of length c, you can conclude that $a^2 + b^2 = c^2$. In this problem, the legs of the triangle are $\overline{XY}$ and $\overline{YZ}$ and the hypotenuse is $\overline{XZ}$.

$$a^2 + b^2 = c^2$$
$$(XY)^2 + (YZ)^2 = (XZ)^2$$

This problem states that each leg has length l. Thus, $l = XY = YZ$.

$$(l)^2 + (l)^2 = (XZ)^2$$
$$l^2 + l^2 = (XZ)^2$$
$$2l^2 = (XZ)^2$$

To calculate the length of XZ in terms of l, solve the equation for XZ.

$$\sqrt{2l^2} = \sqrt{(XZ)^2}$$
$$\sqrt{2} \cdot \sqrt{l^2} = XZ$$
$$\sqrt{2} \cdot l = XZ$$

Therefore, the length of the hypotenuse is $\sqrt{2}$ times the length of the leg, l.

> This may feel familiar. It's a geometry theorem that is often written $H = \sqrt{2} \cdot L$, where H is the hypotenuse length and L is the leg length.

Note: Problems 3.1–3.3 refer to the diagram in Problem 3.1, in which XY = YZ.

3.3 Given $XZ = 7$, calculate the lengths of the legs.

According to Problem 3.2, the hypotenuse of a 45°–45°–90° triangle is equal to $\sqrt{2}$ times the length of a leg.

$$(\text{length of the hypotenuse}) = \sqrt{2} \cdot (\text{length of a leg})$$

Let l represent the length of a leg of triangle XYZ; note that $XZ = 7$ is the length of the hypotenuse.

$$XZ = \sqrt{2} \cdot (l)$$
$$7 = \sqrt{2} \cdot l$$
$$\frac{7}{\sqrt{2}} = l$$

> Multiply the numerator and denominator by the offending square root: $\sqrt{2}$.

The length of the leg, l, is a fraction with an irrational denominator. Traditionally, radicals within a denominator are eliminated in a process called "rationalizing the denominator."

$$l = \frac{7}{\sqrt{2}}\left(\frac{\sqrt{2}}{\sqrt{2}}\right) = \frac{7\sqrt{2}}{\sqrt{2}\sqrt{2}} = \frac{7\sqrt{2}}{\sqrt{(2\cdot2)}} = \frac{7\sqrt{2}}{\sqrt{4}} = \frac{7\sqrt{2}}{2}$$

You conclude that the lengths of the legs are $\dfrac{7\sqrt{2}}{2}$ $\left(\text{or } \dfrac{7}{2}\sqrt{2}\right)$.

3.4 Calculate the length of the legs of an isosceles right triangle if its hypotenuse has length $5\sqrt{6}$.

> In other words, a 45°–45°–90° triangle

Recall that the hypotenuse of an isosceles right triangle is $\sqrt{2}$ times the length of a leg. Let l represent the length of the leg and $h = 5\sqrt{6}$ represent the length of the hypotenuse in the following equation.

$$h = \sqrt{2} \cdot l$$
$$5\sqrt{6} = \sqrt{2} \cdot l$$

Solve for l.

$$\frac{5\sqrt{6}}{\sqrt{2}} = \frac{\sqrt{2} \cdot l}{\sqrt{2}}$$
$$5 \cdot \frac{\sqrt{6}}{\sqrt{2}} = \frac{\sqrt{2}}{\sqrt{2}} \cdot l$$
$$5 \cdot \sqrt{\frac{6}{2}} = \sqrt{\frac{2}{2}} \cdot l$$
$$5\sqrt{3} = \sqrt{1} \cdot l$$
$$5\sqrt{3} = l$$

Both legs of the isosceles right triangle have length $5\sqrt{3}$.

Note: Problems 3.5–3.8 apply the properties of 45°–45°–90° triangles to the coordinate plane in order to generate coordinates on a unit circle.

3.5   Draw a unit circle on the coordinate plane with central angle θ in standard

position, given $m\angle\theta = \dfrac{\pi}{4}$.

A unit circle has radius 1 and is centered at the origin of the coordinate plane. Recall that an angle in standard position has an initial side on the positive x-axis and a vertex that lies on the origin, as explained in Problems 1.3–1.5. You wish to convert the radian angle measure $\pi/4$ into degrees to help visualize where its terminal side lies in the coordinate plane.

$$\frac{\pi}{4}\cdot\frac{180}{\pi} = \frac{\not\pi}{4}\cdot\frac{180}{\not\pi} = \frac{180}{4} = \frac{180\div4}{4\div4} = \frac{45}{1} = 45°$$

> See Problem 1.27 if you can't remember how to convert from radians to degrees. Also, check out Problem 1.17, which helps you visualize common angles in radians.

Draw the unit circle and θ on the same coordinate plane, as illustrated in the following diagram. Notice that the terminal side of angle θ is also a radius of the circle, and recall that a unit circle has radius $r = 1$.

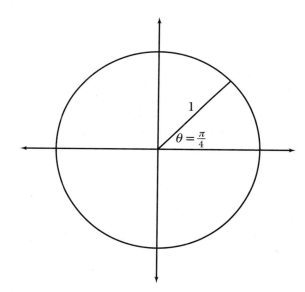

Because you are constructing a 45° angle in the coordinate plane, you can apply the properties of 45°–45°–90° triangles (specifically that the hypotenuse of such a triangle is $\sqrt{2}$ times the length of a leg) in Problems 3.6–3.8.

Note: Problems 3.5–3.8 apply the properties of 45°–45°–90° triangles to the coordinate plane in order to generate coordinates on a unit circle.

3.6 Consider the diagram you created in Problem 3.5, a unit circle with central angle θ that measures $\dfrac{\pi}{4}$ radians. Draw a right triangle in the coordinate plane whose legs represent the horizontal and vertical distances between the origin and the point at which the terminal side of θ intersects the unit circle.

The following diagram specifically examines the first quadrant of the drawing generated in Problem 3.5; the origin is labeled point *A*. Notice that *C* is the point at which the terminal side of θ intersects the unit circle. A vertical segment extends from *C* to the *x*-axis, intersecting it at point *B*.

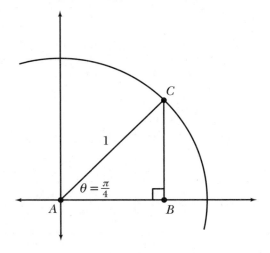

A $\dfrac{\pi}{4} - \dfrac{\pi}{4} - \dfrac{\pi}{2}$ triangle is the same thing as a 45°–45°–90° triangle, but it's written in radians instead of degrees.

Because the *x*-axis is horizontal and $\overline{BC}$ is vertical, they form right angle *ABC*. Therefore, triangle *ABC* is a right triangle, with a right angle measuring $\pi/2$ and two acute angles, each measuring $\pi/4$.

Note that *AB* represents the horizontal distance between the origin and point *C*; similarly, *BC* represents the vertical distance between the origin and point *C*.

Note: Problems 3.5–3.8 apply the properties of 45°–45°–90° triangles to the coordinate plane in order to generate coordinates on a unit circle.

3.7 Calculate the lengths of the sides of the right triangle you identified in Problem 3.6.

> Remember, 45°–45°–90° triangles are automatically isosceles.

The hypotenuse of right triangle *ABC* below is also the radius of the unit circle, so it has length *AC* = 1. Let *l* represent the length of the legs.

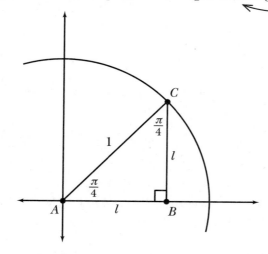

As Problems 3.2–3.3 explain, right triangles with acute angles measuring $\pi/4$ radians (or 45°) have hypotenuses that are $\sqrt{2}$ times the lengths of their legs.

$$h = \sqrt{2} \cdot l$$
$$AC = \sqrt{2} \cdot l$$
$$1 = \sqrt{2} \cdot l$$

Solve for *l*.

$$\frac{1}{\sqrt{2}} = \frac{\sqrt{2} \cdot l}{\sqrt{2}}$$
$$\frac{1}{\sqrt{2}} = \frac{\cancel{\sqrt{2}} \cdot l}{\cancel{\sqrt{2}}}$$
$$\frac{1}{\sqrt{2}} = l$$

Rationalize the denominator by multiplying both the numerator and denominator by $\sqrt{2}$.

$$\left(\frac{1}{\sqrt{2}}\right)\left(\frac{\sqrt{2}}{\sqrt{2}}\right) = l$$

$$\frac{\sqrt{2}}{\sqrt{2 \cdot 2}} = l$$

$$\frac{\sqrt{2}}{\sqrt{4}} = l$$

$$\frac{\sqrt{2}}{2} = l$$

You conclude that $AB = CB = \sqrt{2}/2$.

Note: Problems 3.5–3.8 apply the properties of 45°–45°–90° triangles to the coordinate plane in order to generate coordinates on a unit circle.

3.8   Identify the coordinates of the point at which the terminal side of θ intersects the unit circle.

The diagram below summarizes the conclusions drawn in Problems 3.5–3.7. Of specific interest are the lengths calculated in Problem 3.7, which represent the horizontal and vertical distances between the origin and point C.

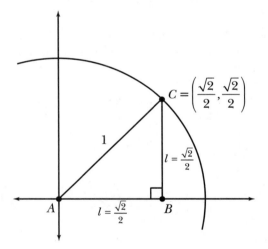

Because $AB = \sqrt{2}/2$, point C is $\sqrt{2}/2$ units right of the origin; therefore, the x-coordinate of point C is $\sqrt{2}/2$. Similarly, the y-coordinate of point C is $\sqrt{2}/2$ because $BC = \sqrt{2}/2$.

$$\text{Point } C = \left(\frac{\sqrt{2}}{2}, \frac{\sqrt{2}}{2}\right)$$

3.9   Identify the coordinates of point *P* in the diagram below.

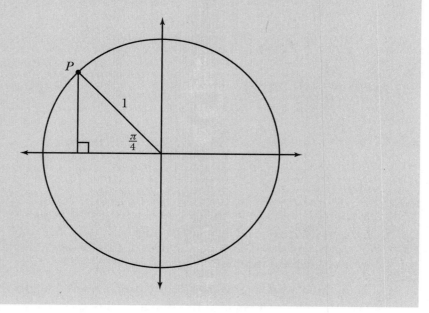

Like triangle *ABC* in Problems 3.6–3.8, this right triangle has two acute angles that measure $\pi/4$ and a hypotenuse of length 1 (because it is a radius of the unit circle). Therefore, the legs of the right triangle have the same lengths as the legs in triangle *ABC*: $\sqrt{2}/2$.

Note that the *x*-coordinate of point *P* must be negative, because *P* is in the second quadrant; you must travel *left* from the origin to reach *P*. Hence, the coordinates of *P* are $(-\sqrt{2}/2, \sqrt{2}/2)$.

> The signs of the coordinates tell you what direction to travel from the origin in order to reach the point:
> +x coordinate: go right
> −x coordinate: go left
> +y coordinate: go up
> −y coordinate: go down.

30°–60°–90° Triangles

$(S = \frac{1}{2}H, L = \sqrt{3} \cdot S)$

Note: Problems 3.10–3.12 refer to the diagram below.

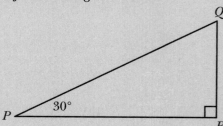

3.10   Given $m\angle P = 30°$, calculate $m\angle Q$.

The interior angles of a triangle have a sum of 180°.

$$m\angle P + m\angle Q + m\angle R = 180°$$

Note that $m\angle R = 90°$, because R is a right angle. Substitute the measures of the two known angles into the previous equation and solve for $m\angle Q$.

$$30° + m\angle Q + 90° = 180°$$
$$(30° + 90°) + m\angle Q = 180°$$
$$120° + m\angle Q = 180°$$
$$m\angle Q = 180° - 120°$$
$$m\angle Q = 60°$$

Note: Problems 3.10–3.12 refer to the diagram in Problem 3.10.

3.11 Let h represent the length of the hypotenuse (PQ), and calculate QR in terms of h.

According to Problem 3.10, the angles of triangle PQR have the following measures: 30°, 60°, and 90°. Such triangles are described as 30°–60°–90° triangles, and their sides (like the sides of 45°–45°–90° triangles) have lengths with well-defined relationships.

> In other words, if you know the length of one side of a 30°-60°-90° triangle, you can quickly calculate the lengths of the other two sides without the use of trig functions.

Recall from geometry that the relative sizes of interior angles relate to the lengths of the sides opposite those angles. More specifically, smaller angles within a figure have shorter opposite sides, and vice versa. In this triangle, $(m\angle P) < (m\angle Q)$, so the side opposite P is shorter than the side opposite Q. The hypotenuse of the triangle is the longest side, as it is opposite the largest angle, right angle R.

> QR < PR

According to a geometric theorem, the shorter leg of a 30°–60°–90° triangle—the leg opposite the 30° angle—is exactly half the length of the hypotenuse. In other words, if s represents the length of the shorter leg and h represents the length of the hypotenuse, then the following equation is true.

$$s = \frac{1}{2}h$$

The shortest side of this triangle has length $s = QR$.

$$QR = \frac{1}{2}h$$

The answer $QR = h/2$, which expresses the product as a single fraction, is also correct.

Note: Problems 3.10–3.12 refer to the diagram in Problem 3.10.

3.12 Let h represent the length of the hypotenuse (PQ), and calculate PR in terms of h.

As Problem 3.11 explains, the length of the shorter leg (s) of a 30°–60°–90° triangle is exactly half the length of the hypotenuse (h). Let l represent the length of the remaining leg of the triangle. Because l is opposite a 60° angle, it is longer than s (which is opposite a 30° angle) but shorter than h (which is opposite a 90° angle). Additionally, l is exactly $\sqrt{3}$ times as long as s.

$$l = \sqrt{3} \cdot s$$

The length of the longer leg in this triangle is $l = PR$. According to Problem 3.11, $s = QR = (1/2) h$.

$$PR = \sqrt{3} \cdot \left(\frac{1}{2} h \right)$$

$$PR = \frac{\sqrt{3}}{2} h$$

You may also correctly conclude that $PR = \dfrac{\sqrt{3} \cdot h}{2}$.

Note: Problems 3.13–3.14 refer to the diagram below.

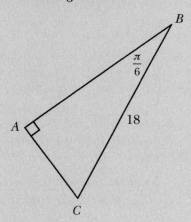

3.13 Calculate AC.

This diagram expresses the angle in radians. Recall that multiplying an angle by $180/\pi$ converts it from radians to degrees.

$$\frac{\pi}{6} \cdot \frac{180}{\pi} = \frac{180 \, \pi}{6 \, \pi} = \frac{180}{6} = 30°$$

Right triangle ABC is a 30°–60°–90° right triangle whose smallest acute angle is B, which measures $\pi/6 = 30°$. Therefore, the side opposite B is the shorter of the two legs in the triangle. According to Problem 3.11, the shorter leg s is half the length of the hypotenuse h.

$$s = \frac{1}{2}h$$

$$AC = \frac{1}{2}(BC)$$

Substitute the length of the hypotenuse into the equation, as indicated on the diagram: $BC = 18$.

$$AC = \frac{1}{2}(18)$$

$$AC = \frac{18}{2}$$

$$AC = 9$$

Note: Problems 3.13–3.14 refer to the diagram in Problem 3.14.

3.14 Calculate AB and verify your answer using a trigonometric function.

As Problem 3.13 explains, ABC is a 30°–60°–90° triangle; its hypotenuse has length $BC = 18$ and its short leg has length $AC = 9$. According to Problem 3.12, the longer leg l of a 30°–60°–90° triangle is $\sqrt{3}$ times as long as the shorter leg s.

$$l = \sqrt{3} \cdot s$$

$$AB = \sqrt{3} \cdot AC$$

$$AB = \sqrt{3} \cdot 9$$

$$AB = 9\sqrt{3}$$

The problem directs you to verify your answer using a trigonometric function. Consider Problems 2.44–2.45, in which you applied trigonometric functions to calculate lengths and angle measurements of right triangles given other lengths and angles.

In this problem, you are given $m\angle B = \pi/6$ and hypotenuse length $BC = 18$. Notice that AB represents the length of the side adjacent to angle B, so you should apply the cosine function.

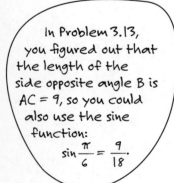

In Problem 3.13, you figured out that the length of the side opposite angle B is $AC = 9$, so you could also use the sine function: $\sin\frac{\pi}{6} = \frac{9}{18}$.

$$\cos B = \frac{\text{length of side adjacent to } \angle B}{\text{length of the hypotenuse}}$$

$$\cos\frac{\pi}{6} = \frac{AB}{BC}$$

$$\cos\frac{\pi}{6} = \frac{AB}{18}$$

Multiply both sides of the equation by 18 to solve for AB.

$$18\left(\cos\frac{\pi}{6}\right) = AB$$

Use a calculator to compute a decimal equivalent of AB, and notice that the decimal value is exactly equal to $9\sqrt{3}$, the length of AB that you just calculated.

$$18\left(\cos\frac{\pi}{6}\right) \approx 15.58845723$$

$$9\sqrt{3} \approx 15.58845723$$

In Problems 3.24–3.30, you learn the EXACT value of $\cos\frac{\pi}{6}$. If you come back to this problem, you can prove these values are equal without having to compare decimals.

Note: Problems 3.15–3.16 refer to the diagram below.

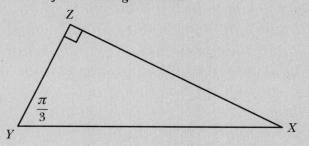

3.15 Given $YZ = 4\sqrt{3}$, calculate the lengths of the other sides of the triangle.

Note that $\pi/3 = 60°$, so XYZ is a 30°–60°–90° triangle with a hypotenuse of length $h = XY$, a short leg with length $s = YZ$, and a long leg with length $l = XZ$. Because XYZ is a 30°–60°–90° triangle, s is half the length of h.

$$s = \frac{1}{2}h$$

$$YZ = \frac{1}{2}(XY)$$

$$4\sqrt{3} = \frac{1}{2}(XY)$$

Solve for XY.

$$2\left(4\sqrt{3}\right) = \left(\frac{\cancel{2}}{1}\right)\left(\frac{1}{\cancel{2}}\right)(XY)$$

$$8\sqrt{3} = XY$$

Recall that the long leg of the triangle is $\sqrt{3}$ times as long as the short leg.

$$l = \sqrt{3} \cdot s$$
$$XZ = \sqrt{3} \cdot YZ$$
$$XZ = \sqrt{3}\left(4\sqrt{3}\right)$$
$$XZ = 4\sqrt{3 \cdot 3}$$
$$XZ = 4\sqrt{9}$$
$$XZ = 4 \cdot 3$$
$$XZ = 12$$

You conclude that the side lengths of triangle XYZ are $YZ = 4\sqrt{3}$, $XZ = 12$, and $XY = 8\sqrt{3}$.

Note: Problems 3.15–3.16 refer to the diagram in Problem 3.15.

3.16 Given $XZ = 6\sqrt{15}$, calculate the lengths of the other sides of the triangle.

As Problem 3.15 explains, triangle XYZ is a 30°–60°–90° right triangle with a hypotenuse of length $h = XY$, a short leg with length $s = YZ$, and a long leg with length $l = XZ$. Recall that the length of the long leg is $\sqrt{3}$ times the length of the short leg.

$$l = \sqrt{3} \cdot s$$
$$XZ = \sqrt{3} \cdot YZ$$

Substitute $XZ = 6\sqrt{15}$ into the equation and solve for YZ.

$$6\sqrt{15} = \sqrt{3} \cdot YZ$$
$$\frac{6\sqrt{15}}{\sqrt{3}} = \frac{\cancel{\sqrt{3}} \cdot YZ}{\cancel{\sqrt{3}}}$$
$$6\sqrt{\frac{15}{3}} = YZ$$
$$6\sqrt{5} = YZ$$

The short leg of triangle ABC is half the length of the hypotenuse.

$$s = \frac{1}{2}h$$
$$YZ = \frac{1}{2}(XY)$$
$$6\sqrt{5} = \frac{1}{2}(XY)$$

Solve for XY.

$$2\left(6\sqrt{5}\right) = \left(\frac{\cancel{2}}{1}\right)\left(\frac{1}{\cancel{2}}\right)(XY)$$
$$12\sqrt{5} = XY$$

The lengths of the sides of triangle XYZ are $YZ = 6\sqrt{5}$, $XZ = 6\sqrt{15}$, and $XY = 12\sqrt{5}$.

Note: Problems 3.17–3.18 refer to the diagram below.

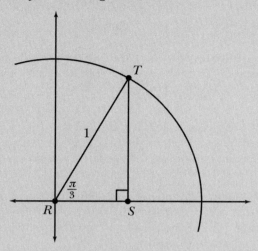

3.17 Given right triangle *RST* and an arc of the unit circle with radius *RT* = 1, calculate *RS* and *RT*, the lengths of the legs of the right triangle.

Like triangle *XYZ* in Problems 3.15–3.16, *RST* is a 30°–60°–90° triangle. Its hypotenuse has length *RT* = 1, the long leg has length $l = ST$, and the short leg has length $s = RS$.

> The longer leg is opposite the larger angle:
> $m\angle R = \dfrac{\pi}{3} = 60°$ and
> $m\angle T = \dfrac{\pi}{6} = 30°.$

The short leg of a 30°–60°–90° triangle is half the length of the hypotenuse.

$$s = \frac{1}{2}h$$

$$RS = \frac{1}{2}(RT)$$

$$RS = \frac{1}{2}(1)$$

$$RS = \frac{1}{2}$$

The long leg of a 30°–60°–90° triangle is $\sqrt{3}$ times the length of the short leg.

$$l = \sqrt{3}\cdot s$$

$$ST = \sqrt{3}\,(RS)$$

$$ST = \sqrt{3}\left(\frac{1}{2}\right)$$

$$ST = \frac{\sqrt{3}}{1}\left(\frac{1}{2}\right)$$

$$ST = \frac{\sqrt{3}}{2}$$

Note: Problems 3.17–3.18 refer to the diagram in Problem 3.17.

3.18 Identify the coordinates of point *T*.

The *x*- and *y*-coordinates of a point are, respectively, the signed horizontal and vertical distances from the origin to that point. Note that these are *signed* distances—the signs of the coordinates indicate direction on the coordinate plane.

Points with positive *x*-coordinates are right of the origin, and points with negative *x*-coordinates are left of the origin. Points with positive *y*-coordinates are above the origin, and points with negative *y*-coordinates are below the origin.

According to Problem 3.17, the horizontal and vertical distances between the origin and point *T* are, respectively, $RS = 1/2$ and $ST = \sqrt{3}/2$. Point *T* lies in the first quadrant, which is both right of and above the origin, so the *x*- and *y*-coordinates of *T* are both positive.

$$T = (+RS, +ST) = \left(\frac{1}{2}, \frac{\sqrt{3}}{2} \right)$$

3.19 Given right triangle *XYZ* and an arc of the unit circle with radius *XZ* = 1, as illustrated below, identify the coordinates of point *Z*.

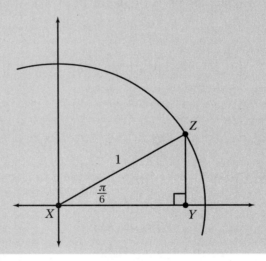

Like Problem 3.18, this problem explores a 30°–60°–90° (or π/6–π/3–π/2 radian) triangle with a hypotenuse of length 1. The legs of *XYZ* have the same lengths as the legs of triangle *RST* in Problem 3.18—the short leg has length $YZ = 1/2$ and the long leg has length $XY = \sqrt{3}/2$. Note that, in triangle *XYZ*, the long leg represents the horizontal distance to *Z*, and the short leg represents the vertical distance. Point *Z* lies in the first quadrant, so its *x*- and *y*-coordinates are positive.

This is point *T* from Problem 3.18, with the *x*- and *y*-coordinates reversed.

$$Z = (+XY, +YZ) = \left(\frac{\sqrt{3}}{2}, \frac{1}{2} \right)$$

3.20 Given an arc of the unit circle with radius 1 and a point *M* on the circle, as illustrated below, identify the coordinates of *M*.

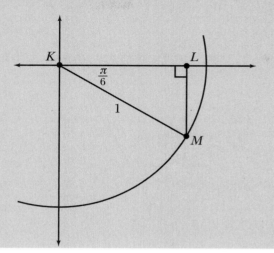

The triangle in this diagram is a copy of the triangle in Problem 3.19, reflected across the *x*-axis from the first to the fourth quadrant. All $\pi/6-\pi/3-\pi/2$ radian triangles with a hypotenuse of 1 have a long leg that measures $\sqrt{3}/2$ and a short leg that measures $1/2$.

Furthermore, if the central angle is $\pi/6$, the length of the long leg is always equal to the *x*-coordinate (and the short leg's length is the *y*-coordinate). Note that *M* lies in the fourth quadrant, right of the origin (so its *x*-coordinate is positive) and below the origin (so its *y*-coordinate is negative).

$$M = (+KL, -LM) = \left(\frac{\sqrt{3}}{2}, -\frac{1}{2}\right)$$

3.21 Given an arc of the unit circle with radius 1 and a point *N* on the circle, as illustrated below, identify the coordinates of *N*.

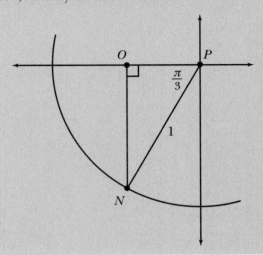

This triangle is a copy of triangle *RST* (from Problems 3.17–3.18), a 30°–60°–90° (or π/6–π/3–π/2 radian) triangle with a central angle of π/3 and a hypotenuse of 1. Therefore, its short leg represents the horizontal distance between the origin and *N*; the long leg represents the vertical distance. Because *N* lies in the third quadrant, which is left of and below the origin, both the *x*- and *y*-coordinates are negative.

$$N = (-OP, -NO) = \left(-\frac{1}{2}, -\frac{\sqrt{3}}{2} \right)$$

Cosine and Sine in the First Quadrant
As easy as $\sqrt{1}, \sqrt{2}, \sqrt{3}$

3.22 Describe the relationship between the coordinates of points on the unit circle and the angles in standard position whose terminal sides intersect those points. Justify your answer using the solution to Problem 3.19.

The *x*-coordinate of the intersection point is the cosine of the angle, and the *y*-coordinate is the sine of the angle. For example, consider the following diagram, the solution to Problem 3.19.

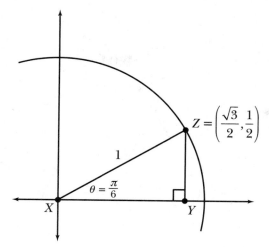

The angle $\theta = \pi/6$ is in standard position, and its terminal side intersects the unit circle at point $Z = \left(\sqrt{3}/2, 1/2\right)$. Therefore, the length of the side adjacent to θ is $XY = \sqrt{3}/2$, and the length of the side opposite θ is $YZ = 1/2$. The hypotenuse of the triangle is $XZ = 1$. Apply trigonometric ratios to calculate $\cos\theta$ and $\sin\theta$.

$$\cos\theta = \frac{\text{length of side adjacent to } \theta}{\text{length of hypotenuse}}$$

$$= \frac{XY}{XZ}$$

$$= \frac{\sqrt{3}/2}{1}$$

$$= \frac{\sqrt{3}}{2}$$

$$\sin\theta = \frac{\text{length of side opposite } \theta}{\text{length of hypotenuse}}$$

$$= \frac{YZ}{XZ}$$

$$= \frac{1/2}{1}$$

$$= \frac{1}{2}$$

The cosine is the adjacent side (which is always the horizontal side for an angle in standard position) divided by 1, so it's basically equal to the adjacent side, the x-coordinate. (Anything divided by 1 is equal to itself.) Same goes for the sine, which equals the length of the vertical side, the y-coordinate.

Notice that $\cos\theta$ is equal to the x-coordinate of point Z, and $\sin\theta$ is equal to the y-coordinate: $\cos\pi/6 = \sqrt{3}/2$ and $\sin\pi/6 = 1/2$.

3.23 Draw a graph in the first quadrant that includes the following elements:

   I.   A unit circle

   II. The angles $0, \dfrac{\pi}{6}, \dfrac{\pi}{4}, \dfrac{\pi}{3},$ and $\dfrac{\pi}{2}$ in standard position

   III. The coordinates of the points at which the angles intersect the unit circle

In Problems 3.8, 3.18, and 3.19, you generate the coordinates for angles $\pi/4, \pi/3,$ and $\pi/6$; add these to a graph of the unit circle. The remaining angles, 0 and $\pi/2$, represent the positive x- and y-axes. Recall that a unit circle is centered at the origin and has a radius of 1. Therefore, a unit circle intersects the positive x-axis at point $(1,0)$ and the positive y-axis at point $(0,1)$, as illustrated in the following diagram.

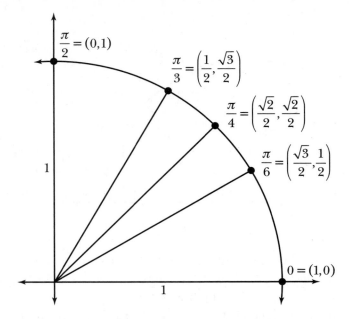

Here's another way to memorize the angles and their points. It involves the denominators of the angles and cosine values:

* $\cos\dfrac{\pi}{3} = \dfrac{1}{2}$, and $1 + 2 = 3$

* $\cos\dfrac{\pi}{4} = \dfrac{\sqrt{2}}{2}$, and $2 \cdot 2 = 4$

* $\cos\dfrac{\pi}{6} = \dfrac{\sqrt{3}}{2}$, and $3 \cdot 2 = 6$

The coordinates follow a very predictable pattern. If you begin at the positive y-axis and proceed clockwise, the x-coordinates of the points are: 0, 1, $\sqrt{2}$, and $\sqrt{3}$. (Because $0 = \sqrt{0}$ and $1 = \sqrt{1}$, you could even conclude that the x-coordinates are $\sqrt{0}$, $\sqrt{1}$, $\sqrt{2}$, and $\sqrt{3}$.) It is important that you are able to recite all of these coordinates from memory, given the corresponding angle.

3.24 Use the diagram you generated in Problem 3.23 to calculate $\cos\dfrac{\pi}{4}$.

In Problem 3.23, you generate the first quadrant of the trigonometric unit circle. Notice that the angle $\theta = \pi/4$ intersects the circle at point $\left(\sqrt{2}/2,\ \sqrt{2}/2\right)$. According to Problem 3.22, points on the unit circle represent trigonometric values of angles. Specifically, the terminal side of angle θ in standard position intersects the unit circle at point ($\cos\theta$, $\sin\theta$). Therefore, the x-coordinate of the intersection point is the cosine of the angle.

$$\cos\frac{\pi}{4} = \frac{\sqrt{2}}{2}$$

Because the x- and y-coordinates are the same for $\theta = \pi/4$, you can also conclude that $\sin\pi/4 = \sqrt{2}/2$.

3.25   Use the diagram you generated in Problem 3.23 to calculate $\sin\dfrac{\pi}{3}$.

The angle $\pi/3$, in standard position, intersects the unit circle at point $\left(1/2, \sqrt{3}/2\right)$. The y-coordinate of this point represents the sine of the angle.

$$\sin\frac{\pi}{3} = \frac{\sqrt{3}}{2}$$

3.26   Use the diagram you generated in Problem 3.23 to calculate $\cos\dfrac{\pi}{2}$.

According to the diagram, a right angle (measuring $\pi/2$ radians) in standard position intersects the unit circle at point $(0,1)$. The x-coordinate of this intersection point represents the cosine of the angle.

$$\cos\frac{\pi}{2} = 0$$

3.27   Use the diagram you generated in Problem 3.23 to calculate $\sin^{-1}\dfrac{1}{2}$.

This problem asks you to identify the angle θ in the first quadrant that has a sine of $1/2$. Note that many angles have this sine value, but only one of those angles is between 0 and $\pi/2$ radians, lying within the first quadrant.

Recall that the sine value of an angle is equal to the y-coordinate on the unit circle. Because the terminal side of $\theta = \pi/6$ intersects the unit circle at point $\left(\sqrt{3}/2, 1/2\right)$, you can conclude that $\sin \pi/6 = 1/2$.

$$\sin^{-1}\frac{1}{2} = \frac{\pi}{6}$$

3.28   Use the diagram you generated in Problem 3.23 to calculate $\cos^{-1} 0$.

This problem asks the following question: "Of the angles listed in the diagram you generated in Problem 3.23, which has a corresponding point on the unit circle whose x-coordinate is 0?" Notice that the terminal side of $\theta = \pi/2$ intersects the circle at point $(0,1)$. Therefore, $\cos^{-1} 0 = \pi/2$.

3.29 Use the diagram you generated in Problem 3.23 to calculate $\tan\dfrac{\pi}{4}$.

Recall that the tangent of an angle is equal to the quotient of its sine and cosine values.

$$\tan\theta = \frac{\sin\theta}{\cos\theta}$$

$$\tan\frac{\pi}{4} = \frac{\sin(\pi/4)}{\cos(\pi/4)}$$

As Problem 3.24 explains, $\sin\pi/4 = \cos\pi/4 = \sqrt{2}/2$. Substitute these values into the equation above.

$$\tan\frac{\pi}{4} = \frac{\sqrt{2}/2}{\sqrt{2}/2}$$

$$\tan\frac{\pi}{4} = 1$$

Any non-zero number divided by itself is equal to 1, even a weird number like $\dfrac{\sqrt{2}}{2}$.

3.30 Use the diagram you generated in Problem 3.23 to calculate $\sec\dfrac{\pi}{3}$.

The secant function is the reciprocal of the cosine function, so begin by calculating $\cos\pi/3$. The terminal side of $\theta = \pi/3$ intersects the unit circle at point $\left(1/2, \sqrt{3}/2\right)$; the x-coordinate of this point is the cosine of the angle.

$$\cos\frac{\pi}{3} = \frac{1}{2}$$

The secant of an angle is equal to the reciprocal of its cosine value.

$$\sec\frac{\pi}{3} = 2$$

The reciprocal of $\dfrac{1}{2}$ is $\dfrac{2}{1}$, or just 2.

Common Angles on the Unit Circle
And their cosine/sine values

3.31 Expand the drawing you created in Problem 3.23 to include the following elements:

  I.   A unit circle passing through all four quadrants of the coordinate plane

  II. The terminal sides of $\dfrac{2\pi}{3}, \dfrac{3\pi}{4}, \dfrac{5\pi}{6}, \pi, \dfrac{7\pi}{6}, \dfrac{5\pi}{4}, \dfrac{4\pi}{3}, \dfrac{3\pi}{2}, \dfrac{5\pi}{3}, \dfrac{7\pi}{4}, \dfrac{11\pi}{6}$, and 2π in standard position

  III. The coordinates of the points at which the angles intersect the unit circle

Although a unit circle is simply a circle with a radius of 1, the term "unit circle" often refers to the drawing described in this problem. This diagram, duplicated in Appendix B for reference, lists the most commonly used angles in trigonometry, as well as their cosine and sine values.

> Most teachers expect you to memorize this unit circle, including all of the angles and their coordinates.

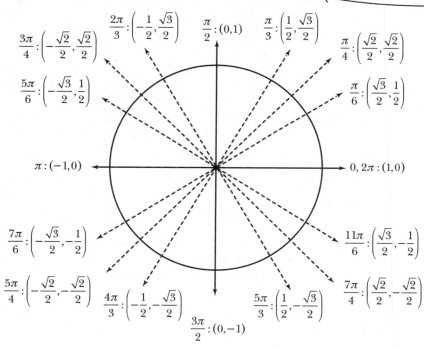

Problem 3.23 generates a single quadrant of the unit circle, but a clear pattern exists for the points in the other three quadrants. Though the signs of the coordinates change, the coordinates themselves remain constant:

• Angles with a denominator of 3 are paired with the point $\left(\pm\dfrac{1}{2}, \pm\dfrac{\sqrt{3}}{2}\right)$

• Angles with a denominator of 4 are paired with the point $\left(\pm\dfrac{\sqrt{2}}{2}, \pm\dfrac{\sqrt{2}}{2}\right)$

• Angles with a denominator of 6 are paired with the point $\left(\pm\dfrac{\sqrt{3}}{2}, \pm\dfrac{1}{2}\right)$

• Angles with a denominator of 1 or 2 have x- and y-coordinates of 0 or ± 1

To help you generate the angles in the unit circle, consider the following observations:

- Angles in the first quadrant have a numerator of 1
- The numerators in the second quadrant are exactly one less than the denominators
- The numerators in the third quadrant are exactly one more than the denominators
- The numerators in the fourth quadrant are exactly one less than twice the denominators

It is very important that you can reproduce the entire unit circle from memory and recall coordinates given an angle (or vice versa). Most instructors expect you to identify the cosine and sine values of these angles without the use of a calculator.

3.32 The diagram below indicates that both cosine and sine are positive for angles whose terminal sides lie in the first quadrant. Determine whether cosine and sine are positive or negative in the remaining three quadrants to complete the diagram, and justify your answer.

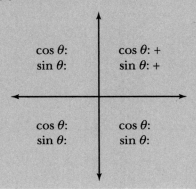

Recall that the cosine and sine of an angle on the unit circle are, respectively, the x- and y-coordinates of the corresponding points. Therefore, cosine is positive whenever a point is right of the origin, in quadrants I and IV. Cosine is negative in quadrants II and III.

Sine is positive whenever a point is above the origin, in quadrants I and II; sine is negative in quadrants III and IV.

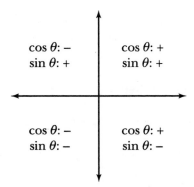

3.33 Identify the signs of the tangent function in each of the four quadrants of the coordinate plane. Justify your answer.

Tangent is equal to the quotient of sine and cosine. To determine the sign of tangent in each quadrant, divide the sign of the sine function by the sign of the cosine function in that quadrant. For example, in quadrant I, both the sine and cosine are positive, and a positive number divided by a positive number is equal to a positive number.

$$\text{Quadrant I:} \quad \tan\theta = \frac{\sin\theta}{\cos\theta} = \frac{+}{+} = +$$

Follow the same procedure for the remaining three quadrants.

$$\text{Quadrant II:} \quad \tan\theta = \frac{\sin\theta}{\cos\theta} = \frac{+}{-} = -$$

$$\text{Quadrant III:} \quad \tan\theta = \frac{\sin\theta}{\cos\theta} = \frac{-}{-} = +$$

$$\text{Quadrant IV:} \quad \tan\theta = \frac{\sin\theta}{\cos\theta} = \frac{-}{+} = -$$

> Tangent is positive whenever cosine and sine have matching signs. In other words, when they are both positive (quadrant I) or both negative (quadrant III).

Note: Problems 3.34–3.35 explain how to identify trigonometric values based on the unit circle diagram you generated in Problem 3.31.

3.34 Assuming $\theta = \dfrac{7\pi}{6}$ is an angle in standard position, in what quadrant does its terminal side lie?

To identify the quadrant in which an angle's terminal side lies, note that each quadrant of the coordinate plane represents 90° or $\pi/2$ radians. Given an angle measured in radians, examine the number multiplied by π:

- If $0 < \theta < 0.5$, the terminal side lies in the first quadrant
- If $0.5 < \theta < 1$, the terminal side lies in the second quadrant
- If $1 < \theta < 1.5$, the terminal side lies in the third quadrant
- If $1.5 < \theta < 2$, the terminal side lies in the fourth quadrant

The number multiplied by π is $7/6 = 1.1\overline{6}$, which belongs to the interval $1 < \theta < 1.5$. Therefore, $\pi = 7\pi/6$ lies in the third quadrant.

Note: Problems 3.34–3.35 explain how to identify trigonometric values based on the unit circle diagram you generated in Problem 3.31.

3.35 Calculate $\cos \dfrac{7\pi}{6}$.

According to Problem 3.31, angles of the unit circle with a denominator of 6 are associated with the coordinate $\left(\pm\sqrt{3}/2, \ \pm 1/2 \right)$. Problem 3.34 explains that the terminal side of $\theta = 7\pi/6$ lies in the third quadrant, in which the x- and y-coordinates of points are both negative. Recall that the cosine of an angle is the x-coordinate of the associated point.

$$\cos \frac{7\pi}{6} = -\frac{\sqrt{3}}{2}$$

If you memorize the unit circle, you don't need to go through all of this logic to answer the question—you just write $-\dfrac{\sqrt{3}}{2}$. This explanation is just meant to step you through the thought process that happens behind the scenes.

Note: Problems 3.36–3.37 explain how to identify trigonometric values based on the unit circle diagram you generated in Problem 3.31.

3.36 Assuming $\theta = \dfrac{3\pi}{2}$ is an angle in standard position, in what quadrant does its terminal side lie?

As Problem 3.34 explains, you can use the number multiplied by π in a radian angle measurement to identify the quadrant in which the terminal side lies: $3/2 = 1.5$. Notice that this angle does not lie within the bounds of the third quadrant (1 and 1.5), nor does it lie within the bounds of the fourth quadrant (1.5 and 2). Instead, it is the boundary of the third and fourth quadrants, the negative y-axis. Angles like $\theta = 3\pi/2$, whose terminal sides lie on the x- or y-axes, are called "quadrantals."

Note: Problems 3.36–3.37 explain how to identify trigonometric values based on the unit circle diagram you generated in Problem 3.31.

3.37 Calculate $\sin \dfrac{3\pi}{2}$.

As the diagram in Problem 3.31 illustrates, the angle $\theta = 3\pi/2$ in standard position intersects the unit circle at point $(0,-1)$. The y-coordinate of this point represents the sine value of θ.

$$\sin \frac{3\pi}{2} = -1$$

3.38 Calculate $\sin\dfrac{3\pi}{4}$ without the use of a trigonometric table or a graphing calculator.

The angle $\theta = 3\pi/4$ intersects the unit circle at point $\left(-\sqrt{2}/2, \sqrt{2}/2\right)$. The y-coordinate of this point represents the sine value of the angle.

$$\sin\frac{3\pi}{4} = \frac{\sqrt{2}}{2}$$

Note that the sines of angles terminating in the second quadrant, like the y-coordinates of points in the second quadrant, are positive.

> Don't keep looking back at the unit circle you made in Problem 3.31. At this point, you need to have those angles, their cosines, and their sines memorized.

3.39 Calculate $\cos\dfrac{2\pi}{3}$ without the use of a trigonometric table or a graphing calculator.

The terminal side of angle $\theta = 2\pi/3$ intersects the unit circle at point $\left(-1/2, \sqrt{3}/2\right)$; the cosine of θ is the x-coordinate of this point.

$$\cos\frac{2\pi}{3} = -\frac{1}{2}$$

3.40 Calculate $\cos\dfrac{11\pi}{6}$ without the use of a trigonometric table or a graphing calculator.

The terminal side of angle $\theta = 11\pi/6$ intersects the unit circle at point $\left(\sqrt{3}/2, -1/2\right)$; the cosine of θ is the x-coordinate of this point.

$$\cos\frac{11\pi}{6} = \frac{\sqrt{3}}{2}$$

Ensure that you are reporting the correct signs in your final answer. Because $\theta = 11\pi/6$ lies in the fourth quadrant, its cosine is positive and its sine is negative.

3.41 Calculate $\sin\dfrac{4\pi}{3}$ without the use of a trigonometric table or a graphing calculator.

The terminal side of angle $\theta = 4\pi/3$ intersects the unit circle at point $\left(-1/2, -\sqrt{3}/2\right)$ in the third quadrant. Therefore, $\sin(4\pi/3) = -\sqrt{3}/2$.

See the margin note next to Problem 2.28.

3.42 Calculate $\cot \dfrac{5\pi}{6}$ without the use of a trigonometric table or a graphing calculator.

Recall that the cosine of an angle is equal to the quotient of its cosine and sine values.

$$\cot \theta = \frac{\cos \theta}{\sin \theta}$$

$$\cot \frac{5\pi}{6} = \frac{\cos(5\pi/6)}{\sin(5\pi/6)}$$

The terminal side of $\pi = 5\pi/6$ intersects the unit circle in the second quadrant at point $\left(-\sqrt{3}/2, 1/2\right)$. The x-coordinate of the point represents its cosine value, and the y-coordinate represents its sine value.

$$\cot \frac{5\pi}{6} = \frac{-\sqrt{3}/2}{1/2}$$

Multiply the numerator and denominator by 2 to reduce the complex fraction.

$$\cot \frac{5\pi}{6} = \frac{\left(-\dfrac{\sqrt{3}}{\cancel{2}}\right)\left(\dfrac{\cancel{2}}{1}\right)}{\left(\dfrac{1}{\cancel{2}}\right)\left(\dfrac{\cancel{2}}{1}\right)}$$

$$= \frac{\left(-\dfrac{\sqrt{3}}{1}\right)}{\left(\dfrac{1}{1}\right)}$$

$$= \frac{-\sqrt{3}}{1}$$

$$= -\sqrt{3}$$

3.43 Calculate $\csc \pi$ without the use of a trigonometric table or a graphing calculator.

The cosecant function is defined as the reciprocal of the sine function.

$$\csc \pi = \frac{1}{\sin \pi}$$

The terminal side of $\theta = \pi$, which lies along the negative x-axis, intersects the unit circle at point $(-1,0)$. Therefore, $\sin \pi = 0$. Substitute this value into the equation above.

$$\csc \pi = \frac{1}{0}$$

Division by zero is an undefined operation. Therefore, you conclude that $\csc \pi$ is undefined.

No cosecant value exists for π or any multiple of π. You'll see this more clearly when you graph cosecant at the end of Chapter 6— there are big linear asymptotes where the graph doesn't exist.

3.44 Calculate tan 0 without the use of a trigonometric table or a graphing calculator.

Note that the "0" in this problem represents the angle measurement 0 radians, which describes an angle whose initial and terminal sides overlap on the positive *x*-axis. The tangent function is equal to the quotient of the sine and cosine functions.

$$\tan\theta = \frac{\sin\theta}{\cos\theta}$$
$$\tan 0 = \frac{\sin 0}{\cos 0}$$

The terminal side of angle $\theta = 0$ lies along the positive *x*-axis and intersects the unit circle at point (1,0). Therefore, $\sin\theta = 0$ and $\cos\theta = 1$.

$$\tan 0 = \frac{0}{1}$$
$$\tan 0 = 0$$

The tangent of an angle measuring 0 radians is equal to 0.

3.45 Calculate $\sin^{-1} 1$ without the use of a trigonometric table or a graphing calculator.

Inverse trigonometric functions provide a trigonometric value and ask you to identify the angle associated with that value. For example, $\sin^{-1} 1$ asks this question: "What angle θ has a sine value of 1?" Recall that the sine of an angle is equal to its *y*-coordinate on the unit circle.

Consider the diagram you created in Problem 3.31; only one point on that diagram has a *y*-coordinate of +1, the point that corresponds with central angle $\theta = \pi/2$. Therefore, $\sin^{-1} 1 = \pi/2$.

Chapter 4

TRIGONOMETRIC VALUES OF GENERAL ANGLES

Think outside the unit circle

In Chapter 3, you familiarize yourself with the most commonly used angles in trigonometry. However, applications in the real world are rarely limited to angles that follow precise, predictable patterns. In this chapter, you explore strategies that analyze angles of any measurement. You begin by calculating reference angles and coterminal angles, which allow you to express angle measurements more manageably.

Did you wonder why the trig tables in Appendix A only covered angles between 0° and 90°? Was it because people in the olden days couldn't conceive of anything bigger than a right angle? Nope. Here's the real reason (drumroll please): No matter how big and ugly an angle is, you can find an angle between 0° and 90° that has the same cosine and sine values (and therefore the same tangent, cotangent, secant, and cosecant values as well). There's no need for a trig table that's 32 pages long when you can settle for one that's a quarter the size. Genius!

Why bother rewriting angles at all, you ask? It allows you to apply everything you learned about right triangle trig (in Chapters 2 and 3) to angles of any shape and size.

Reference Angles
Shrink rays that turn any angles into acute angles

4.1 Given acute angle θ in the diagram below, draw reference angles in the second, third, and fourth quadrants that have the same measure as θ.

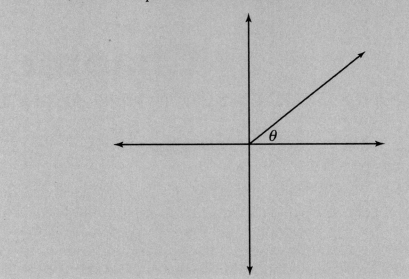

A reference angle is an acute angle formed by a terminal side and the *x*-axis. Reference angles are not necessary in the first quadrant, but each of the other quadrants requires them. The diagram below illustrates the reference angles for the second, third, and fourth quadrants, each with an angle measure of θ.

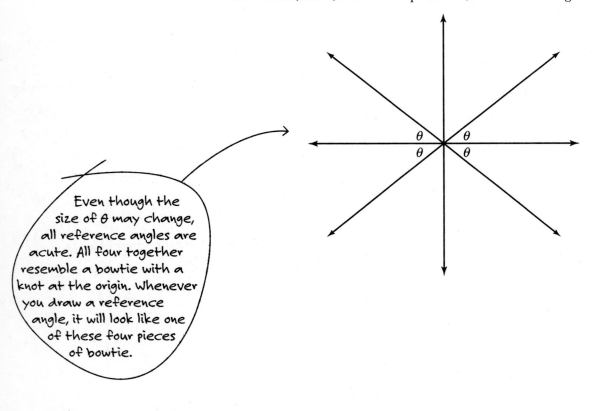

Even though the size of θ may change, all reference angles are acute. All four together resemble a bowtie with a knot at the origin. Whenever you draw a reference angle, it will look like one of these four pieces of bowtie.

Note: Problems 4.2–4.5 refer to angle α in the diagram below.

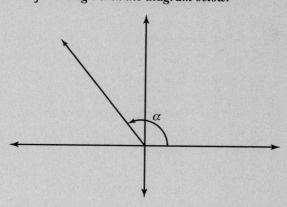

4.2 Draw β, the reference angle for α.

The terminal side of α lies in the second quadrant. Recall that a reference angle is formed by the terminal side of an angle and the *x*-axis, in this case the negative *x*-axis. In the diagram below, β is the reference angle for α.

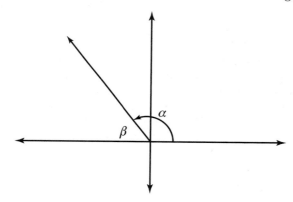

Note: Problems 4.2–4.5 refer to the diagram in Problem 4.2. Note that β is the reference angle for α.

4.3 Given $\sin \beta = \dfrac{4}{5}$, calculate $\sin \alpha$.

The sine function is defined in terms of a right triangle, equal to the length of the leg opposite an angle divided by the length of the hypotenuse. Draw a vertical line segment extending from the terminal side of β to the negative *x*-axis, forming a right triangle as demonstrated in the following diagram. Because sin β = 4/5, the length of the vertical segment (the leg opposite β) is 4, and the length of the hypotenuse is 5.

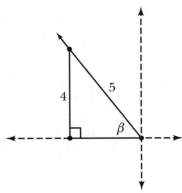

You may be wondering why the x- and y-axes are dotted in this diagram. Forget, for the moment, that β is in the second quadrant. It is a reference angle, and is therefore not required to be in standard position. You can treat the right triangle as you would any triangle not confined to a coordinate plane. Its lengths are positive.

The purpose of a reference angle is to provide the values you need for the original angle. Only when you refer to the actual angle do you need to concern yourself with the quadrants of the coordinate plane and the signs of the functions.

Because $\sin \beta = 4/5$, and β is the reference angle for α, you can conclude that $\sin \alpha = \pm 4/5$. Angles and their reference angles have trigonometric values that are either equal or are opposites.

★ ★ ★
This is very important.

Now it is time to calculate the sine value of α, the original angle.

$$\sin \alpha = \frac{\text{signed length of leg opposite } \beta}{\text{length of the hypotenuse}}$$

In the second quadrant, where angles α and β terminate, you travel left and up from the origin. Therefore, x-values are negative and y-values are positive. The side opposite β (the reference angle) has a signed length of +4, because you travel upward from the x-axis to reach the terminal side. The hypotenuse is always considered a positive value.

$$\sin \alpha = \frac{+4}{+5} = \frac{4}{5}$$

Therefore, $\sin \alpha = 4/5$, which is exactly equal to the given value, $\sin \beta$.

Note: Problems 4.2–4.5 refer to the diagram in Problem 4.2. Note that β is the reference angle for α.

4.4 Given $\cos \beta = \dfrac{3}{5}$, calculate $\cos \alpha$.

The length of the opposite side is still 4, but that's not needed in this problem.

In Problem 4.3, you draw a vertical segment from the terminal side to the x-axis in order to create a right triangle. You can use that triangle again in this problem. Because $\cos \beta = 3/5$ (and cosine is the quotient of the adjacent leg and the hypotenuse), the length of the adjacent leg is 3 and the length of the hypotenuse is 5.

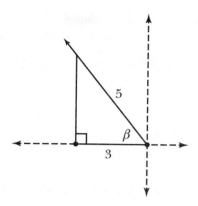

Recall that the trigonometric values of an angle and its reference angle are either equal or opposites, so $\cos\beta = 3/5$. You determine the sign of $3/5$ when you consider the quadrant in which the terminal side of α and β lie.

$$\cos\alpha = \frac{\text{signed length of leg adjacent to }\beta}{\text{length of the hypotenuse}}$$

The terminal sides of α and β lie in the second quadrant. As a result, the side adjacent to β (which is horizontal) extends left of the origin. Therefore, it has a negative signed length. Remember that the length of the hypotenuse is always considered positive. ◄

> Here's a quick guide to signed lengths for each quadrant:
>
> I: horizontal and vertical: +
> II: horizontal = –, vertical = +
> III: horizontal and vertical: –
> IV: horizontal = +, vertical = –
>
> Seem familiar? They are also the signs of the x- and y-coordinates for each quadrant.

$$\cos\alpha = \frac{-3}{+5}$$
$$= -\frac{3}{5}$$

You conclude that $\cos\alpha$ is the opposite of $\cos\beta$.

Note: Problems 4.2–4.5 refer to the diagram in Problem 4.2. Note that β is the reference angle for α.

4.5   Given $\tan\beta = \dfrac{4}{3}$, calculate $\tan\alpha$.

Begin this problem in the same way you began Problems 4.3–4.4, drawing a vertical line from the terminal side toward the *x*-axis to form a right triangle.

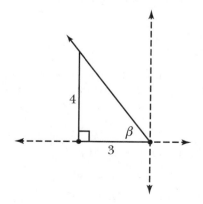

Angle α (and its reference angle β) terminate in the second quadrant, where horizontal signed lengths are negative and vertical signed lengths are positive.

$$\tan\alpha = \frac{\text{signed length of leg opposite }\beta}{\text{signed length of leg adjacent to }\beta}$$

$$\tan\alpha = \frac{+4}{-3}$$

$$\tan\alpha = -\frac{4}{3}$$

Note: Problems 4.6–4.7 refer to the diagram below, in which θ is an angle in standard position that measures 135°.

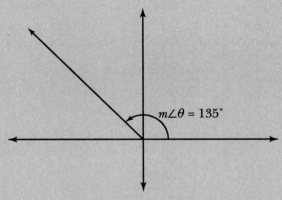

4.6 Draw α, the reference angle for θ, and calculate $m\angle\alpha$ in degrees.

The reference angle for θ is formed by the negative x-axis and the terminal side of θ, as illustrated below.

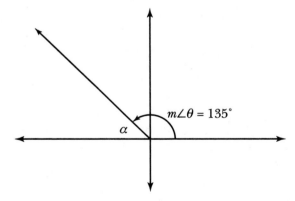

Notice that θ and α are supplementary—together they form a straight angle whose initial side is the positive x-axis and whose terminal side is the negative x-axis. Therefore, the sum of their measures is 180°.

$$m\angle\alpha + m\angle\theta = 180°$$

Substitute $m\angle\theta = 135°$ into the equation and solve for $m\angle\alpha$.

$$m\angle\alpha + 135° = 180°$$

$$m\angle\alpha = 180° - 135°$$

$$m\angle\alpha = 45°$$

Note: Problems 4.6–4.7 refer to the diagram in Problem 4.6, in which θ is an angle in standard position that measures 135°.

4.7 Generalize your solution to Problem 4.6 in order to create a formula that computes the reference angle α for any angle θ in standard position, such that $90° < m\angle\theta < 180°$. Construct the equivalent formula for angles measured in radians, as well.

In Problem 4.6, you recognize that an angle in standard position terminating in the second quadrant has a supplementary reference angle. Therefore, the measure of the reference angle is the difference of 180° and the original angle. In Problem 4.6, angle θ in standard position has a measure of 135°, and its reference angle α has measure 180° − 135°, as illustrated below.

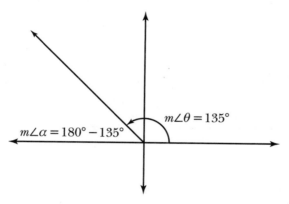

Given any angle θ in standard position such that $90° < m\angle\theta < 180°$, you can conclude (A) the angle terminates in the second quadrant, and (B) the measure of its reference angle is equal to the difference of 180° and $m\angle\theta$.

$$m\angle\alpha = 180° - m\angle\theta$$

If angle θ is measured in radians, the measure of its reference angle α is computed according to the following formula.

$$m\angle\alpha = \pi - m\angle\theta$$

You have to specify that the angle is between 90° and 180° for this formula to work, because all kinds of angles terminate in the second quadrant. For example, a 495° angle in standard position also terminates in the second quadrant, and 180° − 495° = −315° is definitely NOT its reference angle.

These formulas are equivalent because 180° = π.

Note: Problems 4.8–4.9 refer to an angle θ in standard position and its reference angle α. Note that $m\angle\theta = 250°$.

4.8 Draw angles α and θ in the coordinate plane and calculate the measure of α in degrees.

Angle θ measures 250°. An angle in standard position that terminates at the negative y-axis has a measure of 270°. Therefore, the terminal side of θ lies 270° − 250° = 20° clockwise from the negative y-axis, as illustrated in the following diagram. Reference angle α is formed by the terminal side of θ and the negative x-axis. Unlike the angle/reference angle pair examined in Problems 4.6–4.7, θ and its reference angle overlap.

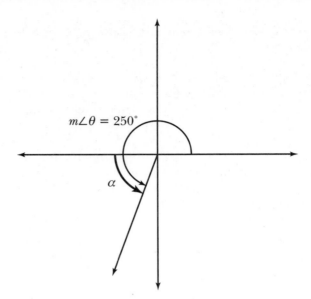

$m\angle\theta = 250°$

α

Notice that θ passes through quadrants I and II before it overlaps angle α in the third quadrant. Therefore, to calculate the measure of α, you subtract the measure of the angle whose initial side is the positive x-axis and whose terminal side is the negative x-axis.

$$m\angle\alpha = m\angle\theta - 180°$$
$$m\angle\alpha = 250° - 180°$$
$$m\angle\alpha = 70°$$

> According to Problem 4.6, the angle containing the first two quadrants measures 180°.

Note: Problems 4.8–4.9 refer to an angle θ in standard position and its reference angle α. Note that $m\angle\theta = 250°$.

4.9 Generalize your solution to Problem 4.8 in order to create a formula that computes the reference angle α for any angle θ in standard position, such that $180° < m\angle\theta < 270°$. Construct the equivalent formula for angles measured in radians, as well.

Any angle whose measure falls between 180° and 270° has a terminal side that lies in the third quadrant. As Problem 4.8 demonstrates, the angle θ and its reference angle α overlap in the third quadrant. Therefore, the measure of reference angle α is exactly 180° less than the measure of angle θ.

$$m\angle\alpha = m\angle\theta - 180°$$

To calculate the measure of reference angle α in radians, express 180° as π.

$$m\angle\alpha = m\angle\theta - \pi$$

Note: Problems 4.10–4.11 refer to an angle θ in standard position and its reference angle α. Note that $m\angle\theta = \dfrac{5\pi}{3}$.

4.10 Draw angles α and θ in the coordinate plane and calculate the measure of α in radians.

Angle θ terminates in the fourth quadrant. (You may wish to convert the angle to degrees if you have trouble visualizing radian angle measurements.) Its reference angle α is formed by the terminal side of θ and the positive *x*-axis, as illustrated below.

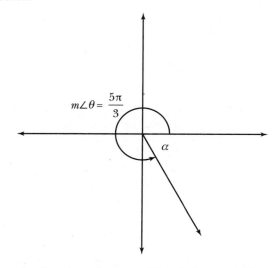

Notice that angles θ and α, when combined, form 1 full rotation in the coordinate plane. Therefore, the sum of their measures is 2π (or 360°).

$$m\angle\theta + m\angle\alpha = 2\pi$$

You are given $m\angle\theta = 5\pi/3$. Substitute this value into the equation and solve for $m\angle\alpha$.

$$\frac{5\pi}{3} + m\angle\alpha = 2\pi$$

$$m\angle\alpha = 2\pi - \frac{5\pi}{3}$$

Express both terms using the least common denominator 3 and simplify the expression.

$$m\angle\alpha = \frac{2\pi}{1} - \frac{5\pi}{3}$$

$$= \frac{2\pi}{1} \cdot \frac{3}{3} - \frac{5\pi}{3}$$

$$= \frac{6\pi}{3} - \frac{5\pi}{3}$$

$$= \frac{1\pi}{3}$$

$$= \frac{\pi}{3}$$

Note: Problems 4.10–4.11 refer to an angle θ in standard position and its reference angle α. Note that $m\angle\theta = \dfrac{5\pi}{3}$.

4.11 Generalize your solution to Problem 4.10 in order to create a formula that computes the reference angle α for any angle θ in standard position, such that $\dfrac{3\pi}{2} < m\angle\theta < 2\pi$. Construct the equivalent formula for angles measured in degrees, as well.

$270° < \theta < 360°$

As Problem 4.10 demonstrates, the measure of reference angle α is equal to the difference of 2π and $m\angle\theta$. To express this formula in degrees, convert 2π into $360°$.

$$m\angle\alpha = 2\pi - m\angle\theta \qquad m\angle\alpha = 360° - m\angle\theta$$

4.12 Given angle θ in standard form that measures 165°, calculate the measure of its reference angle α.

Because $90° < 165° < 180°$, the terminal side of θ lies in the second quadrant. According to Problem 4.7, the measure of reference angle α for a second-quadrant angle θ is computed according to the following formula.

$$m\angle\alpha = 180° - m\angle\theta$$

Substitute the given measure of θ into the equation to calculate the measure of reference angle α.

$$m\angle\alpha = 180° - 165°$$
$$= 15°$$

4.13 Given angle θ in standard form that measures 321°, calculate the measure of its reference angle α.

Because $270° < 321° < 360°$, the terminal side of θ lies in the fourth quadrant. According to Problem 4.11, the measure of reference angle α for a fourth-quadrant angle θ is computed according to the following formula.

$$m\angle\alpha = 360° - m\angle\theta$$

Substitute the given measure of θ into the equation to calculate the measure of reference angle α.

$$m\angle\alpha = 360° - 321°$$
$$= 39°$$

4.14 Given angle θ in standard form that measures $\dfrac{7\pi}{6}$, calculate the measure of its reference angle α.

Note that the terminal side of θ lies in the third quadrant, because $\pi < 7\pi/6 < 3\pi/2$. According to Problem 4.9, the measure of reference angle α for a third-quadrant angle θ is computed according to the following formula.

$$m\angle\alpha = m\angle\theta - \pi$$

Substitute the given measure of θ into the equation to calculate the measure of reference angle α.

$$\alpha = \frac{7\pi}{6} - \pi$$
$$= \frac{7\pi}{6} - \left(\frac{\pi}{1}\right)\left(\frac{6}{6}\right)$$
$$= \frac{7\pi}{6} - \frac{6\pi}{6}$$
$$= \frac{7\pi - 6\pi}{6}$$
$$= \frac{1\pi}{6}$$
$$= \frac{\pi}{6}$$

Problem 3.34 explains how to figure out where an angle's terminal side lies. In fact, it uses this exact angle as an example. Make sure you understand this trick before you move on to Problem 4.15.

4.15 Given angle θ in standard form that measures $\dfrac{11\pi}{13}$, calculate the measure of its reference angle α.

The quotient of the numerator and denominator (excluding π) is equal to $0.\overline{846153}$, so the terminal side of θ lies in the second quadrant. Apply the appropriate reference angle formula to calculate the measure of reference angle α.

$$\alpha = \pi - \theta$$
$$= \pi - \frac{11\pi}{13}$$
$$= \left(\frac{\pi}{1}\right)\left(\frac{13}{13}\right) - \frac{11\pi}{13}$$
$$= \frac{13\pi}{13} - \frac{11\pi}{13}$$
$$= \frac{2\pi}{13}$$

The book is using the trick from Problem 3.34.

Coterminal Angles

Two angles that end at the same ray

4.16 If angles α and β are in standard position on the coordinate plane and are coterminal, what can you conclude about α and β? Can you conclude that α and β have the same measure? Why or why not?

If two angles in standard position are coterminal, their terminal sides coincide. In other words, their terminal sides lie on the same ray. Recall that the angles are in standard position, so their initial sides coincide as well; both initial sides lie along the positive x-axis. Although angles α and β have a great deal in common, you cannot conclude that they have the same measure.

Just because two angles begin and end at the same ray, they may differ in the direction or in the number of rotations about the origin. For instance, an angle in standard position measuring 0 radians and an angle in standard position measuring 2π radians are coterminal, but the latter angle has a significantly larger measure.

Note: Problems 4.17–4.18 refer to angles α and β in standard position, such that $m\angle\alpha = \dfrac{\pi}{2}$ and $m\angle\beta = -\dfrac{3\pi}{2}$.

4.17 Graph angles α and β to visually verify that they are coterminal.

Angle α is positive, so the angle is measured clockwise from its initial side at the positive x-axis. Note that an angle in standard position measuring ($\pi/2$ or 90°) terminates on the positive y-axis. Angle β is negative, so it is measured counterclockwise from the positive x-axis.

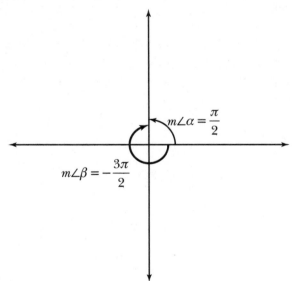

Angles α and β share the same terminal side (the positive y-axis), so the angles are coterminal.

Note: Problems 4.17–4.18 refer to angles α and β in standard position, such that $m\angle\alpha = \dfrac{\pi}{2}$ and $m\angle\beta = -\dfrac{3\pi}{2}$.

4.18 Verify arithmetically that α and β are coterminal.

Coterminal angle measures are calculated by adding or subtracting rotations in the coordinate plane. Because α and β are measured in radians, adding 2π (which represents one rotation in radians) to an angle measure or subtracting 2π from an angle measure produces a coterminal angle measure.

In the context of this problem, adding 2π to $m\angle\beta$ or subtracting 2π from $m\angle\alpha$ produces the measure of the other angle.

$$m\angle\alpha - 2\pi = \frac{\pi}{2} - 2\pi \qquad\qquad m\angle\beta + 2\pi = -\frac{3\pi}{2} + 2\pi$$

$$= \frac{\pi}{2} - \left(\frac{2\pi}{1}\right)\left(\frac{2}{2}\right) \qquad\qquad = -\frac{3\pi}{2} + \left(\frac{2\pi}{1}\right)\left(\frac{2}{2}\right)$$

$$= \frac{\pi}{2} - \frac{4\pi}{2} \qquad\qquad = -\frac{3\pi}{2} + \frac{4\pi}{2}$$

$$= -\frac{3\pi}{2} \qquad\qquad = \frac{\pi}{2}$$

$$= m\angle\beta \qquad\qquad = m\angle\alpha$$

Because the sum of one angle measure and either 2π or -2π produces the other angle measure, angles α and β are coterminal.

> In other words, if you add 360° to an angle measure you get one coterminal angle. Add another rotation, 2(360°) = 720°, to get another coterminal angle. Add 360° multiplied by 3 to get yet another coterminal angle. You could also multiply 360° by –1, –2, –3, etc.

Note: Problems 4.19–4.20 refer to angles α and β in standard position, such that $m\angle\alpha = 225°$ and $m\angle\beta = 585°$.

4.19 Verify arithmetically that α and β are coterminal.

If two angles in standard position are coterminal, then their measures differ by multiples of 1 revolution. Adding 1 revolution (360°) to the measure of angle α produces the measure of β. Thus, α and β are coterminal angles.

$$m\angle\alpha + 360° = 225° + 360°$$

$$= 585°$$

Alternatively, you could subtract 360° from $m\angle\beta$ to verify that the result is $m\angle\alpha$, also proving the angles are coterminal.

$$m\angle\beta - 360° = 585° - 360°$$

$$= 225°$$

> Adding or subtracting 360° means looping around the coordinate plane and ending up where you started. That's why the angles have overlapping terminal sides—they rotate 360° and return to the place they began.

Note: Problems 4.19–4.20 refer to angles α and β in standard position, such that m∠α = 225° and m∠β = 585°.

4.20   Graph angles α and β to visually verify that they are coterminal.

Angle α has a terminal side that lies in the third quadrant. It has a reference angle with measure 225° − 180° = 45°, so its terminal side forms a 45° angle with the negative *x*-axis. Angle β is formed by rotating once fully around the coordinate plane (360°) and then traveling 585° − 360° = 225° to terminate on the same ray as angle α.

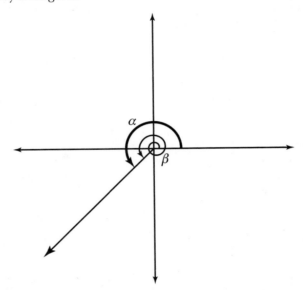

4.21   Given an angle θ in standard position with a measure of 320°, calculate the measures of one positive and one negative coterminal angle.

Coterminal angles have measures that differ by multiples of 1 rotation (360°). Therefore, adding 360° to the measure of θ produces a coterminal angle.

$$m\angle\theta + 360° = 320° + 360°$$
$$= 680°$$

Subtract 360° from the measure of θ to produce a negative coterminal angle.

$$m\angle\alpha - 360° = 320° - 360°$$
$$= -40°$$

Thus, angles in standard position that measure −40°, 320°, and 680° are coterminal.

4.22 Given an angle θ in standard position with a measure of $\dfrac{9\pi}{8}$, calculate the measures of one positive and one negative coterminal angle.

Coterminal angles have radian measures that differ by multiples of 1 rotation (2π radians). Add and subtract 2π from $9\pi/8$ to produce the measures of two coterminal angles, one positive and one negative.

$$m\angle\theta + 2\pi = \frac{9\pi}{8} + 2\pi \qquad\qquad m\angle\theta - 2\pi = \frac{9\pi}{8} - 2\pi$$

$$= \frac{9\pi}{8} + \left(\frac{2\pi}{1}\right)\left(\frac{8}{8}\right) \qquad\qquad = \frac{9\pi}{8} - \left(\frac{2\pi}{1}\right)\left(\frac{8}{8}\right)$$

$$= \frac{9\pi}{8} + \frac{16\pi}{8} \qquad\qquad = \frac{9\pi}{8} - \frac{16\pi}{8}$$

$$= \frac{25\pi}{8} \qquad\qquad = -\frac{7\pi}{8}$$

Thus, angles in standard position that measure $-7\pi/8$, $9\pi/8$, and $25\pi/8$ are coterminal.

4.23 Given an angle θ in standard position with a measure of $-\dfrac{21\pi}{5}$, calculate the measures of one positive and one negative coterminal angle.

Unlike Problems 4.21–4.22, adding 2π (one rotation) to this angle does not produce a positive result, as demonstrated below.

$$-\frac{21\pi}{5} + 2\pi = -\frac{21\pi}{5} + \left(\frac{2\pi}{1}\right)\left(\frac{5}{5}\right)$$

$$= -\frac{21\pi}{5} + \frac{10\pi}{5}$$

$$= -\frac{11\pi}{5}$$

> That's because $\dfrac{21\pi}{5}$ is larger than one rotation: $2\pi = \dfrac{10\pi}{5}$ and $\dfrac{21\pi}{5} > \dfrac{10\pi}{5}$.

Although the result is not positive, by adding 2π, you identify a negative coterminal angle measurement. To continue the search for a positive coterminal angle, add 2π to this result.

$$-\frac{11\pi}{5} + 2\pi = -\frac{11\pi}{5} + \frac{10\pi}{5}$$

$$= -\frac{\pi}{5}$$

The result is still negative (and represents yet another negative coterminal angle), so add 2π to this result.

$$-\frac{\pi}{5}+2\pi=-\frac{\pi}{5}+\frac{10\pi}{5}$$
$$=\frac{9\pi}{5}$$

$-\frac{\pi}{5}$ is also a valid negative coterminal angle.

Therefore, angles in standard position with measures –21π/5, –11π/5, and 9π/5 are coterminal.

4.24 Given an angle θ in standard position with a measure of 1,691°, calculate the measure of coterminal angle α, such that $0° \le m\angle\alpha \le 360°$.

In Problem 4.23, you must repeatedly add 1 rotation to negative coterminal angle measures until the result is positive. In this problem, you must subtract 1 rotation (360°) repeatedly until the difference is greater than or equal to 0° but less than or equal to 360°.

$$1{,}691° - 360° = 1{,}331°$$
$$1{,}331° - 360° = 971°$$
$$971° - 360° = 611°$$
$$611° - 360° = 251°$$

Therefore, $m\angle\alpha = 251°$ and α lies within the interval bounded below by 0° and above by 360°.

4.25 Evaluate cos 840° without the use of a graphing calculator or a trigonometric table.

It doesn't matter how many times you whirl around the origin, like a pointer on a board game spinner. It only matters where that pointer lands on the unit circle. Each time you subtract 360°, you ignore 1 rotation.

Apply the technique described in Problem 4.24 to identify an angle coterminal to 840° but lying within the interval bounded below by 0° and above by 360°. In other words, subtract 360° from 840°; if the difference is greater than 360°, subtract 360° from the result and repeat the process as necessary.

$$840° - 360° = 480°$$
$$480° - 360° = 120°$$

Because 840° and 120° are coterminal angles, they have equivalent trigonometric values: cos 840° = cos 120°. The problem indicates that you cannot use a graphing calculator to evaluate the function, so the angle is likely on the unit circle. Calculate the radian equivalent of 120°.

$$\frac{120}{1}\cdot\frac{\pi}{180}=\frac{120\pi}{180}$$
$$=\frac{(120\div60)\pi}{180\div60}$$
$$=\frac{2\pi}{3}$$

If you need to review the unit circle, refer to Appendix B.

Indeed, the angle belongs to the unit circle. Assuming you have memorized the unit circle—as directed in Chapter 3—you should remember that cos 2π/3 = –1/2. You conclude that cos 840° = –1/2.

4.26 Evaluate sin −675° without the use of a graphing calculator or a trigonometric table.

In Problem 4.23, you repeatedly add 1 rotation to coterminal angle measurements until the result is positive. Apply the same technique here, repeatedly adding 360°, until the result is positive.

$$-675° + 360° = -315°$$
$$-315° + 360° = 45°$$

Because −675° and 45° are coterminal angle measures, their trigonometric values are equal: sin −675° = sin 45°. Note that 45° is equivalent to $\pi/4$ radians, and according to the unit circle, $\sin \pi/4 = \sqrt{2}/2$. You conclude that $\sin(-675°) = \sqrt{2}/2$.

4.27 Given an angle θ in standard position with a measure of $-\dfrac{38\pi}{7}$, calculate the measure of coterminal angle α, such that $0 \leq m\angle\alpha \leq 2\pi$.

This problem, like Problem 4.26, requires you to repeatedly add 1 revolution to the angle measurement until the result is positive. Note that 1 revolution is equal to 2π radians, and the denominator of the given angle measurement is 7. Therefore, you should repeatedly add $2\pi = 14\pi/7$ to the angle measurements.

$$-\frac{38\pi}{7} + \frac{14\pi}{7} = -\frac{24\pi}{7}$$
$$-\frac{24\pi}{7} + \frac{14\pi}{7} = -\frac{10\pi}{7}$$
$$-\frac{10\pi}{7} + \frac{14\pi}{7} = \frac{4\pi}{7}$$

Thus, $m\angle\alpha = 4\pi/7$. Because $0 \leq 4\pi/7 \leq 14\pi/7$, α is coterminal to θ and lies in the interval bounded below by 0 and above by 2π.

4.28 Evaluate $\cos\dfrac{10\pi}{3}$ without the use of a graphing calculator or a trigonometric table.

Calculate the angle coterminal to $10\pi/3$ that lies between 0 and 2π by subtracting one revolution (2π) from the angle.

$$\frac{10\pi}{3} - 2\pi = \frac{10\pi}{3} - \left(\frac{2\pi}{1}\right)\left(\frac{3}{3}\right)$$
$$= \frac{10\pi}{3} - \frac{6\pi}{3}$$
$$= \frac{4\pi}{3}$$

Because $10\pi/3$ and $4\pi/3$ are coterminal, they have the same trigonometric values.

$$\cos\frac{10\pi}{3} = \cos\frac{4\pi}{3}$$
$$= -\frac{1}{2}$$

4.29 Evaluate $\cot\left(-\dfrac{43\pi}{6}\right)$ without the use of a graphing calculator or a trigonometric table.

Calculate the angle coterminal to $-43\pi/6$ that lies on the unit circle by repeatedly adding $2\pi = 12\pi/6$ until the result is positive.

$$-\frac{43\pi}{6} + \frac{12\pi}{6} = -\frac{31\pi}{6}$$
$$-\frac{31\pi}{6} + \frac{12\pi}{6} = -\frac{19\pi}{6}$$
$$-\frac{19\pi}{6} + \frac{12\pi}{6} = -\frac{7\pi}{6}$$
$$-\frac{7\pi}{6} + \frac{12\pi}{6} = \frac{5\pi}{6}$$

Because $-43\pi/6$ and $5\pi/6$ are coterminal, they have the same trigonometric values.

$$\cot\left(-\frac{43\pi}{6}\right) = \cot\frac{5\pi}{6}$$
$$= \frac{\cos(5\pi/6)}{\sin(5\pi/6)}$$
$$= \frac{-\sqrt{3}/2}{1/2}$$

Multiply the numerator and denominator by 2 to reduce the complex fraction.

$$\frac{\left(-\frac{\sqrt{3}}{\cancel{2}}\right)\left(\frac{\cancel{2}}{1}\right)}{\left(\frac{1}{\cancel{2}}\right)\left(\frac{\cancel{2}}{1}\right)} = \frac{\left(-\frac{\sqrt{3}}{1}\right)}{\left(\frac{1}{1}\right)}$$
$$= \frac{-\sqrt{3}}{1}$$
$$= -\sqrt{3}$$

Thus, $\cot\left(-\dfrac{43\pi}{6}\right) = -\sqrt{3}$.

Angles Beyond the Unit Circle

Calculating trig values you didn't memorize

Note: Problems 4.30–4.32 refer to an angle θ in standard position whose terminal side lies in the third quadrant, such that $\tan\theta = \dfrac{1}{3}$.

4.30 Construct a right triangle in the third quadrant using θ's reference angle α. Include the signed lengths of the legs in your diagram.

A third-quadrant reference angle is formed by the terminal side of an angle and the negative *x*-axis. Do not worry about precisely calculating the size of θ or its reference angle α in your diagram. It is more important to label the triangle correctly than to precisely predict the size of the angles. ◄

> *Just make sure the terminal side is in the third quadrant.*

Recall that the tangent ratio relates the opposite leg and the adjacent leg, so the length of the leg opposite α is 1 and the length of the adjacent leg is 3.

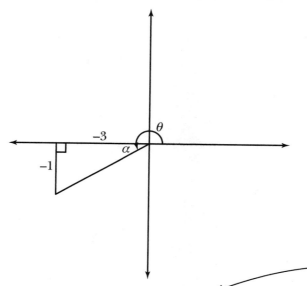

Notice that you must travel *left* and *down* from the origin to reach the end of the hypotenuse in the diagram. Therefore, the horizontal leg and the vertical leg both have negative signed lengths in the diagram. However, $\tan\theta$ is positive because it is equal to the quotient of the two negative lengths, as demonstrated below.

> *This is true for any third-quadrant reference angle right triangle. Points in the third quadrant have negative x-coordinates (negative horizontal lengths) and negative y-coordinates (negative vertical lengths).*

$$\tan\theta = \frac{\text{signed length of side opposite } \alpha}{\text{signed length of side adjacent to } \alpha}$$

$$= \frac{-1}{-3}$$

$$= \frac{1}{3}$$

Note: Problems 4.30–4.32 refer to an angle θ in standard position whose terminal side lies in the third quadrant, such that $\tan \theta = \dfrac{1}{3}$.

4.31 Calculate sin θ.

In Problem 4.30, you construct a right triangle and indicate the signed lengths of its legs: −3 and −1. Substitute these lengths into the Pythagorean theorem to calculate the length of the hypotenuse of that right triangle. Note that it does not matter whether you substitute the lengths of the legs (which are positive values) or the signed lengths of the legs (which are, in this case, negative values), because the Pythagorean theorem squares them. In both cases, the result will be positive.

> The sides really have lengths of +1 and +3. The negatives are only there to indicate direction, so most teachers leave them out of the Pythagorean theorem.

$$a^2 + b^2 = c^2$$
$$1^2 + 3^2 = c^2$$
$$1 + 9 = c^2$$
$$10 = c^2$$
$$\sqrt{10} = c$$

Update the diagram from Problem 4.30 to include the length of the hypotenuse.

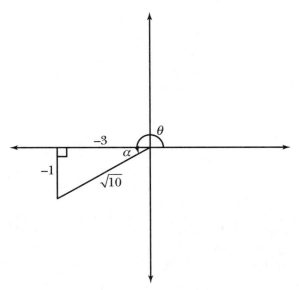

Apply the sine ratio. Substitute signed lengths into the following formula, remembering that the length of the hypotenuse is always considered a positive value.

$$\sin \theta = \frac{\text{signed length of side opposite } \alpha}{\text{length of the hypotenuse}}$$
$$= \frac{-1}{+\sqrt{10}}$$

To rationalize the denominator, multiply the numerator and denominator by $\sqrt{10}$.

$$= \frac{-1}{\sqrt{10}} \cdot \frac{\sqrt{10}}{\sqrt{10}}$$

$$= \frac{-1\sqrt{10}}{\sqrt{100}}$$

$$= -\frac{\sqrt{10}}{10}$$

Note: Problems 4.30–4.32 refer to an angle θ in standard position whose terminal side lies in the third quadrant, such that tan θ = $\frac{1}{3}$.

4.32   Calculate cos θ.

In Problem 4.30, you construct a right triangle and determine the signed lengths of its legs; in Problem 4.31, you compute the length of its hypotenuse. Refer to that right triangle when you calculate cos θ.

$$\cos\theta = \frac{\text{signed length of leg adjacent to } \alpha}{\text{length of the hypotenuse}}$$

$$= \frac{-3}{+\sqrt{10}}$$

Rationalize the denominator to complete the problem.

$$= \frac{-3}{\sqrt{10}} \cdot \frac{\sqrt{10}}{\sqrt{10}}$$

$$= \frac{-3\sqrt{10}}{\sqrt{100}}$$

$$= -\frac{3\sqrt{10}}{10}$$

Note: Problems 4.33–4.35 refer to an angle θ in standard position whose terminal side lies in the second quadrant, such that sin θ = $\frac{3}{4}$.

4.33   Construct a right triangle in the second quadrant using θ's reference angle α. Include the signed lengths of the legs in your diagram.

Second-quadrant reference angles are bounded above by an angle's terminal side and below by the negative x-axis. You are given sin θ = 3/4. Recall that the sine function is the ratio of the opposite leg length to the length of the hypotenuse. Therefore, the side opposite reference angle α has length 3 and the length of the hypotenuse is 4, as illustrated in the following diagram.

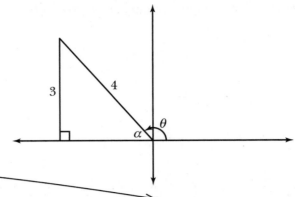

However, the horizontal leg is LEFT of the origin, so it is going to have a negative signed length (as you'll see when you calculate it in Problem 4.34).

Notice that the signed length of the vertical leg is 3. It is positive because it lies above the origin, in the second quadrant.

> *Note: Problems 4.33–4.35 refer to an angle θ in standard position whose terminal side lies in the second quadrant, such that sin θ = $\frac{3}{4}$.*

4.34 Calculate tan θ.

In Problem 4.33, you create a right triangle based on angle α, the reference angle for θ. The vertical leg of the triangle has length 3 and the hypotenuse has length 4. Apply the Pythagorean theorem to calculate the length of the horizontal leg.

$$a^2 + b^2 = c^2$$
$$3^2 + b^2 = 4^2$$
$$9 + b^2 = 16$$
$$b^2 = 16 - 9$$
$$b^2 = 7$$
$$b = \sqrt{7}$$

Recall that horizontal lengths in the second quadrant (as well as the third quadrant) have negative signed lengths. Complete the diagram you began in Problem 4.33, filling in the length of the horizontal leg: $-\sqrt{7}$.

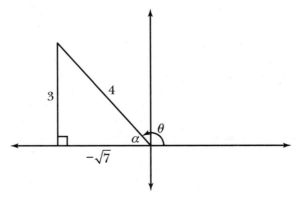

To calculate $\tan \theta$, use the signed lengths in the diagram.

$$\tan \theta = \frac{\text{signed length of leg opposite } \alpha}{\text{signed length of leg adjacent to } \alpha}$$

$$= \frac{+3}{-\sqrt{7}}$$

$$= -\frac{3}{\sqrt{7}}$$

Rationalize the denominator.

$$= -\frac{3}{\sqrt{7}} \cdot \frac{\sqrt{7}}{\sqrt{7}}$$

$$= -\frac{3\sqrt{7}}{\sqrt{49}}$$

$$= -\frac{3\sqrt{7}}{7}$$

Note: Problems 4.33–4.35 refer to an angle θ in standard position whose terminal side lies in the second quadrant, such that $\sin \theta = \frac{3}{4}$.

4.35 Calculate $\sec \theta$.

The secant function is the reciprocal of the cosine function, which divides the length of the adjacent leg by the length of the hypotenuse. Therefore, secant does the opposite—it divides the length of the hypotenuse by the length of the adjacent leg. Consider the diagram you completed in Problem 4.34, which states that the signed length of the leg adjacent to α is $-\sqrt{7}$ and the length of the hypotenuse is 4.

$$\sec \theta = \frac{\text{length of the hypotenuse}}{\text{signed length of leg adjacent to } \alpha}$$

$$= \frac{4}{-\sqrt{7}}$$

$$= -\frac{4}{\sqrt{7}}$$

Rationalize the denominator.

$$= -\frac{4}{\sqrt{7}} \cdot \frac{\sqrt{7}}{\sqrt{7}}$$

$$= -\frac{4\sqrt{7}}{\sqrt{49}}$$

$$= -\frac{4\sqrt{7}}{7}$$

This is worth repeating in case you haven't quite wrapped your head around it: You can calculate the trig values of an angle (like θ in this problem) if you calculate the corresponding trig values of its reference angle (like α) using SIGNED lengths.

Note: Problems 4.36–4.39 refer to an angle θ in standard position such that $\cos \theta = -\dfrac{5}{8}$ and $\sin \theta < 0$.

4.36 Identify the quadrant in which the terminal side of θ lies.

The problem lists a negative value for cos θ and states that sin θ < 0, so sin θ is negative as well. As explained in Problem 3.32, cosine and sine values are both negative in the third quadrant. Therefore, the terminal side of θ lies in the third quadrant.

Note: Problems 4.36–4.39 refer to an angle θ in standard position such that $\cos \theta = -\dfrac{5}{8}$ and $\sin \theta < 0$.

4.37 Construct a right triangle using θ's reference angle α. Include the signed lengths of the legs in your diagram.

According to Problem 4.36, the terminal side of θ lies in the third quadrant, so the right triangle should be located there. The cosine function is defined as the quotient of the adjacent leg's length divided by the length of the hypotenuse.

$$\cos \theta = \frac{\text{signed length of leg adjacent to } \alpha}{\text{length of the hypotenuse}} = \frac{-5}{+8}$$

In the diagram below, the signed length of the leg adjacent to reference angle α is −5, and the length of the hypotenuse is 8.

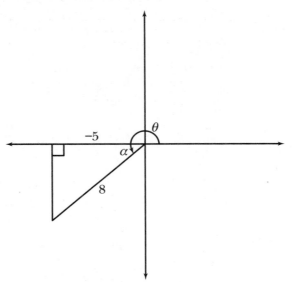

Note: Problems 4.36–4.39 refer to an angle θ in standard position such that $\cos \theta = -\dfrac{5}{8}$ and $\sin \theta < 0$.

4.38 Calculate $\cot \theta$.

In Problem 4.37, you construct a right triangle in the third quadrant based on α, the reference angle for θ. You are given the lengths of the hypotenuse and the horizontal leg adjacent to α, but you need to apply the Pythagorean theorem to calculate the length of the vertical leg opposite α.

$$a^2 + b^2 = c^2$$
$$5^2 + b^2 = 8^2$$
$$25 + b^2 = 64$$
$$b^2 = 64 - 25$$
$$b^2 = 39$$
$$b = \sqrt{39}$$

The vertical leg of this right triangle extends below the origin, so its signed length is $-\sqrt{39}$, as illustrated below.

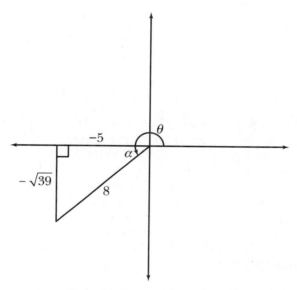

Now that you know the length of the adjacent leg, you can calculate $\cot \theta$.

$$\cot \theta = \frac{\text{signed length of the leg adjacent to } \alpha}{\text{signed length of the leg opposite } \alpha}$$
$$= \frac{-5}{-\sqrt{39}}$$
$$= \frac{5}{\sqrt{39}}$$

> A negative divided by a negative is equal to a positive.

Rationalize the denominator.

$$= \frac{5}{\sqrt{39}} \cdot \frac{\sqrt{39}}{\sqrt{39}}$$

$$= \frac{5\sqrt{39}}{39}$$

Note: Problems 4.36–4.39 refer to an angle θ in standard position such that cos θ = $-\dfrac{5}{8}$ and sin θ < 0.

4.39 Calculate csc θ.

The cosecant function is defined as the reciprocal of the sine function, so it is the quotient of the lengths of the hypotenuse and the opposite leg.

$$\csc \theta = \frac{\text{length of the hypotenuse}}{\text{signed length of leg opposite } \alpha}$$

$$= -\frac{8}{\sqrt{39}}$$

Rationalize the denominator.

$$= -\frac{8}{\sqrt{39}} \cdot \frac{\sqrt{39}}{\sqrt{39}}$$

$$= -\frac{8\sqrt{39}}{39}$$

Note: Problems 4.40–4.43 refer to an angle θ in standard position such that tan θ = $-\dfrac{7}{4}$ and sin θ < 0.

4.40 Identify the quadrant in which the terminal side of θ lies.

You are given a negative value for tan θ, and according to Problem 3.33, tangent is negative for angles that terminate in either the second or the fourth quadrant. You are also given sin θ < 0, which (according to Problem 3.32) is true for angles in the third and fourth quadrants. Therefore, the terminal side of θ must lie in the fourth quadrant to satisfy both of these conditions.

Note: Problems 4.40–4.43 refer to an angle θ in standard position such that tan θ = $-\dfrac{7}{4}$ and sin θ < 0.

4.41 Construct a right triangle using θ's reference angle α. Include the signed lengths of the legs in your diagram.

As Problem 4.40 explains, the terminal side of θ lies in the fourth quadrant. The reference angle α is bounded below by that terminal side and above by the positive x-axis. Notice that the horizontal leg of the right triangle extends right of the origin, so its signed length is positive; the vertical leg extends below the origin, so its signed length is negative.

Recall that tangent is equal to the quotient of the opposite and adjacent legs. Thus, the signed length of the vertical leg opposite α is –7, and the signed length of the horizontal leg adjacent to α is +4, as illustrated below.

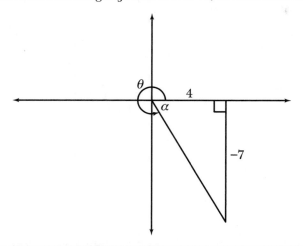

Note: Problems 4.40–4.43 refer to an angle θ in standard position such that $\tan \theta = -\dfrac{7}{4}$ and $\sin \theta < 0$.

4.42 Calculate $\cot \theta$.

You are given $\tan \theta = -7/4$. Recall that cotangent is the reciprocal of tangent. Thus, $\cot \theta = -4/7$.

Note: Problems 4.40–4.43 refer to an angle θ in standard position such that $\tan \theta = -\dfrac{7}{4}$ and $\sin \theta < 0$.

4.43 Calculate $\cos \theta$.

In Problem 4.41, you create a right triangle in the fourth quadrant and list the signed lengths of its legs. Expand the diagram, calculating the length of the hypotenuse according to the Pythagorean theorem.

$$a^2 + b^2 = c^2$$
$$4^2 + 7^2 = c^2$$
$$16 + 49 = c^2$$
$$65 = c^2$$
$$\sqrt{65} = c$$

The leg adjacent to reference angle α has signed length 4, the leg opposite α has signed length –7, and the length of the hypotenuse is $\sqrt{65}$, as illustrated in the following diagram.

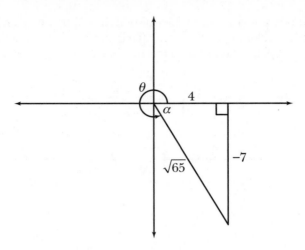

Use this diagram to calculate $\cos \theta$.

$$\cos \theta = \frac{\text{signed length of leg adjacent to } \alpha}{\text{length of the hypotenuse}}$$

$$= \frac{4}{\sqrt{65}}$$

$$= \frac{4}{\sqrt{65}} \cdot \frac{\sqrt{65}}{\sqrt{65}}$$

$$= \frac{4\sqrt{65}}{65}$$

Note: Problems 4.44–4.45 refer to an angle θ in standard position such that $\cos \theta = -\dfrac{3\sqrt{2}}{10}$ and $\sin \theta > 0$.

4.44 Calculate the following: $\sec \theta$, $\sin \theta$, and $\csc \theta$.

To calculate $\sec \theta$, take the reciprocal of the given cosine value and rationalize the denominator.

$$\sec \theta = -\frac{10}{3\sqrt{2}}$$

$$= -\frac{10}{3\sqrt{2}}\left(\frac{\sqrt{2}}{\sqrt{2}}\right)$$

$$= -\frac{10\sqrt{2}}{3\sqrt{4}}$$

$$= -\frac{10\sqrt{2}}{3 \cdot 2}$$

$$= -\frac{10\sqrt{2}}{6}$$

Simplify the fraction to lowest terms.

$$\sec\theta = -\frac{(10\div 2)\sqrt{2}}{(6\div 2)}$$

$$= -\frac{5\sqrt{2}}{3}$$

To calculate $\sin\theta$ (and its reciprocal $\sec\theta$), you must construct a right triangle based on α, the reference angle for θ. Given $\cos\theta = -\dfrac{3\sqrt{2}}{10}$, you can create a right triangle with an adjacent side of length $3\sqrt{2}$ and a hypotenuse of length 10.

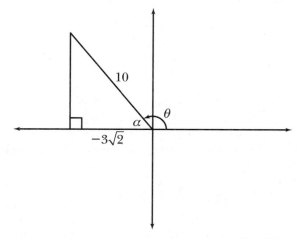

Note that θ terminates in the second quadrant, because the problem states that $\cos\theta < 0$ and $\sin\theta > 0$. Furthermore, the signed horizontal length is negative because it extends left of the origin.

Apply the Pythagorean theorem to calculate the length of the leg opposite α.

$$a^2 + b^2 = c^2$$

$$\left(3\sqrt{2}\right)^2 + b^2 = 10^2$$

$$(9)\left(\sqrt{4}\right) + b^2 = 100$$

$$9(2) + b^2 = 100$$

$$18 + b^2 = 100$$

$$b^2 = 100 - 18$$

$$b^2 = 82$$

$$b = \sqrt{82}$$

Note that the signed length of the vertical leg is $+\sqrt{82}$, because it extends above the origin. Now calculate $\sin\theta$.

$$\sin\theta = \frac{\text{signed length of leg opposite }\alpha}{\text{length of the hypotenuse}}$$

$$= \frac{\sqrt{82}}{10}$$

The reciprocal of $\sin\theta$ is equal to $\csc\theta$.

$$\csc\theta = \frac{10}{\sqrt{82}}$$

$$= \frac{10}{\sqrt{82}}\left(\frac{\sqrt{82}}{\sqrt{82}}\right)$$

$$= \frac{10\sqrt{82}}{82}$$

Reduce the fraction to lowest terms.

$$\csc\theta = \frac{(10 \div 2)\sqrt{82}}{(82 \div 2)}$$

$$= \frac{5\sqrt{82}}{41}$$

Thus, $\sec\theta = -\dfrac{5\sqrt{2}}{3}$, $\sin\theta = \dfrac{\sqrt{82}}{10}$, and $\csc\theta = \dfrac{5\sqrt{82}}{41}$.

Note: Problems 4.44–4.45 refer to an angle θ in standard position such that $\cos\theta = -\dfrac{3\sqrt{2}}{10}$ and $\sin\theta > 0$.

4.45 Calculate $\tan\theta$ and $\cot\theta$.

In Problem 4.44, you construct a right triangle in the second quadrant based on α, the reference angle for θ. You also determine the signed lengths of the legs of the right triangle. The diagram below summarizes the information.

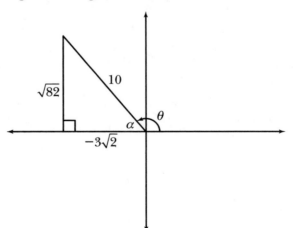

Because $\tan\theta = \dfrac{\sin\theta}{\cos\theta}$, and you figured out what $\sin\theta$ and $\cos\theta$ were in Problem 4.44, you could divide the sine by the cosine to get the answer.

Use the right triangle above to calculate $\tan\theta$.

$$\tan\theta = \frac{\text{signed length of leg opposite } \alpha}{\text{signed length of leg adjacent to } \alpha}$$

$$= \frac{+\sqrt{82}}{-3\sqrt{2}}$$

$$= -\frac{\sqrt{41 \cdot 2}}{3\sqrt{2}}$$

$$= -\frac{\sqrt{41} \cdot \cancel{\sqrt{2}}}{3 \cdot \cancel{\sqrt{2}}}$$

$$= -\frac{\sqrt{41}}{3}$$

Calculate the reciprocal of $\tan\theta$ and rationalize the denominator to compute $\cot\theta$.

$$\cot\theta = -\frac{3}{\sqrt{41}}$$

$$= -\frac{3}{\sqrt{41}}\left(\frac{\sqrt{41}}{\sqrt{41}}\right)$$

$$= -\frac{3\sqrt{41}}{41}$$

Thus, $\tan\theta = -\dfrac{\sqrt{41}}{3}$ and $\cot\theta = -\dfrac{3\sqrt{41}}{41}$.

Chapter 5
GRAPHING SINE AND COSINE FUNCTIONS

Get on the right wavelength

In the preceding chapters, you explore sine and cosine as numeric values based on the lengths of the sides of a right triangle. However, this does not provide a holistic picture of the trigonometric functions themselves. Graphing the functions and transforming those graphs are visually effective ways to better understand sine and cosine as periodic functions.

The graphs of cosine and sine are "periodic," which means that they repeat over and over as you travel horizontally across the graph. More specifically, they look like waves that wiggle above and below the x-axis to a maximum height of $y = 1$ and a minimum height of $y = -1$.

In this chapter, you graph sine and cosine, but you don't stop there. You also find out what happens to the graphs when you tweak them a little. For example, once you know what the graph of sin x looks like, you'll be able to quickly sketch the graphs of $3 \cdot \sin x$ and sin 2x.

However, before you even think about sine and cosine, it's worth spending some time with less complicated periodic graphs to see how they work.

Periodic Functions
Graphs that repeat and sort of look like a heartbeat

Problems 5.1–5.5 refer to the graph of periodic function f(x) below.

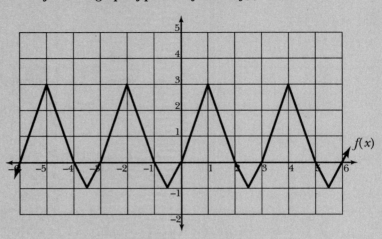

5.1 Identify the domain of $f(x)$.

The domain of a function is the collection of values for which the function is defined. Given a graph of a function, the most straightforward method to determine its domain is to imagine a series of vertical lines on the coordinate plane. Any vertical line drawn on the plane that intersects the graph represents an x-value for which the function is defined (and, therefore, a value in the domain of the function).

Consider the graph of $f(x)$ in this problem. If you were to draw a vertical line at $x = 1$, it would intersect the graph at height $y = 3$. Therefore, the function is defined at $x = 1$, and $f(1) = 3$. In fact, any vertical line drawn on the coordinate plane intersects this graph, so the domain of $f(x)$ is all real numbers.

The correct answer in set notation, if required by your instructor, is "$\{x \mid x$ is a real number$\}$." Note that the small vertical line between the x's in set notation is read "such that." Therefore, you read the answer as "the set of all x, such that x is a real number." Written in interval notation, the domain is $(-\infty, \infty)$.

> The collection of all the numbers you can plug into the function, usually in place of x or θ.

> Interval notation indicates a set of possible values between two boundaries. In this case the boundaries are −∞ and ∞ (the ends of the real number line), so all of the real numbers fall within the boundaries.

Problems 5.1–5.5 refer to the graph of periodic function f(x) in Problem 5.1.

5.2 Identify the range of $f(x)$.

In Problem 5.1, vertical lines in the coordinate plane help you determine the domain of $f(x)$, the set of possible inputs for a function. Similarly, horizontal lines that intersect the graph of a function represent members of the range— the set of outputs for the function.

Consider the given graph of $f(x)$. If you were to draw a horizontal line along the x-axis, which has equation $f(x) = 0$ (or $y = 0$), it would intersect the graph repeatedly. The function has a vertical height of 0 at $x = -6$, $x = -4$, $x = -3$, $x = -1$, $x = 0$, $x = 2$, $x = 3$, $x = 5$, and $x = 6$. Keep in mind that $f(x)$ is a periodic function, so the repetitive pattern of its graph will continue in both the positive and negative directions along the x-axis. It will intersect the graph infinitely many times over the entire domain of $f(x)$.

Horizontal lines as high as $f(x) = 3$ and as low as $f(x) = -1$ intersect the graph, but anything outside of those bounds will not. For instance, the horizontal line $f(x) = 4$ that is 4 units above the x-axis lies a full unit above the peaks of $f(x)$; it does not intersect the graph, so $f(x) = 4$ does not belong to the domain.

Because the horizontal lines between (and including) $f(x) = -1$ and $f(x) = 3$ all intersect the graph, the domain of $f(x)$ is $-1 \le f(x) \le 3$. In set notation, the range is $\{f(x) \mid -1 \le f(x) \le 3\}$, and in interval notation, the solution is $[-1,3]$.

> Parentheses in interval notation indicate that the boundaries are excluded, and brackets mean the boundaries are included. For example, [-1,3] means "all the numbers between -1 and 3, INCLUDING -1 and 3."

Problems 5.1–5.5 refer to the graph of periodic function f(x) in Problem 5.1.

5.3 Identify the minimum and maximum function values of $f(x)$.

Problem 5.2 states that the range of $f(x)$ is $-1 \le f(x) \le 3$, because the function has a maximum height of 3 and a minimum height of -1.

Problems 5.1–5.5 refer to the graph of periodic function f(x) in Problem 5.1.

5.4 Calculate the amplitude of $f(x)$.

The "amplitude" of a function is defined as the difference of the function's maximum and minimum values, divided by two. According to Problem 5.3, those values are 3 and -1, respectively.

$$\text{amplitude} = \frac{(\text{maximum} - \text{minimum})}{2}$$
$$= \frac{3 - (-1)}{2}$$
$$= \frac{3 + 1}{2}$$
$$= \frac{4}{2}$$
$$= 2$$

Visually, the amplitude of a periodic function represents the distance a function stretches from the horizontal line that is equidistant from the function's maximum and minimum values—the "middle" of the function, as illustrated below.

> Think of the max and min as the top and bottom of a big rectangle that stretches across the graph. If you divide that rectangle in half, the amplitude is the distance from the "middle" to the top and bottom of the rectangle.

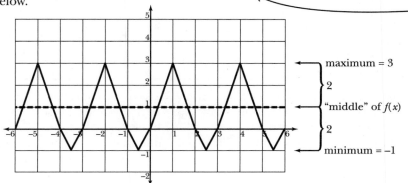

maximum = 3

2

"middle" of $f(x)$

2

minimum = −1

Problems 5.1–5.5 refer to the graph of periodic function f(x) in Problem 5.1.

5.5 Calculate the period of *f(x)*.

Recall that a periodic function infinitely repeats the same function values in the coordinate plane. The smallest horizontal distance over which the function does not repeat is called the period of the function. In other words, the period is the number of units, along the *x*-axis, it takes for the function to repeat itself.

Consider the graph of *f(x)*. It intersects the *x*-axis at *x* = 0, rises to a height of 3 at *x* = 1, returns to the *x*-axis at *x* = 2, dips below the *x*-axis until approximately *x* = 2.5, and then returns to the *x*-axis at *x* = 3. The function repeats the same pattern again between *x* = 3 and *x* = 6. In fact, the same pattern is repeated left of the *y*-axis, between *x* = −6 and *x* = −3 and again between *x* = −3 and *x* = 0.

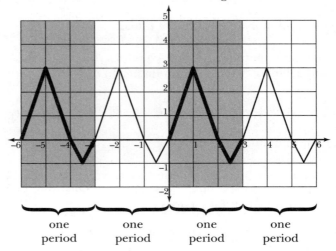

one period   one period   one period   one period

The values of *f(x)* repeat every 3 units along the *x*-axis, so the period of *f(x)* is 3. More formally, the period is equal to the difference of two *x*-values marking the beginning and end of one complete iteration of the graph. For instance, the rightmost full iteration of *f(x)* begins at *x* = 3 and ends at *x* = 6. Therefore, the period is 6 − 3 = 3.

Problems 5.6–5.9 refer to the graph of periodic function g(x) below.

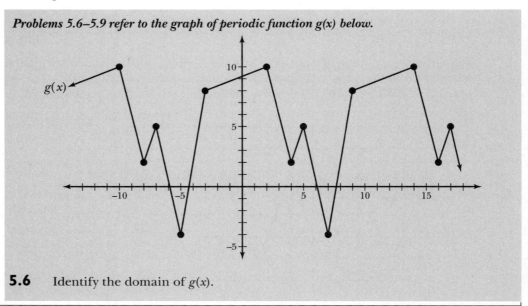

5.6 Identify the domain of *g(x)*.

As Problem 5.1 explains, any vertical line that intersects the graph of $g(x)$ represents a member of the domain. Any vertical line drawn on this coordinate plane will intersect the graph, so the domain of $g(x)$ is all real numbers. In set notation, the domain of $g(x)$ is $\{x \mid x$ is a real number$\}$; in interval notation, the domain is $(-\infty, \infty)$.

Problems 5.6–5.9 refer to the graph of periodic function g(x) in Problem 5.6.

5.7   Identify the range of $g(x)$. ←

> Different instructors require you to report the answer different ways, so the book lists the three most commonly used formats. You don't have to use all three—just figure out which your instructor prefers.

The function reaches a maximum height of 10 (at $x = -10$, $x = 2$, and $x = 14$) and a minimum height of -4 (at $x = -5$ and $x = 7$). All horizontal lines between $g(x) = 10$ and $g(x) = -4$ also intersect the graph, so the range of $g(x)$ is $-4 \leq g(x) \leq 10$. In set notation, the range is $\{g(x) \mid -4 \leq g(x) \leq 10\}$, and in interval notation, the range is $[-4, 10]$.

Problems 5.6–5.9 refer to the graph of periodic function g(x) in Problem 5.6.

5.8   Calculate the amplitude of $g(x)$.

According to Problem 5.7, the maximum and minimum values of $g(x)$ are 10 and -4, respectively. Substitute these values into the amplitude formula presented in Problem 5.4.

$$\begin{aligned}
\text{amplitude} &= \frac{(\text{maximum} - \text{minimum})}{2} \\
&= \frac{10 - (-4)}{2} \\
&= \frac{10 + 4}{2} \\
&= \frac{14}{2} \\
&= 7
\end{aligned}$$

Problems 5.6–5.9 refer to the graph of periodic function g(x) in Problem 5.6.

5.9   Calculate the period of $g(x)$.

To identify the period, locate points on the graph that have the same height and determine whether or not they represent a point after which the graph repeats. For example, the graph of $g(x)$ reaches its maximum height at $x = 2$ and again at $x = 14$. In both cases, the graph immediately drops to a height of $g(x) = 2$ exactly 2 units to the right: $g(4) = g(16) = 2$. The pattern is further confirmed when you move 1 unit to the right; both graphs jump 3 units higher: $g(5) = g(17) = 5$.

In the following diagram, one period of $g(x)$ is shaded, representing the horizontal distance between two points at which $g(x)$ reaches its maximum value. You are not required to use the maximum values of a periodic graph to mark the "beginning" and "end" of one period. You may choose to mark the beginning and end of a period using any two corresponding points, such as minimum points or x-intercepts, and your final answer will be the same.

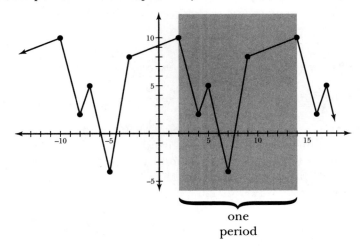

one
period

One period of $g(x)$ is bounded on the left by the vertical line $x = 2$ and is bounded on the right by the vertical line $x = 14$. The period is equal to the difference between these x values.

$$\text{period of } g(x) = 14 - 2$$
$$= 12$$

Note: Problems 5.10–5.12 refer to the graph of periodic function $h(x)$ below.

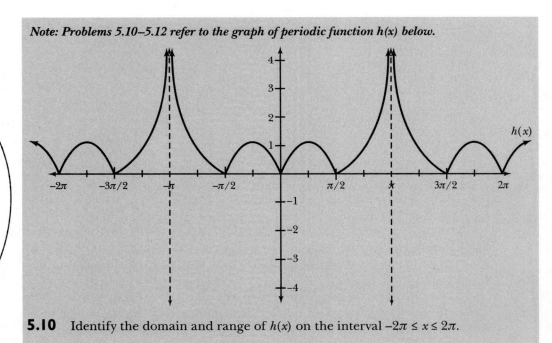

Asymptotes are lines that a function approaches but never reaches; they are usually drawn as dotted lines. While they are not technically part of the function, they definitely shape its graph.

5.10 Identify the domain and range of $h(x)$ on the interval $-2\pi \le x \le 2\pi$.

The graph has vertical asymptotes at $x = -\pi$ and $x = \pi$, so vertical lines drawn at these x-values will not intersect the graph. Whereas $h(x)$ gets infinitely close to the asymptotes as you approach those dotted vertical lines from the right and the left, the graph never actually intersects them. Hence, the graph is undefined at $x = -\pi$ and $x = \pi$. The domain of $h(x)$ is all real numbers, excluding $-\pi$ and π. In set notation, the domain is $\{x \mid x$ is a real number such that $-2\pi \le x \le 2\pi$, $x \ne -\pi$, and $x \ne \pi\}$. In interval notation, the domain is $[-2\pi, -\pi) \cup (-\pi, \pi) \cup (\pi, 2\pi]$.

The problem only asks you to identify the domain for the portion of $h(x)$ displayed in this diagram ($-2\pi \le x \le 2\pi$), but $h(x)$ is periodic. All of the function values (and the undefined values) will repeat infinitely.

Horizontal lines drawn on the coordinate plane will intersect $h(x)$ at and above the x-axis, so the range of $h(x)$ is all real numbers greater than or equal to 0. In set notation, the range is $\{h(x) \mid h(x) \ge 0\}$. In interval notation, the range is $[0, \infty)$.

Note: Problems 5.10–5.12 refer to the graph of periodic function h(x) in Problem 5.10.

5.11   Calculate the amplitude of $h(x)$.

The minimum value of the function is $h(x) = 0$, but the function has no finite maximum value. Its graph increases without bound as it approaches the vertical asymptotes. Therefore, amplitude is not defined for $h(x)$.

Note: Problems 5.10–5.12 refer to the graph of periodic function h(x) in Problem 5.10.

5.12   Calculate the period of $h(x)$.

One full period of the graph occurs between $x = -\pi$ and $x = \pi$. To compute the period, calculate the difference of the x-values.

$$\text{period} = \pi - (-\pi)$$
$$= \pi + \pi$$
$$= 2\pi \ \leftarrow$$

Remember, $h(x)$ is undefined at π, at the asymptote. If you add the period 2π to (or subtract it from) that undefined x-value, you get other x-values at which $h(x)$ is undefined. For example, $\pi + 2\pi = 3\pi$, so $h(x)$ is also undefined at 3π.

Transforming Periodic Graphs
Move, stretch, squish, and flip graphs

Note: Problems 5.13–5.17 refer to the graph of periodic function f(x) below.

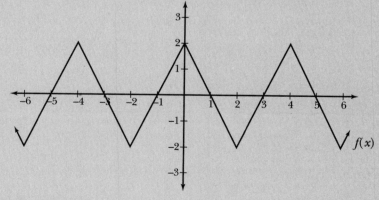

5.13   Graph $f(x) - 1$.

It is helpful to identify key points on a graph before you attempt to transform it. For example, (–6,–2), (–4,2), (–2,–2), (0,2), (2,–2), (4,2), and (6,–2) may be considered key points on the graph of $f(x)$ because they are the points at which $f(x)$ changes direction.

When you transform a graph, think about the way in which the transformation affects the individual points you've identified. For example, to graph $f(x) - 1$, you subtract 1 from each of the $f(x)$, or y, coordinates of the points. The transformed key points on the graph of $f(x) - 1$ are (–6,–3), (–4,1), (–2,–3), (0,1), (2,–3), (4,1), and (6,–3), as illustrated below.

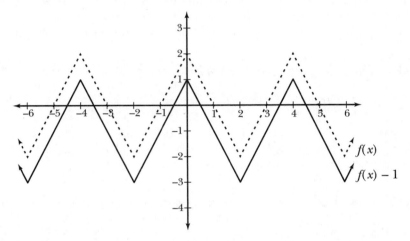

Subtracting a constant from a function shifts its graph vertically downward the corresponding number of units.

Note: Problems 5.13–5.17 refer to the graph of periodic function f(x) in Problem 5.13.

5.14 Graph $f(x) + 2$.

In other words, a real number (not a variable).

As Problem 5.13 explains, subtracting a constant from a function shifts its graph downward the corresponding number of units. Similarly, adding a constant to a function shifts its graph upwards. Thus, the graph of $f(x) + 2$ is simply the graph of $f(x)$ shifted up 2 units on the coordinate plane.

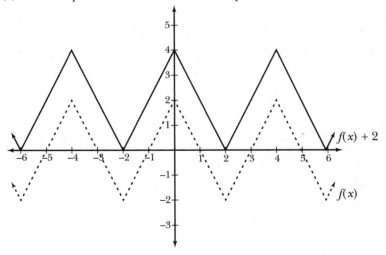

Note: Problems 5.13–5.17 refer to the graph of periodic function f(x) in Problem 5.13.

5.15 Graph $f(x + 1)$.

Problems 5.13–5.14 explore the transformations that occur when constants are added to, or subtracted from, a function. This problem is slightly different, because it adds 1 to the *input* of the function, rather than the output. Accordingly, the graph will shift in a horizontal direction rather than vertically.

Think about how adding 1 to the input of a function affects its graph. If you substitute $x = 0$ into $f(x)$, the output is 2, so $f(0) = 2$. You will get the same output (2) from the function $f(x + 1)$ if you evaluate it for $x = -1$.

Because the graph of f(x) passes through the point (x,f(x)) = (0,2).

$$f((-1)+1) = f(-1+1)$$
$$= f(0)$$
$$= 2$$

Because 1 is added to each input value, $f(x + 1)$ reaches $f(x)$'s function values exactly 1 unit sooner than $f(x)$. Interpreted graphically, this means that $f(x + 1)$ is the graph of $f(x)$ shifted 1 unit to the left, as illustrated below.

A horizontal shift is also called a "phase shift."

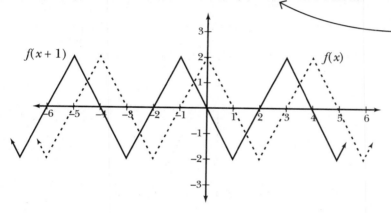

Note: Problems 5.13–5.17 refer to the graph of periodic function f(x) in Problem 5.13.

5.16 Graph $f(x - 3)$.

As Problem 5.15 explains, adding a constant to the input of a function—in other words, adding it inside the parentheses rather than outside—shifts the graph to the left. Similarly, subtracting a constant from the input of a function shifts the graph to the right. Therefore, the graph of $f(x - 3)$ is the graph of $f(x)$ shifted 3 units to the right.

In the following diagram, corresponding key points on the graph are compared. The y-intercept of the original, dotted graph of $f(x)$ is $(0,2)$. This point is shifted 3 units to the right in the graph of $f(x - 3)$, and the corresponding coordinate on the new graph is $(3,2)$.

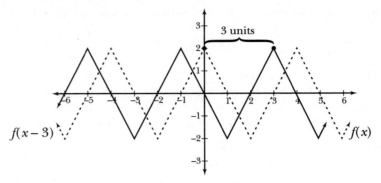

Many students errantly interpret $f(x-3)$ as a leftward horizontal shift, because they equate a negative number with leftward movement in the coordinate plane. Remember, adding within parentheses results in a leftward shift and subtracting within parentheses results in a rightward shift.

Note: Problems 5.13–5.17 refer to the graph of periodic function f(x) in Problem 5.13.

5.17 Graph $f(x-1) + 2$.

The graph of $f(x-1) + 2$ undergoes two transformations from the base graph of $f(x)$. Subtracting 1 within parentheses results in a horizontal shift 1 unit to the right. Adding 2 results in a vertical shift of 2 units upward, as illustrated below.

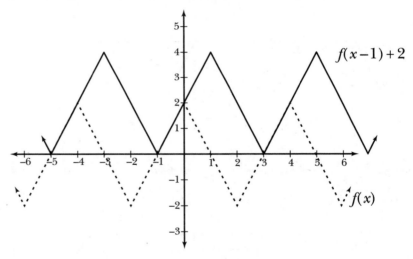

Consider key point $(0,2)$ on the original graph of $f(x)$. Shifting this point 1 unit to the right and 2 units up produces new point $(0 + 1, 2 + 2) = (1,4)$ on the graph of $f(x-1) + 2$.

Note: Problems 5.18–5.21 refer to the graph of periodic function g(x) below.

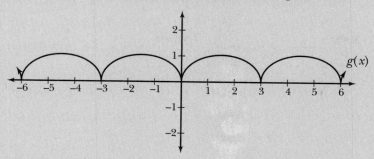

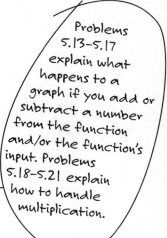

Problems 5.13–5.17 explain what happens to a graph if you add or subtract a number from the function and/or the function's input. Problems 5.18–5.21 explain how to handle multiplication.

5.18 Graph 2g(x).

Multiplying a function by 2 stretches the graph of the function vertically by a factor of 2. In other words, its maximum and minimum values stretch twice as far from the x-axis. Consider the graph of g(x); its maximum value is 1 and its minimum value is 0. Once transformed, the new graph 2g(x) will have a maximum value of $1 \cdot 2 = 2$, but its minimum value remains unchanged: $0 \cdot 2 = 0$.

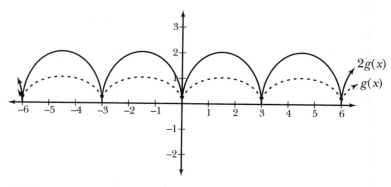

Much like adding a constant to a function shifts its graph vertically (as explained in Problems 5.13–5.14), multiplying a function by a constant stretches its graph vertically.

Note: Problems 5.18–5.21 refer to the graph of periodic function g(x) in Problem 5.18.

5.19 Graph $\frac{1}{3}g(x)$.

Multiplying a function by a positive real number less than 1 will compress the graph, rather than stretch it. The original graph of g(x) has maximum and minimum values of 1 and 0, respectively. Multiply each by 1/3 to calculate the maximum and minimum values of (1/3)(g(x)): (1/3)(1) = 1/3 and (1/3)(0) = 0.

Note that the scale of the following graph is slightly modified to better visualize the effects of this transformation. The original, dotted graph of g(x) stretches to a maximum height of 1, but the transformed graph of (1/3)(g(x)) reaches only one-third as high.

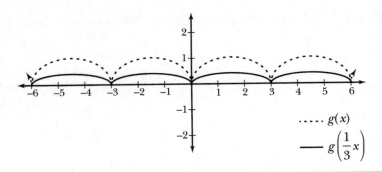

Note: *Problems 5.18–5.21 refer to the graph of periodic function g(x) in Problem 5.18.*

5.20 Graph $g(3x)$.

Multiplying the input of a function—in other words, multiplying inside parentheses instead of outside—either compresses or stretches the graph horizontally. The coefficient of the independent variable (in this case 3 is the coefficient of the independent variable x) identifies the number of periods that will repeat within one period of the original graph.

This transformation is a bit more complex than simple shifts, so it bears some further explanation. To calculate the period of a transformed graph, divide the period of the original graph by the coefficient within parentheses. In this example, the original graph of $g(x)$ has a period of 3. Note that the period is a positive value, so you must take the absolute value of the quotient.

$$\text{period of } g(3x) = \left| \frac{\text{period of } g(x)}{3} \right|$$
$$= \left| \frac{3}{3} \right|$$
$$= |1|$$
$$= 1$$

> So the graph of g(3x) has three humps between x = 0 and x = 3 instead of one hump like g(x).

The period of the transformed graph $g(3x)$ is 1, so rather than requiring a horizontal distance of 3 to complete one period of the graph, $g(3x)$ requires a horizontal distance of only 1. Therefore, three complete periods of the graph appear where only one period of the graph appeared originally, as illustrated below.

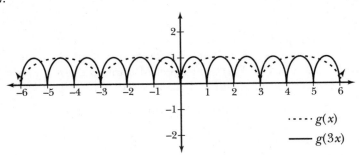

Note: Problems 5.18–5.21 refer to the graph of periodic function g(x) in Problem 5.18.

5.21 Graph $g\left(\dfrac{1}{2}x\right)$.

As Problem 5.20 explains, multiplying the input of a function by a constant either compresses or stretches the graph horizontally. While multiplying by a positive constant greater than 1 results in a compressed graph (as demonstrated in Problem 5.20), multiplying by a positive constant less than 1 elongates the graph.

To compute the period of the transformed graph $g\left(\dfrac{1}{2}x\right)$, divide the period of the original graph $g(x)$ by the coefficient of x.

$$\text{period of } g\left(\frac{1}{2}x\right) = \left|\frac{\text{period of } g(x)}{1/2}\right|$$

$$= \left|\frac{3}{1/2}\right|$$

Multiply the numerator and denominator of the complex fraction by 2 to simplify it.

$$= \left|\frac{3(2)}{(1/2)(2)}\right|$$

$$= \left|\frac{6}{1}\right|$$

$$= 6$$

The transformed graph has a period of 6, so one period is twice as long as a period of $g(x)$, as illustrated below.

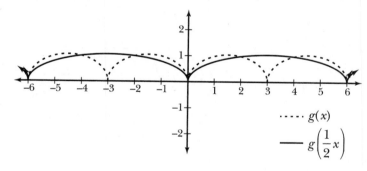

The coefficient of x tells you how much of the new graph will appear in one period of the old graph. In this problem, $g(x)$ has a period of 3, so you know that only 1/2 of the original graph fits in the same horizontal space. In other words, the period of the new graph is twice the period of $g(x)$.

Note: Problems 5.22–5.25 refer to the graph of periodic function h(x) below.

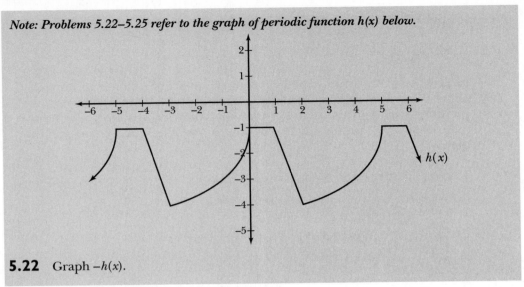

5.22 Graph −h(x).

Multiplying a function by −1 reflects its graph across the *x*-axis. To generate the transformed graph −h(x), multiply the h(x)-coordinates (or *y*-coordinates) of each point on the original graph by −1. For example, key points on the original graph h(x) include (−4,−1), (−3,−4), (0,−1), (1,−1), (2,−4), and (5,−1). Therefore, the graph of −h(x) will contain points (−4,1), (−3,4), (0,1), (1,1), (2,4), and (5,1).

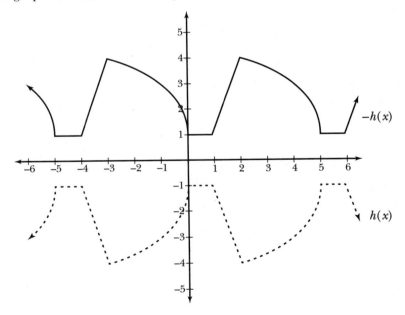

Note: Problems 5.22–5.25 refer to the graph of periodic function h(x) in Problem 5.22.

5.23 Graph h(−x).

Multiplying the input of a function by −1 reflects the graph across the *y*-axis. To generate the transformed graph h(−x), multiply the *x*-coordinate of each point on the original graph of h(x) by −1. For example, the key points on the original graph h(x) include (−4,−1), (−3,−4), (0,−1), (1,−1), (2,−4), and (5,−1); they correspond to the points (4,−1), (3,−4), (0,−1), (−1,−1), (−2,−4), and (−5,−1) on the transformed graph h(−x).

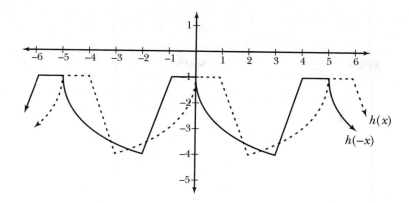

Note: Problems 5.22–5.25 refer to the graph of periodic function h(x) in Problem 5.22.

5.24 Graph $-2h(x)$.

This problem combines two transformations: multiplying a function by -1 (which reflects the graph across the x-axis) and multiplying a function by 2 (which stretches the graph vertically). To generate the graph, multiply the y-coordinate of each point on the original graph of $h(x)$ by -2. For example, the point $(0,-1)$ on $h(x)$ corresponds with the point $(0,2)$ on $-2h(x)$.

In other words, this combines Problem 5.18 (where you stretch by a factor of 2) and Problem 5.22 (where you reflect across the x-axis).

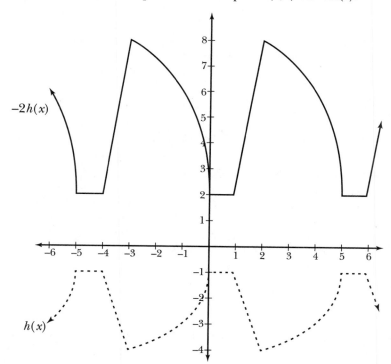

Note: Problems 5.22–5.25 refer to the graph of periodic function h(x) in Problem 5.22.

5.25 Graph $h(-x) + 4$.

This problem requires you to perform two transformations. Because the input of the function is $-x$ rather than x, you must reflect $h(x)$ across the y-axis. Then, because 4 is added to $h(-x)$, you must shift the graph of $h(x)$ up 4 units.

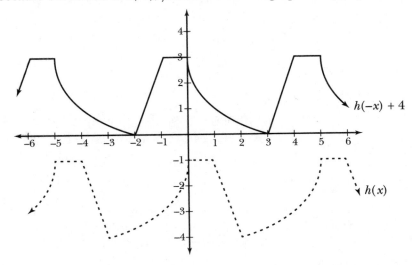

Note that horizontal and vertical shifts should be completed after all reflections, stretches, and compressions have been completed. In other words, address multiplied constants before constants that are added or subtracted.

Sine Functions

Oscillating between –1 and 1, until you transform them

5.26 Graph $\sin \theta$ on the interval $-2\pi \le \theta \le 2\pi$.

The graph of function $f(\theta) = \sin \theta$ is a periodic wave that intersects the horizontal axis at multiples of π. Consider one period of the graph, bounded on the left by $\theta = 0$ and on the right by $\theta = 2\pi$. The following sine values are derived from the trigonometric unit circle: $\sin 0 = \sin 2\pi = 0$, $\sin \pi/2 = 1$, $\sin \pi = 0$, and $\sin 3\pi/2 = -1$. Therefore, the points $(0,0)$, $(\pi/2,1)$, $(\pi,0)$, and $(3\pi/2,-1)$ belong to the graph of $\sin \theta$, as illustrated below.

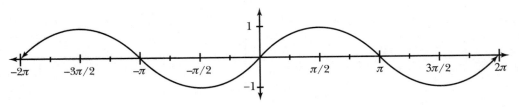

5.27 Graph $f(\theta) = \sin(\theta - \pi)$.

As Problem 5.16 explains, subtracting π from the input of a function shifts the graph of $\sin \theta$ a total of π units to the right.

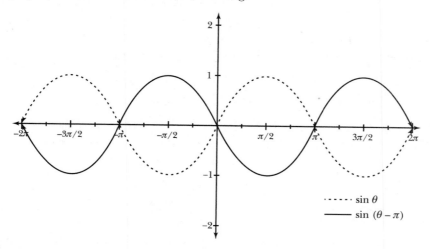

$\cdots\cdots$ $\sin \theta$

———— $\sin (\theta - \pi)$

5.28 Graph $f(\theta) = \sin \theta + 1$.

As Problem 5.14 explains, adding 1 to a function shifts its graph up 1 unit.

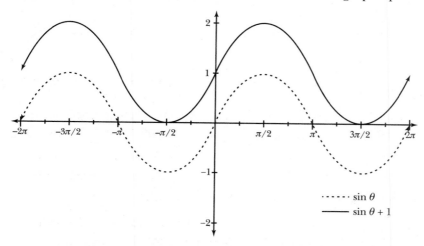

$\cdots\cdots$ $\sin \theta$

———— $\sin \theta + 1$

5.29 Graph $f(\theta) = -\sin \theta$.

As Problem 5.22 explains, multiplying a function by -1 reflects its graph across the x-axis.

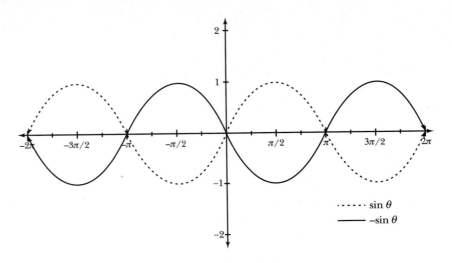

····· sin θ
—— −sin θ

5.30 Graph $f(\theta) = \sin(-\theta)$.

As Problem 5.23 explains, multiplying the input of a function by −1 reflects its graph across the *y*-axis.

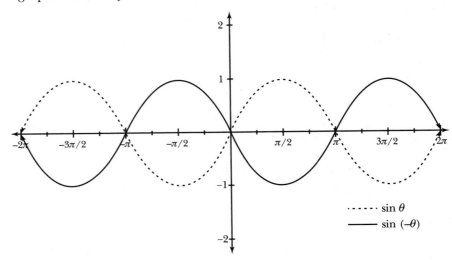

····· sin θ
—— sin (−θ)

This is true for this graph, not for all graphs.

Note that reflecting sin θ across either the *x*- or *y*-axis produces the same graph.

5.31 Graph $f(\theta) = 3\sin\theta$ and calculate the amplitude of $f(\theta)$.

The function $y = \sin\theta$ has an amplitude of 1, because the untransformed graph of sin θ stretches to a maximum vertical height of 1 and a minimum height of −1. According to Problem 5.18, multiplying sin θ by 3 changes the amplitude of the graph, which now reaches maximum and minimum heights of 3 and −3, respectively. Therefore, the amplitude of $f(\theta)$ is 3.

If a trig function is multiplied by a real number, the absolute value of that real number is the amplitude of the graph. In other words, the amplitude of $f(x) = A \cdot \sin x$ is |A|. Amplitude is always a positive value.

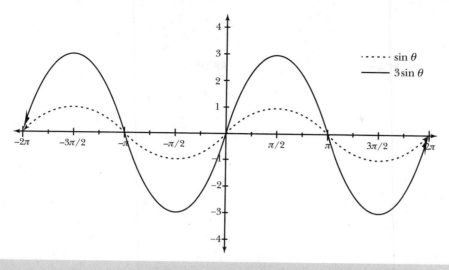

5.32 Graph $f(\theta) = \sin 2\theta$ and calculate the period of $f(\theta)$.

As Problem 5.20 explains, multiplying the input of a function by a real number greater than 1 causes the graph to compress horizontally. In this problem, you compress the graph of $y = \sin \theta$ by a factor of 2 because the input θ is multiplied by 2. In other words, two complete periods of $\sin \theta$ appear between $\theta = 0$ and $\theta = 2\pi$ on the horizontal axis, in the interval where one period of the untransformed graph normally appears.

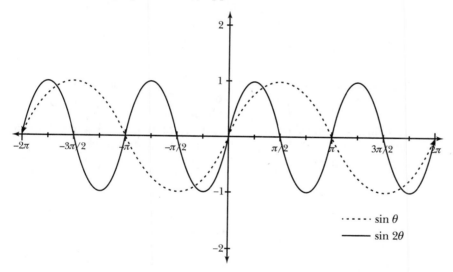

To calculate the period of $f(\theta)$, divide the period of the untransformed graph $y = \sin \theta$ by the coefficient of θ in the transformed graph $f(\theta)$.

If the new graph has equation $f(x) = \sin (Bx)$, then divide by B.

$$\text{period of } f(\theta) = \left| \frac{\text{period of } \sin\theta}{\text{coefficient of } \theta \text{ in } f(\theta)} \right|$$

$$= \left| \frac{2\pi}{2} \right|$$

$$= \left| \frac{\cancel{2}\pi}{\cancel{2}} \right|$$

$$= \left| \frac{\pi}{1} \right|$$

$$= \pi$$

5.33 Graph $f(\theta) = \sin\left(\theta + \dfrac{\pi}{4}\right) - 2.$

Adding $\pi/4$ to the input of the function $y = \sin\theta$ shifts its graph left $\pi/4$ units; subtracting 2 from the function shifts the graph down 2 units.

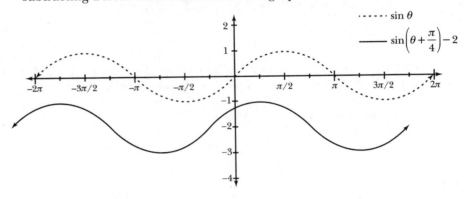

5.34 Graph $f(\theta) = -\dfrac{1}{3}\sin\left(\dfrac{1}{4}\theta\right).$

Transform the graph of $y = \sin\theta$, compressing it to one-third of its original amplitude, reflecting it about the x-axis, and then stretching the graph horizontally, so that only one-fourth of the original graph of $\sin\theta$ appears between 0 and 2π.

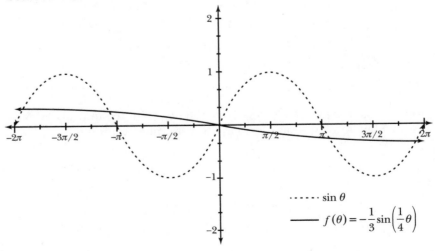

Cosine Functions

The sine graph schooched slightly to the left

5.35 Graph $f(\theta) = \cos\theta$ on the interval $-2\pi \le \theta \le 2\pi$.

According to the trigonometric unit circle, $\cos 0 = \cos 2\pi = 1$, $\cos \pi/2 = 0$, $\cos \pi = -1$, and $\cos 3\pi/2 = 0$. The graph of $f(\theta) = \cos\theta$ is very similar to the graph of $y = \sin\theta$. In fact, as Problem 5.36 explains, the two graphs are equivalent apart from a simple horizontal shift.

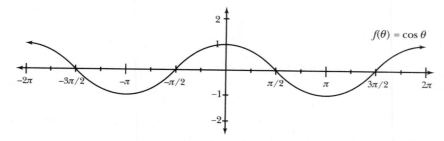

5.36 Graph $f(\theta) = \cos\theta$ and $g(\theta) = \sin\left(\theta + \dfrac{\pi}{2}\right)$ to visually verify that $f(\theta) = g(\theta)$.

The graph of $g(\theta)$ shifts the graph of $y = \sin\theta$ a horizontal distance of $\pi/2$ units to the left. The resulting graph coincides with the graph of $f(\theta) = \cos\theta$.

Overlaps

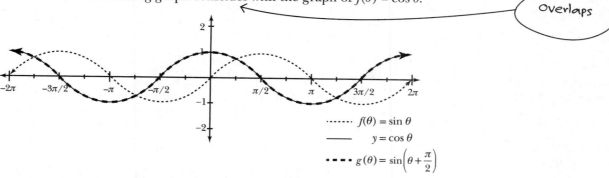

5.37 Graph $f(\theta) = 4\cos\left(\theta - \dfrac{3\pi}{4}\right)$.

Multiplying the cosine function by 4 stretches its amplitude by a factor of 4. Thus, the graph of $f(\theta)$ stretches to a maximum height of 4 and a minimum height of -4. Subtracting $3\pi/4$ from the input of the function shifts the graph $3\pi/4$ units to the right.

Each of the tic marks along the x-axis represents a distance of $\pi/4$, so move the graph three tic marks to the right.

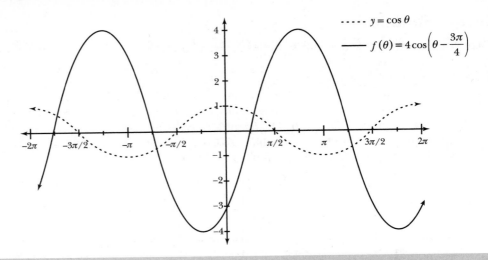

5.38 Graph $f(\theta) = \cos(-3\theta) + 2$.

When graphing a function using numerous transformations, you should complete them in the following order:

1. Horizontal and vertical stretching/compression, which occur when either the function or its input is multiplied by a real number

2. Reflections about the x- or y-axes, which occur when either the function or its input is multiplied by −1

3. Horizontal and vertical shifts, which occur when a real number is added to (or subtracted from) a function or its input

Technically speaking, the first two steps in the list above may be reversed, but horizontal and vertical shifts must always be completed last.

In this problem, the input of $y = \cos \theta$ is multiplied by −3, so the graph of $y = \cos \theta$ is reflected about the y-axis and compressed by a factor of 3. Then, because 2 is added to the function, the graph is shifted upward a distance of 2 units.

The graph of $y = \cos\theta$ is "y-symmetric," because the right and left sides of the graph look like reflections of each other, as if the y-axis were a mirror. When you reflect a y-symmetric graph across the y-axis, it looks exactly the same as it did before you reflected it.

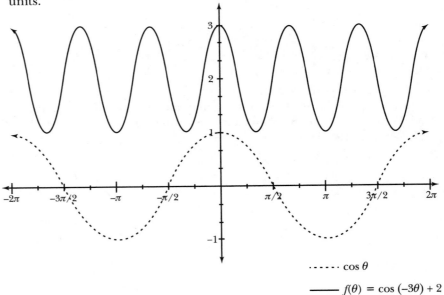

Note: Problems 5.39–5.40 explain how to graph $g(\theta) = \cos\left(\dfrac{1}{2}\theta - \pi\right)$, *which contains a specific pair of transformations: a horizontal shift and a period change.*

5.39 Calculate the θ-coordinates on the graph of $g(\theta)$ that correspond with $\theta = 0$ and $\theta = 2\pi$ on the untransformed graph of $y = \cos\theta$.

If you multiply the input of a function by a real number *and* add a value to (or subtract a value from) the input as well, you have to apply a specific technique to graph this pair of transformations. ⟵

> *In other words, if you graph a function $f(x) = \cos(Bx + C)$, where $B \neq 1$ and $C \neq 0$, then you have to apply this technique. In this problem, $B = 1/2$ and $C = -\pi$.*

The period of the untransformed graph $y = \cos\theta$ is 2π. Recall that one full period of $\cos\theta$ occurs between the horizontal boundaries of $\theta = 0$ and $\theta = 2\pi$. To locate the starting and ending points of one period on the transformed graph $g(\theta)$, set the parenthetical quantity $\left(\dfrac{1}{2}\theta - \pi\right)$ equal to 0 and 2π and solve for θ.

$$\frac{1}{2}\theta - \pi = 0 \qquad\qquad \frac{1}{2}\theta - \pi = 2\pi$$

$$\frac{1}{2}\theta = \pi \qquad\qquad\quad \frac{1}{2}\theta = 3\pi$$

$$2\left(\frac{1}{2}\theta\right) = 2(\pi) \qquad 2\left(\frac{1}{2}\theta\right) = 2(3\pi)$$

$$\theta = 2\pi \qquad\qquad\quad \theta = 6\pi$$

Note: Problems 5.39–5.40 explain how to graph $g(\theta) = \cos\left(\dfrac{1}{2}\theta - \pi\right)$, *which contains a specific pair of transformations: a horizontal shift and a period change.*

5.40 Graph $g(\theta)$.

According to Problem 5.39, the period of $y = \cos\theta$ that normally lies between $\theta = 0$ and $\theta = 2\pi$ lies between $\theta = 2\pi$ and $\theta = 6\pi$ on the transformed graph ⟵ of $g(\theta)$. Note that neither the amplitude of $y = \cos\theta$ nor its vertical position changes.

> *Look at the graph of cosine between $\theta = 0$ and $\theta = 2\pi$. Draw that same shape between the new boundaries, $\theta = 2\pi$ and $\theta = 6\pi$.*

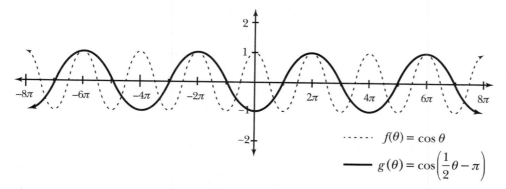

5.41 Graph $f(\theta) = \cos\left(4\theta - \dfrac{\pi}{2}\right)$ and calculate its period.

Apply the technique described in Problems 5.40–5.41. Because one full period of $y = \cos\theta$ lies between $\theta = 0$ and $\theta = 2\pi$, set the parenthetical quantity $\left(4\theta - \dfrac{\pi}{2}\right)$ equal to both values of θ to identify the values of θ that bound one period of $f(\theta)$.

$$4\theta - \frac{\pi}{2} = 0 \qquad\qquad 4\theta - \frac{\pi}{2} = 2\pi$$

$$4\theta = \frac{\pi}{2} \qquad\qquad 4\theta = 2\pi + \frac{\pi}{2}$$

$$\left(\frac{1}{4}\right)(4\theta) = \left(\frac{1}{4}\right)\left(\frac{\pi}{2}\right) \qquad\qquad 4\theta = \frac{4\pi}{2} + \frac{\pi}{2}$$

$$\theta = \frac{\pi}{8} \qquad\qquad 4\theta = \frac{5\pi}{2}$$

$$\left(\frac{1}{4}\right)(4\theta) = \left(\frac{1}{4}\right)\left(\frac{5\pi}{2}\right)$$

$$\theta = \frac{5\pi}{8}$$

The portion of the untransformed graph $y = \cos\theta$ that lies between $\theta = 0$ and $\theta = 2\pi$ will lie between $\theta = \pi/8$ and $\theta = 5\pi/8$ on the transformed graph of $f(\theta)$.

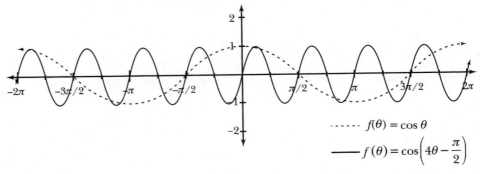

The period of a transformed graph is equal to the absolute value of the original, untransformed period divided by the coefficient of θ.

$$\text{period of } f(\theta) = \left|\frac{\text{period of }\cos\theta}{\text{coefficient of }\theta\text{ in } f(\theta)}\right|$$

$$= \left|\frac{2\pi}{4}\right|$$

$$= \frac{\pi}{2}$$

Note that the difference of the θ-values calculated above is also equal to the period of $f(\theta)$.

$$\text{period of } f(\theta) = \frac{5\pi}{8} - \frac{\pi}{8}$$

$$= \frac{4\pi}{8}$$

$$= \frac{\pi}{2}$$

Chapter 6
GRAPHING OTHER TRIGONOMETRIC FUNCTIONS

Tan, cot, sec, and csc

Chapter 5 explores graphical transformations, the effects that constants have on the graphs of functions depending upon their placement within the function. For example, multiplying by a constant within a function stretches or compresses that function, whereas adding or subtracting a value within the function shifts its graph but does not affect its shape. In this chapter, you are introduced to the graphs of the four remaining trigonometric functions, and you apply the same transformations to these slightly more complex graphs.

Sine and cosine are defined for all real numbers, but that's not true for the other four trig functions. Because tangent, cotangent, secant, and cosecant are all defined as quotients containing sine and/or cosine, graphs in this chapter are a little more complicated. Whenever quotients are involved, you have to worry about dividing by zero, which is not allowed.

Any time the denominator of a trig function could equal zero, you have to create an invisible fence in the graph—like the kind people dig into their yards to keep their dogs from running away. These "fences" are vertical lines called asymptotes, which functions (like dogs) can approach but not cross.

Tangent

Large function values, limited domain

6.1 Graph $f(\theta) = \tan \theta$ on the interval $-2\pi \le \theta \le 2\pi$.

Recall that $\tan \theta$ is defined as the quotient of $\sin \theta$ and $\cos \theta$. Therefore, $\tan \theta$ is equal to 0 whenever $\sin \theta = 0$, but $\tan \theta$ is undefined whenever $\cos \theta = 0$. Points of interest on the graph of $f(\theta)$ include $(0, 0)$, $(\pi/4, 1)$, and $(-\pi/4, -1)$.

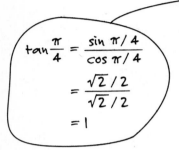

$$\tan \frac{\pi}{4} = \frac{\sin \pi/4}{\cos \pi/4}$$
$$= \frac{\sqrt{2}/2}{\sqrt{2}/2}$$
$$= 1$$

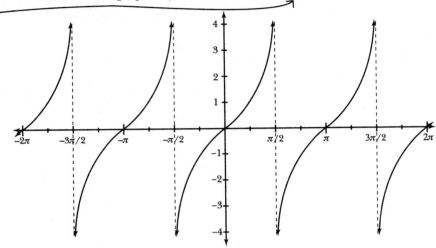

6.2 Based on the graph you created in Problem 6.1, identify the domain, range, and period of $f(\theta) = \tan \theta$.

As Problem 6.1 explains, $f(\theta) = (\sin \theta)/(\cos \theta)$ is undefined whenever its denominator is equal to 0, as indicated by vertical asymptotes on the graph. Therefore, the domain of $f(\theta)$ is all real numbers such that $\theta \ne (k\pi)/2$, where k is an odd number.

In other words, the domain of cosine excludes $\dfrac{1\pi}{2}, \dfrac{3\pi}{2}, \dfrac{5\pi}{2}, \dfrac{7\pi}{2}, \dfrac{9\pi}{2}, \ldots$ as well as the opposites of those numbers.

Apply the technique first described in Problem 5.2 to identify the range of the graph. Note that $f(\theta) = \tan \theta$ has neither a minimum nor a maximum value, and any horizontal line drawn on the coordinate plane intersects the graph. Therefore, the range of $f(\theta)$ is all real numbers.

One full period of the graph appears between consecutive asymptotes. For example, one period of $f(\theta) = \tan \theta$ lies between $\theta = -\pi/2$ and $\theta = \pi/2$. Subtract those θ-values to calculate the period.

$$\text{period of } f(\theta) = \frac{\pi}{2} - \left(-\frac{\pi}{2}\right)$$
$$= \frac{\pi}{2} + \frac{\pi}{2}$$
$$= \frac{2\!\!\!/\,\pi}{2\!\!\!/}$$
$$= \pi$$

6.3   Graph $g(\theta) = 2\tan\theta$.

Multiplying the function $y = \tan\theta$ by 2 doubles each of the function values. Consider the point $(\pi/4, 1)$ on the untransformed graph of $y = \tan\theta$. Multiply its y-coordinate by 2 to generate the corresponding point on the graph of $g(\theta)$. You conclude that $g(\theta)$ passes through the point $(\pi/4, 2)$.

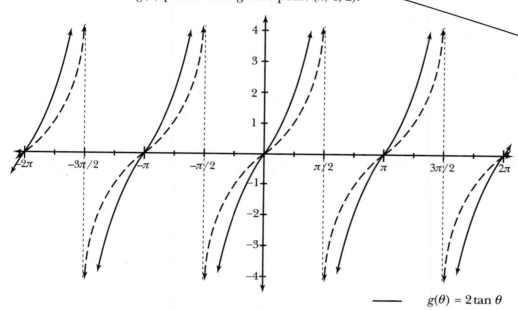

Flip back to Problem 5.18 if you need to review this transformation.

$$\text{———} \quad g(\theta) = 2\tan\theta$$
$$\text{- - -} \quad y = \tan\theta$$

6.4   Graph $h(\theta) = \tan\left(\theta + \dfrac{\pi}{2}\right)$ and identify the domain of $h(\theta)$.

Adding $\pi/2$ to the input of the untransformed function $y = \tan\theta$ shifts its graph $\pi/2$ units to the left.

This is explained in Problem 5.15.

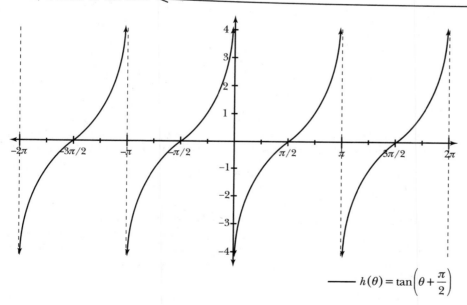

$$\text{———} \quad h(\theta) = \tan\left(\theta + \dfrac{\pi}{2}\right)$$

6.5 Identify the function $k(\theta)$ graphed below, a transformation of the graph $y = \tan \theta$.

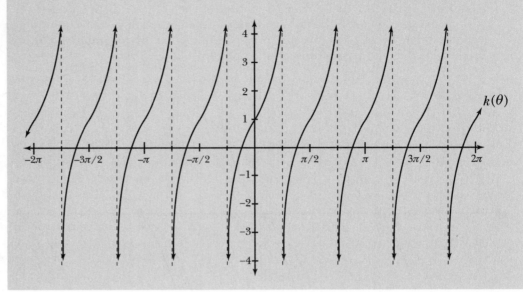

Two transformations change the graph of $y = \tan \theta$ into the graph of $k(\theta)$. First, it is shifted 1 unit up on the coordinate plane, so you must add 1 to $\tan \theta$ to create $k(\theta)$. Second, the period of the graph is compressed horizontally, so you will need to multiply the input by a real number value. It appears that the graph is compressed by a factor of 2, but it is important to verify this observation empirically.

Notice that one full period of the graph lies between two consecutive vertical asymptotes, $\theta = -\pi/4$ and $\theta = \pi/4$, for example. Thus, the period of $k(\theta)$ is the distance between those θ-values.

$$\text{period of } k(\theta) = \frac{\pi}{4} - \left(-\frac{\pi}{4}\right)$$

$$= \frac{\pi}{4} + \frac{\pi}{4}$$

$$= \frac{2\pi}{4}$$

$$= \frac{\pi}{2}$$

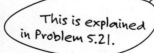

This is explained in Problem 5.21.

The graph is, indeed, compressed horizontally by a factor of 2, because the new period is half the size of the original period $\left(\frac{\pi}{2} = \frac{1}{2} \cdot \pi\right)$. Thus, the input of function $k(\theta)$ should be multiplied by 2.

If you need further evidence, recall that the period of a transformed graph is equal to the absolute value of the untransformed period divided by the coefficient of θ in the transformed function. In the following formula, let x represent the coefficient of θ in $k(\theta)$. Solve the proportion for x.

$$\text{period of } k(\theta) = \frac{\text{period of } \tan\theta}{\text{coefficient of } \theta \text{ in } k(\theta)}$$

$$\frac{\pi}{2} = \frac{\pi}{x}$$

$$\pi x = 2\pi$$

$$\frac{\cancel{\pi} x}{\cancel{\pi}} = \frac{2\cancel{\pi}}{\cancel{\pi}}$$

$$x = 2$$

In conclusion, the graph of $k(\theta)$ is compressed horizontally by a factor of 2 and is shifted up 1 unit. Therefore, $k(\theta) = \tan(2\theta) + 1$.

Cotangent
A reflection of tangent

6.6 Graph $f(\theta) = \cot\theta$ on the interval $-2\pi \le \theta \le 2\pi$ and identify the period of $f(\theta)$.

Recall that $f(\theta) = \cot\theta$ is defined as the quotient of $\cos\theta$ and $\sin\theta$. Therefore, it is equal to 0 (and intersects the x-axis) whenever $\cos\theta = 0$ and is undefined whenever $\sin\theta = 0$.

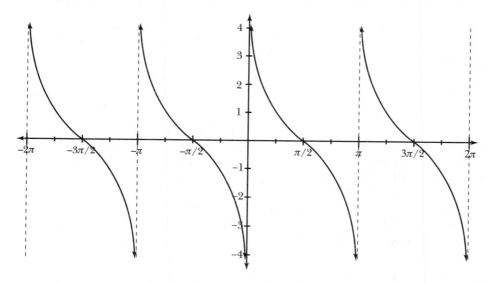

Like the graph of $y = \tan\theta$, the graph of $f(\theta) = \cot\theta$ passes through the point $(\pi/4, 1)$, as demonstrated below.

$$\frac{\sin\pi/4}{\cos\pi/4} = \frac{\cos\pi/4}{\sin\pi/4} = \frac{\sqrt{2}/2}{\sqrt{2}/2} = 1$$

As illustrated in the graph of $\cot\theta$, one period of the graph lies between vertical asymptotes $\theta = 0$ and $\theta = \pi$. Therefore, the period of $f(\theta)$ is the distance between those θ-values.

$$\text{period of } f(\theta) = \pi - 0$$
$$= \pi$$

Thus, the graphs of tangent and cotangent both have the same period, π.

6.7 Based on the graph you created in Problem 6.6, identify the domain and range of $f(\theta) = \cot \theta$.

The graph of $f(\theta) = \cot \theta$ has vertical asymptotes at each multiple of π. Hence, the domain of $\cot \theta$ is all real numbers such that $\theta \neq k\pi$, where k is an integer. Like the graph of $y = \tan \theta$, the graph of $f(\theta) = \cot \theta$ increases and decreases without bound, and any horizontal line drawn on the coordinate plane intersects the graph. Thus, the range of $f(\theta)$ is all real numbers.

> $\cot \theta$ is not defined at $\theta = \ldots, -3\pi, -2\pi, -1\pi, 0\pi, 1\pi, 2\pi, 3\pi, \ldots$.

6.8 Graph $j(\theta) = \cot(-\theta) + 2$.

Compared to the untransformed function $y = \cot \theta$, the input of $j(\theta)$ is multiplied by -1 and 2 is added to the function value. Therefore, to graph $j(\theta)$, you reflect the graph of $y = \cot \theta$ about the y-axis and move it up two units.

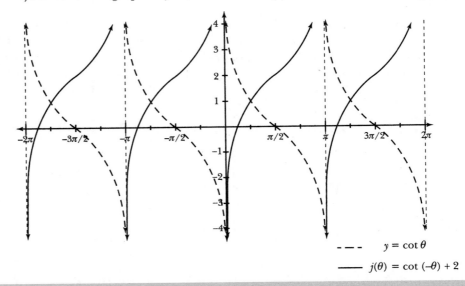

$- - -$ $y = \cot \theta$

$-\!\!-\!\!-$ $j(\theta) = \cot(-\theta) + 2$

6.9 Graph $h(\theta) = \dfrac{1}{3} \cot(2\theta)$.

All of the points on $h(\theta)$ are one-third as high as the corresponding points on the graph of $y = \cot \theta$. Furthermore, two periods of $h(\theta)$ lie in the interval $0 < \theta < \pi$, where only one interval of $y = \cot \theta$ lies.

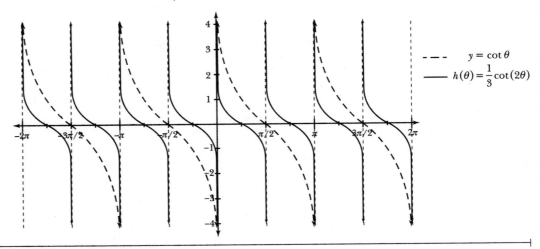

$- - -$ $y = \cot \theta$

$-\!\!-\!\!-$ $h(\theta) = \dfrac{1}{3} \cot(2\theta)$

6.10 Create a function $g(\theta)$, a transformation of $y = \tan \theta$ that coincides with the graph of $f(\theta) = \cot \theta$.

Reflecting the graph of tangent across the x-axis and then moving it to the left $\pi/2$ units produces the graph of $f(\theta) = \cot \theta$. Thus, the graph of $g(\theta) = -\tan (\theta + \pi/2)$ coincides with the graph of $f(\theta)$.

Note that there are numerous correct answers to this problem. You can reflect the graph of $y = \tan \theta$ across either the x- or the y-axis, and you can shift the reflected graph $\pi/2$ units to the left or the right.

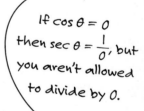

$g(\theta) = -\tan\left(\theta - \dfrac{\pi}{2}\right)$
and
$g(\theta) = \tan\left(\dfrac{\pi}{2} - \theta\right)$
are other correct answers.

Secant

U-shaped pieces that shoot off of the cosine graph

Note: Problems 6.11–6.13 explore the graph of $f(\theta) = \sec \theta$.

6.11 Draw the graph of $y = \cos \theta$ as a dotted curve on the coordinate plane and use the graph to determine the domain and period of $f(\theta) = \sec \theta$.

Notice that $\sec \theta = 1/\cos \theta$ and $\tan \theta = (\sin \theta)/(\cos \theta)$ have the same denominator, $\cos \theta$. They also have the same domain, because they are both undefined only when $\cos \theta = 0$. Problem 6.2 states that the domain of $\tan \theta$ is all real numbers such that $\theta \neq k\pi/2$, where k is an odd number; that is also the domain of $f(\theta) = \sec \theta$.

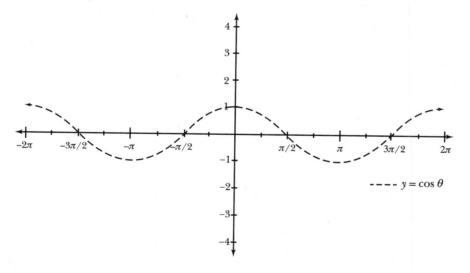

If $\cos \theta = 0$ then $\sec \theta = \dfrac{1}{0}$, but you aren't allowed to divide by 0.

The function values of $f(\theta) = \sec \theta$ are equal to the reciprocals of the corresponding values of $\cos \theta$, so they repeat when the values of $\cos \theta$ repeat. Because $y = \cos \theta$ has period 2π, you conclude that $f(\theta) = \sec \theta$ also has period 2π.

In other words, if $f(x)$ is the reciprocal of periodic function $g(x)$, then $f(x)$ and $g(x)$ have the same period.

Note: Problems 6.11–6.13 explore the graph of $f(\theta) = \sec\theta$.

6.12 Demonstrate how the graph of $f(\theta) = \sec\theta$ can be generated from the graph you created in Problem 6.11.

> If cosine is tiny and positive, secant is a huge positive number; if cosine is tiny and negative, then secant is a huge negative number.

Recall that $\sec\theta$ and $\cos\theta$ are reciprocal functions. Therefore, the larger one function is, the smaller the other is. For example, according to the unit circle, $\cos \pi/3 = 1/2$. The corresponding value of secant is larger: $\sec\dfrac{\pi}{3} = \dfrac{1}{1/2} = 2$. Given an even smaller cosine value, the secant grows even larger: if $\cos\theta = 1/100$, then $\sec\theta = \dfrac{1}{1/100} = 100$.

Graphically speaking, as the graph of $\cos\theta$ gets closer to the x-axis, the graph of $\sec\theta$ increases or decreases without bound. As illustrated in the following graph, $f(\theta) = \sec\theta$ has the same vertical asymptotes as the graph of $y = \tan\theta$; both are undefined at $\theta = k\pi/2$, where k is an odd number. As the graph of $f(\theta) = \sec\theta$ approaches the vertical asymptotes at those values, the function values are very large positive or negative numbers, like the graph of $\tan\theta$.

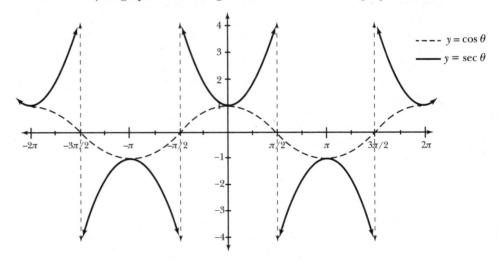

Notice that the graphs of $\cos\theta$ and $\sec\theta$ share point $(0, 1)$. According to the unit circle, $\cos 0 = 1$, and the reciprocal of 1 is 1. Similarly, the graphs share point $(\pi, -1)$ because $\cos\pi = \sec\pi = -1$.

Note: Problems 6.11–6.13 explore the graph of $f(\theta) = \sec\theta$.

6.13 Use the graph you generated in Problem 6.12 to identify the range of $f(\theta)$.

Horizontal lines drawn on the coordinate plane intersect the graph of $f(\theta)$ at height $f(\theta) = 1$ and above, as well as height $f(\theta) = -1$ and below. However, the graph does not exist between heights of -1 and 1. Therefore, the range of $f(\theta)$ is all real numbers excluding any value y such that $-1 < y < 1$. In interval notation, the range of $f(\theta)$ is $(-\infty, -1] \cup [1, \infty)$.

6.14 Graph $g(\theta) = -\sec\theta - 1$.

Begin with the graph of $y = \sec\theta$ and apply two transformations. Multiplying a function by -1 reflects it across the x-axis, and subtracting 1 from it moves the graph down 1 unit.

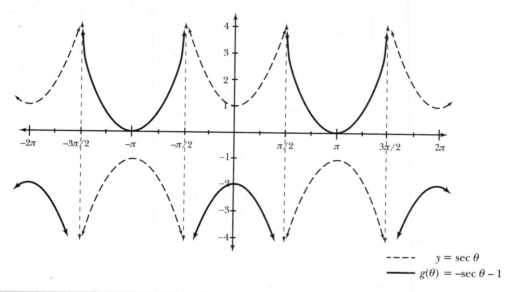

- - - - $y = \sec\theta$
———— $g(\theta) = -\sec\theta - 1$

6.15 Graph $h(\theta) = \sec(3\theta)$.

Consider one period of the untransformed secant graph, beginning at $\theta = 0$ and ending at $\theta = 2\pi$. The graph of $h(\theta)$ will contain three full periods of this shape in the same horizontal interval, as illustrated below.

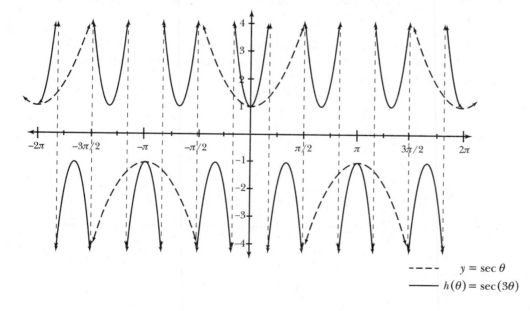

- - - - $y = \sec\theta$
———— $h(\theta) = \sec(3\theta)$

Cosecant

Similar to secant, but based off of sine

Note: Problems 6.16–6.18 explore the graph of $f(\theta) = \csc \theta$.

6.16 Draw the graph of $y = \sin \theta$ as a dotted curve on the coordinate plane and use the graph to determine the domain of $f(\theta) = \csc \theta$.

In Problem 6.11, you determine the domain of $\sec \theta$ by examining the graph of its reciprocal, $\cos \theta$. As that solution demonstrates, $\sec \theta$ is undefined wherever $\cos \theta$ intersects the horizontal axis. Similarly, $f(\theta) = \csc \theta$ is undefined at each value of θ at which $\sin \theta$ intersects the horizontal axis.

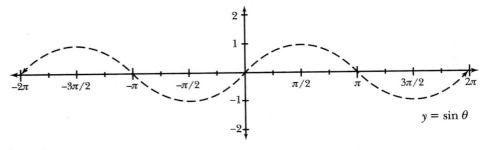

$y = \sin \theta$

> The sine function is equal to zero at positive and negative multiples of $\theta = \ldots, -3\pi, -2\pi, -1\pi, 0\pi, 1\pi, 2\pi, 3\pi, \ldots$.

Thus, the domain of $f(\theta)$ is all real numbers such that $\theta \neq k\pi$, where k is an integer. Note that $f(\theta) = \csc \theta$ has the same domain as $y = \cot \theta$, as $\cot \theta$ and $\csc \theta$ have the same denominator, $\sin \theta$.

Note: Problems 6.16–6.18 explore the graph of $f(\theta) = \csc \theta$.

6.17 Demonstrate how the graph of $f(\theta) = \csc \theta$ can be generated from the graph you created in Problem 6.16.

Consider the solution to Problem 6.12, which explains that small values on a function correspond with large values on the reciprocal function. Thus, whenever the graph of $y = \sin \theta$ approaches the horizontal axis, the graph of $f(\theta)$ increases or decreases without bound.

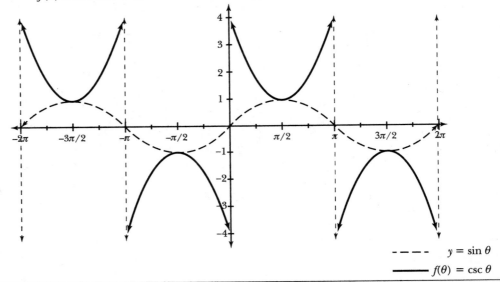

$- - - - \quad y = \sin \theta$

$\underline{\qquad} \quad f(\theta) = \csc \theta$

Note: Problems 6.16–6.18 explore the graph of $f(\theta) = \csc \theta$.

6.18 Use the graph you generated in Problem 6.17 to identify the range of $f(\theta)$.

Like the graph of $y = \sec \theta$, any horizontal line drawn on the coordinate plane intersects the graph of $f(\theta) = \csc \theta$ except horizontal lines between $f(\theta) = -1$ and $f(\theta) = 1$. Thus, the range of $f(\theta)$, like the range of $y = \sec \theta$ identified in Problem 6.13, is all real numbers excluding any value y such that $-1 < y < 1$. In interval notation, the range of $f(\theta)$ is $(-\infty, -1] \cup [1, \infty)$.

6.19 Graph $g(\theta) = \csc\left(\dfrac{1}{2}\theta\right)$.

One full period of $y = \csc \theta$ lies between $\theta = 0$ and $\theta = 2\pi$. Because the input of $g(\theta)$ is multiplied by $1/2$, only half of that period appears in the transformed graph of $g(\theta)$, as illustrated below.

The period of csc θ (like the period of sec θ) is 2π.

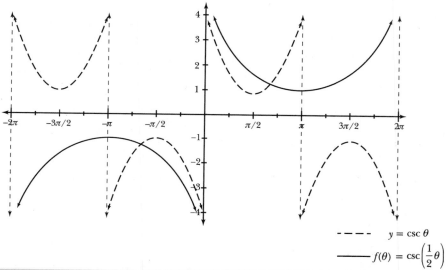

$$- - - - \quad y = \csc \theta$$
$$\text{———} \quad f(\theta) = \csc\left(\dfrac{1}{2}\theta\right)$$

6.20 Graph $h(\theta) = -\csc\left(\theta + \dfrac{\pi}{4}\right)$.

The graph of $h(\theta)$ is the graph of $y = \csc \theta$ reflected across the x-axis and shifted to the left $\pi/4$ units.

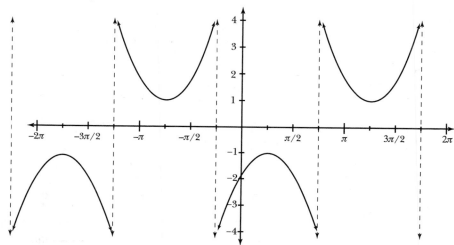

Chapter 7
BASIC TRIGONOMETRIC IDENTITIES

Simplifying trig statements

This chapter is the first of two dedicated to trigonometric identities, equivalency statements that will serve as tools for future chapters and mathematics courses. The preceding chapters focus on a concrete representation of trigonometric functions—how they are defined in terms of right triangles, how they derive their values from the unit circle, and how they are affected graphically with the introduction of real number values. In order to master more advanced trigonometric theorems, however, you must first focus on comparatively more abstract concepts.

The first six chapters of the book may have felt a little like geometry, but the next clump of chapters will feel more like algebra. This chapter marks the beginning of trig identities, formulas you need to memorize. These formulas allow you to manipulate complicated trig expressions and make them much simpler looking and easier to deal with. And don't worry about having to commit even more things to memory—unlike the unit circle, you'll use these formulas so much, the memorization will come naturally.

Each new section in this chapter brings a new batch of identities to memorize. Basically, all of the problems in each section will fall into one of two categories:

1. Simplifying trig expressions (sort of like simplifying fractions to lowest terms, so that what you end up with is nicer looking than what you started with)

2. Verifying identities (where you take an equation that doesn't look like it could possibly be equal and show that it is)

Reciprocal and Cofunction Identities

Cos is reciprocal of sec, cofunction of sin

7.1 List the six cofunction identities and explain how they are related.

> The other two pairs of cofunctions: secant/cosecant, tangent/cotangent

Each of the three trigonometric functions that begin with the prefix "co-" is described as the "cofunction" of the function with the same name, when the prefix is omitted. For example, sine and cosine are cofunctions. If $f(x)$ and $g(x)$ are trigonometric cofunctions, then substituting $\pi/2 - x$ into one of the functions is equivalent to substituting x into its cofunction.

$$f\left(\frac{\pi}{2} - x\right) = g(x) \quad \text{and} \quad g\left(\frac{\pi}{2} - x\right) = f(x)$$

For example, because $\sin x$ and $\cos x$ are cofunctions, the following statements are true.

$$\sin\left(\frac{\pi}{2} - x\right) = \cos x \quad \text{and} \quad \cos\left(\frac{\pi}{2} - x\right) = \sin x$$

Ensure that you state the parenthetical quantity correctly—you must subtract x from $\pi/2$, not vice versa. Apply this pattern to the four remaining cofunctions to complete the set of six cofunction identities.

$$\tan\left(\frac{\pi}{2} - x\right) = \cot x \qquad \cot\left(\frac{\pi}{2} - x\right) = \tan x$$

$$\sec\left(\frac{\pi}{2} - x\right) = \csc x \qquad \csc\left(\frac{\pi}{2} - x\right) = \sec x$$

7.2 Simplify the expression: $\cos\left(\dfrac{\pi}{2} - x\right) \cdot \sin x$.

As Problem 7.1 explains, $\cos\left(\dfrac{\pi}{2} - x\right) = \sin x$, so substitute the equivalent quantity into the expression to simplify it.

$$\cos\left(\frac{\pi}{2} - x\right) \cdot \sin x = (\sin x)(\sin x)$$

$$= \sin^2 x$$

> If an entire trig expression is squared, write the exponent after the name of the function. In other words, $(\cos 5x)(\cos 5x)(\cos 5x) = \cos^3 5x$. If you write "$\cos 5x^3$," it looks like you're plugging $5x^3$ into cosine, not raising $(\cos 5x)$ to the third power.

The product $(\sin x)(\sin x)$ may also be written $(\sin x)^2$, but the notation $\sin^2 x$ is more common. Note that the statement $(\sin x)(\sin x) = \sin x^2$ is incorrect—multiplying $\sin x$ times itself is equal to the square of the entire *quantity* $(\sin x)$, not just the sine of x^2.

7.3 Simplify the expression: $\cot\left(\dfrac{\pi}{2} - x\right) - 5\tan x$.

Apply a cofunction identity to rewrite $\cot(\pi/2 - x)$ as its cofunction, $\tan x$.

$$\cot\left(\frac{\pi}{2} - x\right) - 5\tan x = \tan x - 5\tan x$$

$$= -4\tan x$$

Notice that tan x and $-5\tan x$ are like terms, so they should be combined to simplify the expression; the result is $-4\tan x$.

7.4 Simplify the expression: $\dfrac{\sin(\pi/2 - \theta)}{\cos\theta}$.

Apply the cofunction identity $\sin(\pi/2 - \theta) = \cos\theta$ to rewrite the numerator of the fraction.

$$\frac{\sin(\pi/2 - \theta)}{\cos\theta} = \frac{\cos\theta}{\cos\theta}$$
$$= 1$$

Reducing the fraction $(\cos\theta)/(\cos\theta)$ to lowest terms produces the equivalent value 1. However, you cannot conclude that $(\cos\theta)/(\cos\theta) = 1$ for all values of θ. For instance, substituting $\theta = \pi/2$ into the expression produces an indeterminate value, as demonstrated below.

$$\frac{\cos\pi/2}{\cos\pi/2} = \frac{0}{0}$$

Typically, instructors do not require you to state limitations of simplified expressions, so 1 is an acceptable answer for this problem.

7.5 List the six reciprocal functions and express each in terms of sine and cosine.

Each of the six trigonometric functions can be expressed as the reciprocal of another trigonometric function, as demonstrated below.

$$\sin\theta = \frac{1}{\csc\theta} \quad \cos\theta = \frac{1}{\sec\theta} \quad \tan\theta = \frac{1}{\cot\theta}$$
$$\csc\theta = \frac{1}{\sin\theta} \quad \sec\theta = \frac{1}{\cos\theta} \quad \cot\theta = \frac{1}{\tan\theta}$$

Furthermore, the tangent and cotangent functions may be expressed in terms of sine and cosine.

$$\tan\theta = \frac{1}{\cot\theta} = \frac{\sin\theta}{\cos\theta} \qquad \cot\theta = \frac{1}{\tan\theta} = \frac{\cos\theta}{\sin\theta}$$

7.6 Simplify the expression: $\dfrac{1}{\cos(\pi/2 - \theta)} - 3\csc\theta$.

Apply the cofunction identity $\cos(\pi/2 - \theta) = \sin\theta$ to rewrite the denominator.

$$\frac{1}{\cos(\pi/2 - \theta)} - 3\csc\theta = \frac{1}{\sin\theta} - 3\csc\theta$$

According to a reciprocal identity, $1/\sin\theta = \csc\theta$.

$$= \csc\theta - 3\csc\theta$$
$$= -2\csc\theta$$

7.7 Simplify the expression: $\dfrac{\sin\theta}{\csc\theta}$.

It is helpful to rewrite this expression as a product—the numerator divided by 1 multiplied by the denominator divided into 1.

$$\frac{\sin\theta}{\csc\theta} = \frac{\sin\theta}{1} \cdot \frac{1}{\csc\theta}$$

Recall that $1/\csc\theta = \sin\theta$.

$$= \frac{\sin\theta}{1} \cdot \sin\theta$$
$$= \sin^2\theta$$

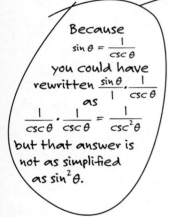

Because $\sin\theta = \dfrac{1}{\csc\theta}$ you could have rewritten $\dfrac{\sin\theta}{1} \cdot \dfrac{1}{\csc\theta}$ as $\dfrac{1}{\csc\theta} \cdot \dfrac{1}{\csc\theta} = \dfrac{1}{\csc^2\theta}$ but that answer is not as simplified as $\sin^2\theta$.

7.8 Rewrite the expression as a single fraction: $\sec\theta + \tan\theta$.

Express $\sec\theta$ and $\tan\theta$ in terms of sine and cosine.

$$\sec\theta + \tan\theta = \frac{1}{\cos\theta} + \frac{\sin\theta}{\cos\theta}$$

The fractions have a common denominator, so they may be combined.

$$= \frac{1 + \sin\theta}{\cos\theta}$$

7.9 Simplify the expression: $\sec x\,(\cos x - 1)$.

Express $\sec x$ in terms of cosine using a reciprocal identity.

$$\sec x\,(\cos x - 1) = \frac{1}{\cos x}(\cos x - 1)$$

Apply the distribution property, multiplying each of the terms within the parenthetical quantity by $1/\cos x$.

$$= \frac{1}{\cos x}(\cos x) + \frac{1}{\cos x}(-1)$$
$$= \frac{\cos x}{\cos x} - \frac{1}{\cos x}$$

Note that $\cos x/\cos x = 1$ and $1/\cos x = \sec x$.

$$= 1 - \sec x$$

7.10 Simplify the expression: $\tan^2\left(\dfrac{\pi}{2} - x\right) \cdot \sec^3 x$.

According to a cofunction identity, $\tan(\pi/2 - x) = \cot x$. Therefore, $\tan^2(\pi/2 - x) = \cot^2 x$.

$$\tan^2\left(\frac{\pi}{2} - x\right) \cdot \sec^3 x = \cot^2 x \cdot \sec^3 x$$

Rewrite the expression $\cot^2 x \cdot \sec^3 x$ in terms of sine and cosine to simplify.

$$= \frac{\cos^2 x}{\sin^2 x} \cdot \frac{1}{\cos^3 x}$$

$$= \frac{\cancel{\cos x} \cdot \cancel{\cos x}}{\cancel{\cos x} \cdot \cancel{\cos x} \cdot \cos x \cdot \sin^2 x}$$

$$= \frac{1}{\cos x \cdot \sin^2 x}$$

Apply reciprocal identities to eliminate the fraction.

$$= \frac{1}{\cos x} \cdot \frac{1}{\sin^2 x}$$

$$= \sec x \cdot \csc^2 x$$

> This answer is pretty good— it's probably okay if you stopped here. However, an answer without fractions is usually better than one with fractions in it.

7.11 Simplify the expression: $3 \cot \theta \left[\dfrac{\sin^2 \theta}{4 \sin (\pi / 2 - \theta)} \right]$.

Express $\cot \theta$ in terms of sine and cosine, and apply a cofunction identity to rewrite the denominator within brackets.

$$3 \cot \theta \left[\frac{\sin^2 \theta}{4 \sin (\pi / 2 - \theta)} \right] = 3 \left(\frac{\cos \theta}{\sin \theta} \right) \left[\frac{\sin^2 \theta}{4 \cos \theta} \right]$$

$$= \frac{3 \cdot \cos \theta \cdot \sin^2 \theta}{4 \cdot \sin \theta \cdot \cos \theta}$$

$$= \frac{3 \cdot \cancel{\cos \theta} \cdot \cancel{\sin \theta} \cdot \sin \theta}{4 \cancel{\sin \theta} \cdot \cancel{\cos \theta}}$$

$$= \frac{3 \sin \theta}{4}$$

$$= \frac{3}{4} \sin \theta$$

7.12 Simplify the expression: $\sec x \left[3 \cot x + \csc^2 \left(\dfrac{\pi}{2} - x \right) + \sin x \right]$.

Express $\sec x$ and $\cot x$ in terms of sine and cosine, and apply a cofunction identity to rewrite $\csc^2 (\pi/2 - x)$ as $\sec^2 x$.

$$\sec x \left[3 \cot x + \csc^2 \left(\frac{\pi}{2} - x \right) + \sin x \right] = \frac{1}{\cos x} \left[3 \cdot \frac{\cos x}{\sin x} + \sec^2 x + \sin x \right]$$

$$= \frac{1}{\cos x} \left[3 \cdot \frac{\cos x}{\sin x} + \frac{1}{\cos^2 x} + \frac{\sin x}{1} \right]$$

Apply the distributive property.

$$= \left(\frac{1}{\cos x} \right) \left(3 \cdot \frac{\cos x}{\sin x} \right) + \left(\frac{1}{\cos x} \right) \left(\frac{1}{\cos^2 x} \right) + \left(\frac{1}{\cos x} \right) \left(\frac{\sin x}{1} \right)$$

$$= 3 \cdot \frac{\cancel{\cos x}}{\cancel{\cos x} \cdot \sin x} + \frac{1}{\cos^3 x} + \frac{\sin x}{\cos x}$$

$$= 3 \cdot \frac{1}{\sin x} + \sec^3 x + \tan x$$

$$= 3 \csc x + \sec^3 x + \tan x$$

> $\cos x = \cos^1 x$, so when you multiply $(\cos x)(\cos^2 x)$, it's the same as multiplying $(\cos^1 x)(\cos^2 x)$. Add the exponents when you multiply: $(\cos^1 x)(\cos^2 x) = \cos^{1+2} x = \cos^3 x$.

Negative Identities

What happens when you plug in –x?

7.13 Evaluate each of the six trigonometric functions for –x to generate the negative trigonometric identities.

Consider the graphs of the trigonometric functions that you generate in Chapters 5 and 6. Any graph that is symmetric about the y-axis—in other words, the graph is a reflection of itself across the y-axis, as though that vertical line were a mirror—is described as an "even function." If $f(x)$ is an even function, then $f(-x) = f(x)$. You get the same output when a real number and its opposite are substituted into the function.

Of the six trigonometric functions, only two have y-symmetric graphs and are, therefore, even functions: cos x and its reciprocal sec x. Therefore, cos $(-x)$ = cos x and sec $(-x)$ = sec x. The other four trigonometric functions are origin-symmetric, and are considered "odd functions." If $g(x)$ is an odd function, then $g(-x) = -g(x)$. Hence, substituting x and –x into an odd function produce opposite results: sin$(-x)$ = –sin x, tan $(-x)$ = –tan x, csc $(-x)$ = –csc x, and cot $(-x)$ = –cot x.

> Odd functions do exactly opposite things on either side of the origin. Think of the graph of tangent as an example. Immediately to the right of the origin, the graph shoots up and to the right, but to the left of the origin, the graph shoots down and to the left.

7.14 Simplify the expression: sin $(-x)$ csc $(-x)$.

Both sine and cosecant are odd functions, as explained in Problem 7.13, so sin $(-x)$ = –sin x and csc $(-x)$ = –csc x.

$$\sin (-x) \csc (-x) = (-\sin x)(-\csc x)$$

Apply a reciprocal identity to simplify the expression, and note that the product of two negative values is positive.

$$= +(\sin x)(\csc x)$$
$$= (\sin x)\left(\frac{1}{\sin x}\right)$$
$$= \frac{\sin x}{\sin x}$$
$$= 1$$

7.15 Write the expression as a single fraction: $3\sec (-\theta) + 2$.

Secant is an even function, so sec $(-\theta)$ = sec θ. Apply a reciprocal identity to express the first term as a fraction.

$$3\sec(-\theta) + 2 = 3\sec\theta + 2$$
$$= 3 \cdot \frac{1}{\cos\theta} + 2$$
$$= \frac{3}{\cos\theta} + 2$$

In order to add the terms, they must share a common denominator, so multiply the numerator and denominator of the second term by cos θ.

$$= \frac{3}{\cos\theta} + \frac{2}{1} \cdot \frac{\cos\theta}{\cos\theta}$$

$$= \frac{3}{\cos\theta} + \frac{2\cos\theta}{\cos\theta}$$

$$= \frac{3 + 2\cos\theta}{\cos\theta}$$

7.16 Verify the statement: $\tan x = \dfrac{\sin x}{\cos(-x)}$.

Whereas the preceding problems in this chapter ask you to simplify an expression, this problem directs you to verify a statement—an equation. In order to prove the statement true, you will manipulate one or both sides of the equation until it is absolutely clear that both sides of the equation are equal.

In this problem, you should manipulate the right side of the equation, applying the negative identity cos $(-x)$ = cos x.

$$\tan x = \frac{\sin x}{\cos(-x)}$$

$$\tan x = \frac{\sin x}{\cos x}$$

Because tangent is defined as the quotient of sine and cosine, tan x = sin x/cos x. is an obviously true statement, so you have successfully verified the original statement.

7.17 Verify the statement: $\cos(-x)\tan(-x) = -\cos\left(\dfrac{\pi}{2} - x\right)$.

In order to verify this statement, you will simplify each side of the equation individually. Begin by applying the negative identities cos $(-x)$ = cos x and tan $(-x)$ = $-$tan x to the left side of the equation.

$$(\cos x)(-\tan x) = -\cos\left(\frac{\pi}{2} - x\right)$$

$$-\cos x \tan x = -\cos\left(\frac{\pi}{2} - x\right)$$

Now apply a cofunction identity to the right side of the equation.

$$-\cos x \tan x = -(\sin x)$$

The statement is not yet obviously true. Multiply both sides by -1 and express tan x in terms of sine and cosine. ⬅

$$(-1)(-\cos x \tan x) = (-1)(-\sin x)$$

$$\cos x \tan x = \sin x$$

$$\frac{\cancel{\cos x}}{1} \cdot \frac{\sin x}{\cancel{\cos x}} = \sin x$$

$$\frac{\sin x}{1} = \sin x$$

> Expressing everything in terms of sine and cosine is a very useful trick for verifying trig statements. Remember it.

The final equation ($\sin x = \sin x$) is an obviously true statement that is logically equivalent to the initial equation. Because the final statement is true, you can conclude that the first statement is true as well.

7.18 Simplify the expression: $\cot\left(\theta - \dfrac{\pi}{2}\right)\sin\left(\dfrac{\pi}{2} - \theta\right)$.

Although it may appear so at first glance, $\cot(\theta - \pi/2)$ is not a cofunction identity, because the argument of the trigonometric expression is $\theta - \pi/2$, not $\pi/2 - \theta$. In order to apply the cofunction identity, you must factor -1 out of the terms in order to reverse them, as demonstrated below. Notice that the -1 that is factored out of the expression still appears within the argument of cotangent.

$$\cot\left(\theta - \frac{\pi}{2}\right)\sin\left(\frac{\pi}{2} - \theta\right) = \cot\left[-1\left(\frac{\pi}{2} - \theta\right)\right]\sin\left(\frac{\pi}{2} - \theta\right)$$

According to the negative identities, $\cot(-x) = -\cot x$.

$$= -\cot\left(\frac{\pi}{2} - \theta\right)\sin\left(\frac{\pi}{2} - \theta\right)$$

Apply cofunction identities and express the functions in terms of sine and cosine.

$$= -\tan\theta\cos\theta$$

$$= -\frac{\sin\theta}{\cos\theta}\cdot\frac{\cos\theta}{1}$$

$$= -\sin\theta$$

7.19 Verify the statement: $\sin(-x)\cos(-x)\cot(-x) = \dfrac{\cos x}{\sec x}$.

Apply negative identities to simplify the left side of the equation, and then express the functions in terms of sine and cosine.

$$\sin(-x)\cos(-x)\cot(-x) = \frac{\cos x}{\sec x}$$

$$(-\sin x)(\cos x)(-\cot x) = \frac{\cos x}{\sec x}$$

$$(\sin x)(\cos x)\left(\frac{\cos x}{\sin x}\right) = \frac{\cos x}{\sec x}$$

$$(\cos x)(\cos x) = \frac{\cos x}{\sec x}$$

$$\cos^2 x = \frac{\cos x}{\sec x}$$

Now that you have fully simplified the left side of the equation, simplify the right side. Notice that dividing by $\sec x$ is the same as multiplying by its reciprocal, $\cos x$.

$$\cos^2 x = \frac{\cos x}{1}\cdot\frac{1}{\sec x}$$

$$\cos^2 x = \frac{\cos x}{1}\cdot\cos x$$

$$\cos^2 x = \cos^2 x$$

"Argument" means "the thing you plug into the function." In other words, the argument of sin x is x.

In other words, that -1 you just factored out of the input can now be pulled out of the input and placed in front of the function.

Remember, once you get an obviously true equation, you've verified the original statement.

7.20   Verify the statement: $\dfrac{1}{\cos(-x)+1}+\dfrac{1}{\sec(-x)+1}=1$.

Both cosine and secant are even functions, so according to the negative identities, $\cos(-x)=\cos x$ and $\sec(-x)=\sec x$. Note that adding fractions requires common denominators, so you must multiply the numerator and denominator of the first fraction by $\sec x+1$ and multiply the numerator and denominator of the second fraction by $\cos x+1$.

$$\frac{1}{\cos(-x)+1}+\frac{1}{\sec(-x)+1}=1$$

$$\frac{1}{\cos x+1}+\frac{1}{\sec x+1}=1$$

$$\left(\frac{\sec x+1}{\sec x+1}\right)\left(\frac{1}{\cos x+1}\right)+\left(\frac{\cos x+1}{\cos x+1}\right)\left(\frac{1}{\sec x+1}\right)=1$$

$$\frac{\sec x+1}{(\sec x+1)(\cos x+1)}+\frac{\cos x+1}{(\sec x+1)(\cos x+1)}=1$$

$$\frac{\sec x+\cos x+2}{(\sec x+1)(\cos x+1)}=1$$

Multiply the factors in the denominator using the FOIL method. Note that $(\sec x)(\cos x)=1$ because secant and cosine are reciprocal functions.

$$\frac{\sec x+\cos x+2}{(\sec x)(\cos x)+(\sec x)(1)+(1)(\cos x)+(1)(1)}=1$$

$$\frac{\sec x+\cos x+2}{1+\sec x+\cos x+1}=1$$

$$\frac{\sec x+\cos x+2}{\sec x+\cos x+2}=1$$

$$1=1$$

Any value divided by itself, even something big like $(\sec x + \cos x + 2)$, is equal to 1.

7.21   Write the fraction as a sum or difference of two terms: $\dfrac{1-\cos x}{\sin(-x)}$.

Note that this problem simply asks you to simplify a single expression, not to verify an equation (like Problems 7.19 and 7.20). To begin, apply a negative identity to express $\sin(-x)$ as $-\sin x$.

$$\frac{1-\cos x}{\sin(-x)}=\frac{1-\cos x}{-\sin x}$$

If a fraction contains two or more terms in the numerator and only one term in the denominator, then you can rewrite that fraction as the sum or difference of the individual numerator terms, each divided by the common denominator.

$$=\frac{1}{-\sin x}+\frac{-\cos x}{-\sin x}$$

$$=-\frac{1}{\sin x}+\frac{\cos x}{\sin x}$$

$$=-\csc x+\cot x$$

For example, $\dfrac{1+2-3}{5}=\dfrac{1}{5}+\dfrac{2}{5}-\dfrac{3}{5}$.

7.22 Verify the statement: $\dfrac{\sin x - \tan x}{\sin(-x)} = \sec(-x) - 1$.

Begin by applying negative identities to rewrite sin (−x) and sec (−x), and then use the technique demonstrated in Problem 7.21 to express the single fraction on the left side of the equation as two fractions with a common denominator.

$$\frac{\sin x - \tan x}{-\sin x} = \sec x - 1$$

$$\frac{\sin x}{-\sin x} + \frac{-\tan x}{-\sin x} = \sec x - 1$$

$$-1 + \frac{\tan x}{\sin x} = \sec x - 1$$

If you add 1 to both sides of the equation, and apply a reciprocal identity to rewrite sec x as 1/cos x, the result is a proportion.

$$\left(-1 + \frac{\tan x}{\sin x}\right) + 1 = (\sec x - 1) + 1$$

$$\frac{\tan x}{\sin x} = \sec x$$

$$\frac{\tan x}{\sin x} = \frac{1}{\cos x}$$

Cross-multiply to eliminate the fractions from the equation and simplify.

$$(\tan x)(\cos x) = (\sin x)(1)$$

$$\left(\frac{\sin x}{\cancel{\cos x}}\right)(\cancel{\cos x}) = \sin x$$

$$\sin x = \sin x$$

Pythagorean Identities

For example, $\cos^2 x + \sin^2 x = 1$

7.23 List the three Pythagorean identities, in terms of θ.

Each of the three Pythagorean identities is an equation that contains two squared trigonometric functions, a sum, and the real number 1.

- $\cos^2\theta + \sin^2\theta = 1$
- $1 + \tan^2\theta = \sec^2\theta$
- $1 + \cot^2\theta = \csc^2\theta$

Note that θ, x, or any other variable may be used in the identities; the choice of variable does not affect the validity of the statement. For example, the first identity written in terms of x is equally valid: $\cos^2 x + \sin^2 x = 1$.

7.24 Verify that the Pythagorean identity relating sine and cosine is true for $\theta = \pi/4$.

According to Problem 7.23, $\cos^2\theta + \sin^2\theta = 1$. Substitute $\theta = \pi/4$ into each of the trigonometric functions and square the values to verify that the equation is true.

$$\cos^2\left(\frac{\pi}{4}\right) + \sin^2\left(\frac{\pi}{4}\right) = 1$$

$$\left(\frac{\sqrt{2}}{2}\right)^2 + \left(\frac{\sqrt{2}}{2}\right)^2 = 1$$

$$\frac{\sqrt{4}}{4} + \frac{\sqrt{4}}{4} = 1$$

$$\frac{2}{4} + \frac{2}{4} = 1$$

$$\frac{4}{4} = 1$$

$\cos^2\left(\frac{\pi}{4}\right)$ means the same thing as $\left(\cos\frac{\pi}{4}\right)^2$. Plug in $\theta = \frac{\pi}{4}$ to get $\frac{\sqrt{2}}{2}$ and then square that.

Any non-zero real number divided by itself equals 1. In this case, $4/4 = 1$, so the statement is true.

7.25 Verify that the Pythagorean identity relating secant and tangent is true for $\theta = \pi/6$.

Substitute $\theta = \pi/6$ into the Pythagorean identity $1 + \tan^2\theta = \sec^2\theta$ and verify that the resulting statement is true.

$$1 + \tan^2\theta = \sec^2\theta$$

$$1 + \tan^2\left(\frac{\pi}{6}\right) = \sec^2\left(\frac{\pi}{6}\right)$$

$$1 + \left[\frac{\sin(\pi/6)}{\cos(\pi/6)}\right]^2 = \left[\frac{1}{\cos(\pi/6)}\right]^2$$

$$1 + \frac{(1/2)^2}{\left(\sqrt{3}/2\right)^2} = \frac{1^2}{\left(\sqrt{3}/2\right)^2}$$

$$1 + \frac{1/4}{3/4} = \frac{1}{3/4}$$

Multiply the entire equation by $3/4$ to eliminate the fraction in the denominators.

$$\frac{3}{4}(1) + \frac{3}{4}\left(\frac{1/4}{3/4}\right) = \frac{3}{4}\left(\frac{1}{3/4}\right)$$

$$\frac{3}{4} + \frac{1/4}{1} = 1$$

$$\frac{3}{4} + \frac{1}{4} = 1$$

$$\frac{4}{4} = 1$$

7.26 Simplify the expression: $\dfrac{1-\sin^2 x}{\sec x}$.

Recall that $\cos^2 x + \sin^2 x = 1$. If you subtract $\sin^2 x$ from both sides of that identity, you produce the equally valid identity $\cos^2 x = 1 - \sin^2 x$. Therefore, you can replace $1 - \sin^2 x$ in the numerator of this expression with $\cos^2 x$.

$$\frac{1-\sin^2 x}{\sec x} = \frac{\cos^2 x}{\sec x}$$
$$= \frac{\cos^2 x}{1}\cdot \cos x$$
$$= \cos^3 x$$

> Dividing by sec x is the same thing as multiplying by its reciprocal, cos x. This works just like Problem 7.7, where dividing by cosecant is the same as multiplying by sine.

7.27 Simplify the expression: $\dfrac{\sec^2 \theta - 1}{\tan \theta}$.

According to a Pythagorean identity, $1 + \tan^2\theta = \sec^2\theta$. If you subtract 1 from both sides of that equation, you produce the equally valid identity $\tan^2\theta = \sec^2\theta - 1$. Use this identity to rewrite the numerator of the fraction and simplify.

$$\frac{\sec^2 \theta - 1}{\tan \theta} = \frac{\tan^2 \theta}{\tan \theta}$$
$$= \frac{\cancel{\tan\theta}\cdot \tan\theta}{\cancel{\tan\theta}}$$
$$= \tan \theta$$

7.28 Simplify the expression: $\dfrac{\sin^3 x - \sin x}{\cos x - \cos^3 x}$.

Factor $\sin x$ out of each term in the numerator, and factor $\cos x$ out of each term in the denominator.

$$\frac{\sin^3 x - \sin x}{\cos x - \cos^3 x} = \frac{\sin x\left(\sin^2 x - 1\right)}{\cos x\left(1 - \cos^2 x\right)}$$

If you subtract $\cos^2 x$ and 1 from both sides of the expression $\cos^2 x + \sin^2 x = 1$, the result is $\sin^2 x - 1 = -\cos^2 x$.

$$= \frac{\sin x\left(-\cos^2 x\right)}{\cos x\left(1 - \cos^2 x\right)}$$

> If you need to review factoring, look at Chapter 12 in The Humongous Book of Algebra Problems.

If you subtract $\cos^2 x$ from both sides of the Pythagorean identity $\cos^2 x + \sin^2 x = 1$, the result is $\sin^2 x = 1 - \cos^2 x$. Use this equation to rewrite the parenthetical quantity in the denominator. Then, simplify the fraction.

$$= \frac{\sin x\left(-\cos^2 x\right)}{\cos x\left(\sin^2 x\right)}$$
$$= -\frac{\cancel{\sin x}\cdot \cancel{\cos x}\cdot \cos x}{\cancel{\cos x}\cdot \cancel{\sin x}\cdot \sin x}$$
$$= -\cot x$$

7.29   Verify the statement: $\csc^4\theta - \cot^4\theta = 2\csc^2\theta - 1$.

Factor the difference of perfect squares on the left side of the equation.

$$(\csc^2\theta - \cot^2\theta)(\csc^2\theta + \cot^2\theta) = 2\csc^2\theta - 1$$

Recall that $1 + \cot^2\theta = \csc^2\theta$. If you subtract $\cot^2\theta$ from both sides of that Pythagorean identity, the result is $1 = \csc^2\theta - \cot^2\theta$. Use this statement to simplify the left side of the equation.

$$(1)\ (\csc^2\theta + \cot^2\theta) = 2\csc^2\theta - 1$$

Add and subtract like terms to simplify the expression.

$$\cot^2\theta = \left(2\csc^2\theta - \csc^2\theta\right) - 1 \longleftarrow$$
$$\cot^2\theta = \csc^2\theta - 1$$
$$\cot^2\theta + 1 = \csc^2\theta$$

> *Subtract $\csc^2\theta$ from both sides, then add 1 to both sides. You end up with a Pythagorean identity, so you're done.*

7.30   Simplify the expression: $\dfrac{1}{1-\sin x} + \dfrac{1}{1+\sin x}$.

In order to add the fractions, you need to express them in terms of the least common denominator $(1 - \sin x)(1 + \sin x)$.

$$\frac{1}{1-\sin x} + \frac{1}{1+\sin x} = \left(\frac{1}{1-\sin x}\right)\left(\frac{1+\sin x}{1+\sin x}\right) + \left(\frac{1}{1+\sin x}\right)\left(\frac{1-\sin x}{1-\sin x}\right)$$
$$= \frac{1+\sin x}{(1-\sin x)(1+\sin x)} + \frac{1-\sin x}{(1+\sin x)(1-\sin x)}$$
$$= \frac{1+\sin x + 1 - \sin x}{(1-\sin x)(1+\sin x)}$$

Combine like terms in the numerator and multiply the factors in the denominator.

$$= \frac{(1+1)+(\sin x - \sin x)}{1+\sin x - \sin x - \sin^2 x}$$
$$= \frac{2}{1-\sin^2 x}$$

Note that $1 - \sin^2 x = \cos^2 x$, as explained in Problem 7.26.

$$= \frac{2}{\cos^2 x}$$
$$= 2\sec^2 x$$

7.31   Verify the statement: $(1+\tan\theta)^2 = \dfrac{1+2\sin\theta\cos\theta}{\cos^2\theta}$.

Expand the left side of the equation, and rewrite the single fraction on the right side of the equation as the sum of two fractions with common denominator $\cos^2\theta$.

$$(1+\tan\theta)(1+\tan\theta) = \frac{1}{\cos^2\theta} + \frac{2\sin\theta\cos\theta}{\cos^2\theta}$$

$$1+\tan\theta+\tan\theta+\tan^2\theta = \sec^2\theta + \frac{2\sin\theta\,\cancel{\cos\theta}}{\cos\theta\cdot\cancel{\cos\theta}}$$

$$1+2\tan\theta+\tan^2\theta = \sec^2 + 2\left(\frac{\sin\theta}{\cos\theta}\right)$$

According to a Pythagorean identity, $1+\tan^2\theta = \sec^2\theta$.

$$\left(1+\tan^2\theta\right)+2\tan\theta = \sec^2\theta + 2\tan\theta$$

$$\sec^2\theta + 2\tan\theta = \sec^2\theta + 2\tan\theta$$

Both sides of the equation are equal, so the original statement is verified.

7.32 Verify the identity: $\dfrac{\sec x - \tan x}{\sec x} = \dfrac{\cos^2 x}{1+\sin x}$.

This equation is a proportion, so you can cross-multiply to eliminate the fractions.

$$(\sec x - \tan x)(1+\sin x) = (\sec x)(\cos^2 x)$$

$$(\sec x)(1)+(\sec x)(\sin x)+(-\tan x)(1)+(-\tan x)(\sin x) = (\sec x)(\cos^2 x)$$

$$\frac{1}{\cos x} + \left(\frac{1}{\cos x}\cdot\frac{\sin x}{1}\right) - \frac{\sin x}{\cos x} - \left(\frac{\sin x}{\cos x}\cdot\frac{\sin x}{1}\right) = \frac{1}{\cos x}\left(\frac{\cos^2 x}{1}\right)$$

$$\frac{1}{\cos x} + \frac{\sin x}{\cos x} - \frac{\sin x}{\cos x} - \frac{\sin^2 x}{\cos x} = \frac{\cancel{\cos x}\cdot\cos x}{\cancel{\cos x}}$$

$$\frac{1}{\cos x} - \frac{\sin^2 x}{\cos x} = \cos x$$

Combine the fractions on the left side of the equation and apply the Pythagorean identity $\cos^2 x + \sin^2 x = 1$, rewritten as $1-\sin^2 x = \cos x$.

$$\frac{1-\sin^2 x}{\cos x} = \cos x$$

$$\frac{\cos^2 x}{\cos x} = \cos x$$

$$\cos x = \cos x$$

Like most of the problems in this chapter, there are different ways to verify the statement. For example, you could rewrite $\cos^2 x$ as $(1 - \sin^2 x)$ and then factor that as $(1 + \sin x)(1 - \sin x)$. Then, you can simplify the fraction on the right side of the equation, eliminating $(1 + \sin x)$ from the numerator and denominator.

7.33 Verify the statement: $\dfrac{\cos x}{\tan x} - \csc x = \sin(-x)$.

Dividing by $\tan x$ is equivalent to multiplying by its reciprocal, $\cot x$.

$$\frac{\cos x}{1}\cdot\frac{1}{\tan x} - \csc x = \sin(-x)$$

$$\cos x(\cot x) - \csc x = -\sin x$$

Rewrite the trigonometric functions in terms of sine and cosine and simplify.

$$\frac{\cos x}{1}\left(\frac{\cos x}{\sin x}\right) - \frac{1}{\sin x} = -\sin x$$

$$\frac{\cos^2 x}{\sin x} - \frac{1}{\sin x} = -\sin x$$

$$\frac{\cos^2 x - 1}{\sin x} = -\sin x$$

Note that $\cos^2 x - 1 = -\sin^2 x$.

$$\frac{-\sin^2 x}{\sin x} = -\sin x$$

$$-\sin x = -\sin x$$

Sum and Difference Formulas for Sine and Cosine

Expanding things like sin (x + y)

7.34 Apply an identity to express both of the following expressions as a sum of two products: sin $(x + y)$ and sin $(x - y)$.

Sum and difference identities allow you to rewrite a trigonometric statement whose argument is a sum or a difference. The result is usually not simpler; you transform a single trigonometric statement into a sum or difference of two products. Though these identities do not necessarily *simplify* an expression, they are no less important.

> The argument of sin (x + y) is (x + y).

$$\sin (x + y) = \sin x \cos y + \cos x \sin y \qquad \sin(x - y) = \sin x \cos y - \cos x \sin y$$

Notice that the left-hand formula above contains addition signs on both sides of the equation. In the right-hand formula, both are replaced by subtraction signs. Because the signs match, some textbooks combine the pair of formulas into a single formula, as demonstrated below.

$$\sin (x \pm y) = \sin x \cos y \pm \cos x \sin y$$

> If the argument has addition in it, you substitute an expression that also contains addition. Subtraction works the same way.

Note: Problems 7.35–7.36 demonstrate two different ways to calculate sin 105°, an angle not on the standard unit circle, using two angles that do appear on the standard unit circle.

7.35 Given $45° + 60° = 105°$, apply a sum formula to calculate the exact value of sin 105°.

Because $45° + 60° = 105°$, you can write the expression sin 105° as sin $(45° + 60°)$. Notice that both of the angles in the sum appear in the unit circle $(45° = \pi/4$ radians and $60° = \pi/3$ radians), and you should have these sine and cosine values committed to memory. Apply the sum formula for sine (stated in Problem 7.34).

$$\sin(105°) = \sin(45° + 60°)$$
$$= \sin 45° \cos 60° + \cos 45° \sin 60°$$
$$= \frac{\sqrt{2}}{2} \cdot \frac{1}{2} + \frac{\sqrt{2}}{2} \cdot \frac{\sqrt{3}}{2}$$
$$= \frac{\sqrt{2}}{4} + \frac{\sqrt{6}}{4}$$
$$= \frac{\sqrt{2} + \sqrt{6}}{4}$$

Note: Problems 7.35–7.36 demonstrate two different ways to calculate sin 105°, an angle not on the standard unit circle, using two angles that do appear on the standard unit circle.

7.36 Given $135° - 30° = 105°$, apply a difference formula to calculate sin 105°.

In Problem 7.35, you apply a sum formula to calculate sin 105°. In this problem, you apply a difference formula, and the results are equal. Note that $135° = 3\pi/4$ radians and $30° = \pi/6$ radians.

$$\sin 105° = \sin(135° - 30°)$$
$$= \sin 135° \cos 30° - \cos 135° \sin 30°$$
$$= \frac{\sqrt{2}}{2} \cdot \frac{\sqrt{3}}{2} - \left(-\frac{\sqrt{2}}{2}\right)\left(\frac{1}{2}\right)$$
$$= \frac{\sqrt{6}}{4} + \frac{\sqrt{2}}{4}$$
$$= \frac{\sqrt{6} + \sqrt{2}}{4}$$

7.37 Apply an identity to rewrite each of the following expressions as a sum of two products: cos $(x + y)$ and cos $(x - y)$.

The cosine sum and difference formulas are similar to the sine sum and difference formulas, with two major dissimilarities. Rather than products of different trigonometric functions (like sin x cos y), the cosine formulas contain products of the same function (cos x cos y). Furthermore, the signs within the formulas are opposites; if the argument contains an addition sign, you expand it into an expression containing a subtraction sign.

$$\cos(x + y) = \cos x \cos y - \sin x \sin y \qquad \cos(x - y) = \cos x \cos y + \sin x \sin y$$

Because the two signs in each expression are opposites, some textbooks combine the formulas above into a single formula using the symbols "$\pm$ and $\mp$" to indicate the change in sign.

$$\cos(x \pm y) = \cos x \cos y \mp \sin x \sin y$$

Note: Problems 7.38–7.39 explain how to calculate $\cos -\dfrac{\pi}{12}$, an angle not on the standard unit circle, using two angles that do appear on the standard unit circle.

7.38   Identify two angles on the unit circle that have a difference of $-\dfrac{\pi}{12}$.

Notice that $\dfrac{\pi}{4} - \dfrac{\pi}{3} = -\dfrac{\pi}{12}$, as demonstrated below.

$$\frac{\pi}{4} - \frac{\pi}{3} = \left(\frac{\pi}{4} \cdot \frac{3}{3}\right) - \left(\frac{\pi}{3} \cdot \frac{4}{4}\right)$$

$$= \frac{3\pi}{12} - \frac{4\pi}{12}$$

$$= -\frac{\pi}{12}$$

Note: Problems 7.38–7.39 explain how to calculate $\cos -\dfrac{\pi}{12}$, an angle not on the standard unit circle, using two angles that do appear on the standard unit circle.

7.39   Apply the difference formula for cosine to the angles you identified in Problem 7.38 to calculate $\cos -\dfrac{\pi}{12}$.

Recall that $\cos (x - y) = \cos x \cos y + \sin x \sin y$.

$$\cos\left(-\frac{\pi}{12}\right) = \cos\left(\frac{\pi}{4} - \frac{\pi}{3}\right)$$

$$= \cos \frac{\pi}{4} \cos \frac{\pi}{3} + \sin \frac{\pi}{4} \sin \frac{\pi}{3}$$

$$= \frac{\sqrt{2}}{2} \cdot \frac{1}{2} + \frac{\sqrt{2}}{2} \cdot \frac{\sqrt{3}}{2}$$

$$= \frac{\sqrt{2}}{4} + \frac{\sqrt{6}}{4}$$

$$= \frac{\sqrt{2} + \sqrt{6}}{4}$$

7.40   Compare the trigonometric values calculated in Problems 7.35, 7.36, and 7.39, and justify the similarities or differences.

According to Problems 7.35, 7.36, and 7.39, $\sin 105° = \cos (-\pi/12)$. To explain why this is true, it is helpful to express the degree measure in radians.

$$105 \cdot \frac{\pi}{180} = \frac{105}{180} \cdot \pi$$

$$= \frac{7}{12}\pi$$

Therefore, these problems conclude that $\sin (7\pi/12) = \cos (-\pi/12)$. This is true according to the cofunction identity $\sin (\pi/2 - \theta) = \cos \theta$, if $\theta = -\pi/12$.

$$\sin\left(\frac{\pi}{2} - \theta\right) = \cos\theta$$

$$\sin\left[\frac{\pi}{2} - \left(-\frac{\pi}{12}\right)\right] = \cos\left(-\frac{\pi}{12}\right)$$

$$\sin\left(\frac{6\pi}{12} + \frac{\pi}{12}\right) = \cos\left(-\frac{\pi}{12}\right)$$

$$\sin\left(\frac{7\pi}{12}\right) = \cos\left(-\frac{\pi}{12}\right)$$

While it may be, at first, surprising that sin 105° = cos (−π/12), a cofunction identity easily verifies the equality of the statements.

7.41 Verify the cofunction identity $\cos\left(\frac{\pi}{2} - \theta\right) = \sin\theta$.

Apply the difference formula for cosine to expand the left side of the equation and demonstrate that both sides of the statement are equal.

$$\cos\frac{\pi}{2}\cos\theta + \sin\frac{\pi}{2}\sin\theta = \sin\theta$$

$$(0)\cos\theta + (1)\sin\theta = \sin\theta$$

$$0 + \sin\theta = \sin\theta$$

$$\sin\theta = \sin\theta$$

7.42 Students often mistakenly apply the distributive property to trigonometric arguments, resulting in incorrect claims such as $\sin\left(\frac{\pi}{6} + \frac{\pi}{3}\right) = \sin\frac{\pi}{6} + \sin\frac{\pi}{3}$. Demonstrate that this statement is incorrect.

Do you see the error here? You can't distribute the letters "sin" through parentheses. Those letters don't represent a value—they are just the name of the function.

Evaluate each side of the equation individually and then compare the values to demonstrate they are unequal. Begin with the left side of the equation.

$$\sin\left(\frac{\pi}{6} + \frac{\pi}{3}\right) = \sin\left(\frac{\pi}{6} + \frac{\pi}{3}\cdot\frac{2}{2}\right)$$

$$= \sin\left(\frac{\pi}{6} + \frac{2\pi}{6}\right)$$

$$= \sin\left(\frac{3\pi}{6}\right)$$

$$= \sin\left(\frac{\pi}{2}\right)$$

$$= 1$$

Now simplify the right side of the original statement.

$$\sin\frac{\pi}{6} + \sin\frac{\pi}{3} = \frac{1}{2} + \frac{\sqrt{3}}{2}$$

$$= \frac{1 + \sqrt{3}}{2}$$

The left and right sides of the original statement have different values, so the original statement is incorrect.

7.43 Simplify the expression: $\sin\left(\dfrac{\pi}{4}-\theta\right)-\cos\left(\theta-\dfrac{\pi}{4}\right)$.

Apply the difference formulas for sine and cosine.

$$\sin\left(\frac{\pi}{4}-\theta\right)-\cos\left(\theta-\frac{\pi}{4}\right)=\left[\sin\frac{\pi}{4}\cos\theta-\cos\frac{\pi}{4}\sin\theta\right]-\left[\cos\theta\cos\frac{\pi}{4}+\sin\theta\sin\frac{\pi}{4}\right]$$

$$=\frac{\sqrt{2}}{2}\cos\theta-\frac{\sqrt{2}}{2}\sin\theta-\frac{\sqrt{2}}{2}\cos\theta-\frac{\sqrt{2}}{2}\sin\theta$$

$$=-\frac{\sqrt{2}}{2}\sin\theta-\frac{\sqrt{2}}{2}\sin\theta$$

$$=-2\left(\frac{\sqrt{2}}{2}\sin\theta\right)$$

$$=-\sqrt{2}\sin\theta$$

7.44 Verify the statement: $\sin(x+y)+\sin(x-y)=2\sin x\cos y$.

Expand the expressions on the left side of the equation and simplify.

$$\left[\sin x\cos y+\cos x\sin y\right]+\left[\sin x\cos y-\cos x\sin y\right]=2\sin x\cos y$$

$$\left(\sin x\cos y+\sin x\cos y\right)+\left(\cos x\sin y-\cos x\sin y\right)=2\sin x\cos y$$

$$2\sin x\cos y+0=2\sin x\cos y$$

The two sides of the equation are clearly equal, so the original statement is verified.

7.45 Given an angle w that terminates in the third quadrant and an angle z that terminates in the fourth quadrant, such that $\sin w=-\dfrac{2}{3}$ and $\cos z=\dfrac{1}{7}$, calculate $\sin(w+z)$.

Two angles, w and z, are described in this problem. You are given the quadrant in which those angles terminate, and you are provided one trigonometric value of each angle. In order to solve the problem, however, you will need additional information, as you discover when you expand the expression $\sin(w+z)$.

$$\sin(w+z)=\sin w\cos z+\cos w\sin z$$

While you are given the values of $\sin w$ and $\cos z$, you also need the values of $\cos w$ and $\sin z$ to substitute into the sum formula for sine. Apply the techniques described in Problems 4.36–4.43 (and illustrated in the following diagram), to calculate the missing values.

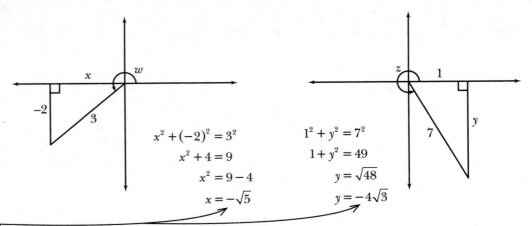

$$x^2 + (-2)^2 = 3^2$$
$$x^2 + 4 = 9$$
$$x^2 = 9 - 4$$
$$x = -\sqrt{5}$$

$$1^2 + y^2 = 7^2$$
$$1 + y^2 = 49$$
$$y = \sqrt{48}$$
$$y = -4\sqrt{3}$$

x and y have negative SIGNED values because x is left of, and y is below, the origin.

According to the diagram, $\cos w = -\dfrac{\sqrt{5}}{3}$ and $\sin z = -\dfrac{4\sqrt{3}}{7}$. Substitute these values, and the original values given by the problem, into the expanded sum formula for sine.

$$\sin(w + z) = \sin w \cos z + \cos w \sin z$$
$$= \left(-\frac{2}{3}\right)\left(\frac{1}{7}\right) + \left(-\frac{\sqrt{5}}{3}\right)\left(-\frac{4\sqrt{3}}{7}\right)$$
$$= -\frac{2}{21} + \frac{4\sqrt{15}}{21}$$
$$= \frac{4\sqrt{15} - 2}{21}$$

Chapter 8
ADVANCED TRIGONOMETRIC IDENTITIES

"Advanced" means "brimming with fractions"

Chapter 7 introduces the concept of trigonometric identities—true statements that are applied to simplify and verify trigonometric expressions and statements. In this chapter, you explore more advanced trigonometric identities, including double-angle, half-angle, power-reducing, product-to-sum, and sum-to-product formulas. Most of the identities in Chapters 7 and 8 are expressed in terms of sine and cosine, but the final section of this chapter explores identities involving the tangent function.

Chapter 7 dealt with trig expressions containing "x." (Remember, x is called the "argument" of the expression cos x.) This chapter deals with more complicated arguments, like cos 2x and sin x/2.

Don't start this chapter unless you've already finished Chapter 7, because you'll see some of those single-angle identities seep through into these problems. Focus your attention on the first and last sections of this chapter (double angles and tangent identities). The rest of the sections discuss identities that are used much less frequently—they each have a particular niche that makes them useful only in very specific circumstances.

Double-Angle Formulas
Ditch the 2s in sin 2x and cos 2y

8.1 Express $\sin 2x$ as a product of single-angle trigonometric functions.

The expression $\sin 2x$ is described as a "double angle," because the argument of the expression is $2x$ rather than simply x. The identity below allows you to rewrite the double-angle expression $\sin 2x$ as the product of two single-angle expressions.

$$\sin 2x = 2\sin x \cos x$$

8.2 Verify your answer to Problem 8.1 by applying the sum formula for sine.

Apply the sum formula for sine to expand $\sin 2x$. Note that $\sin 2x = \sin (x + x)$.

$$\sin 2x = \sin(x + x)$$
$$= \sin x \cos x + \cos x \sin x$$
$$= \sin x \cos x + \sin x \cos x$$
$$= 2\sin x \cos x$$

> This formula comes from Problem 7.34.

8.3 Simplify the expression: $\dfrac{\sin 2x}{\sin x}$.

Apply the double-angle identity $\sin 2x = 2\sin x \cos x$ to expand the numerator, and then simplify the fraction.

$$\frac{\sin 2x}{\sin x} = \frac{2\,\cancel{\sin x}\,\cos x}{\cancel{\sin x}}$$
$$= \frac{2\cos x}{1}$$
$$= 2\cos x$$

> Draw a right triangle in the second quadrant with an adjacent side with signed length –12 and a hypotenuse with length 13. Then use the Pythagorean theorem to calculate the signed length of the opposite side: +5.

8.4 Calculate $\sin 2\theta$, given $\cos\theta = -\dfrac{12}{13}$ and $\dfrac{\pi}{2} < \theta < \pi$.

This problem provides the cosine of θ and indicates that θ terminates in the second quadrant. Apply the technique modeled in Problems 4.30–4.45 (and again in Problem 7.45) to calculate $\sin \theta$. Note that $\sin \theta = +5/13$. Substitute the known values of $\cos \theta$ and $\sin \theta$ to calculate $\sin 2\theta$.

$$\sin 2\theta = 2\sin\theta \cos\theta$$
$$= 2\left(\frac{5}{13}\right)\left(-\frac{12}{13}\right)$$
$$= -\frac{120}{169}$$

8.5 Calculate $\sin 2\alpha$, given $\sin \alpha = -\dfrac{1}{4}$ and $\dfrac{3\pi}{2} < \alpha < 2\pi$.

Apply the technique described in Problem 8.4 to calculate the value of $\cos \alpha$, knowing α terminates in the fourth quadrant and $\sin \alpha = -1/4$. According to the Pythagorean theorem, $\cos \alpha = +\sqrt{15}\,/\,4$. Note that $\cos \alpha > 0$ because angles in the fourth quadrant have positive cosine values and negative sine values. Substitute $\cos \alpha$ and $\sin \alpha$ into the sine double-angle formula to calculate $\sin 2\alpha$.

$$\sin 2\alpha = 2\sin \alpha \cos \alpha$$

$$= 2\left(-\frac{1}{4}\right)\left(\frac{\sqrt{15}}{4}\right)$$

$$= -\frac{2\sqrt{15}}{16}$$

$$= -\frac{\sqrt{15}}{8}$$

8.6 Verify the statement: $(1 - \cos 2x)(1 + \cos 2x) = 4\sin^2 x \cos^2 x$.

Multiply the terms left of the equal sign.

$$1 + \cos 2x - \cos 2x - \cos^2 2x = 4\sin^2 x \cos^2 x$$

$$1 - \cos^2 2x = 4\sin^2 x \cos^2 x$$

According to a Pythagorean identity, $\cos^2 x + \sin^2 x = 1$. This identity is true for any value of x. For example, if you replace x with $2x$, the identity remains valid: $\cos^2 2x + \sin^2 2x = 1$. If you subtract $\cos^2 2x$ from both sides of the new identity, the result is $\sin^2 2x = 1 - \cos^2 2x$. Use this equality statement to rewrite the left side of the equation above.

$$\sin^2 2x = 4\sin^2 x \cos^2 x$$

Note that $\sin^2 2x = (\sin 2x)(\sin 2x)$.

$$(\sin 2x)(\sin 2x) = 4\sin^2 x \cos^2 x$$

$$(2\sin x \cos x)(2\sin x \cos x) = 4\sin^2 x \cos^2 x$$

$$4\sin^2 x \cos^2 x = 4\sin^2 x \cos^2 x$$

8.7 Verify the statement: $\csc 2x = \dfrac{1}{2}\cot x + \dfrac{1}{2}\tan x$.

Recall that $\csc x$ and $\sin x$ are reciprocal functions. Therefore, $\csc 2x = \dfrac{1}{\sin 2x}$. Use this equality statement to rewrite the left side of the statement. Express the right side of the statement in terms of sine and cosine.

$$\frac{1}{\sin 2x} = \frac{1}{2} \cdot \frac{\cos x}{\sin x} + \frac{1}{2} \cdot \frac{\sin x}{\cos x}$$

$$\frac{1}{2\sin x \cos x} = \frac{\cos x}{2\sin x} + \frac{\sin x}{2\cos x}$$

Combine the fractions on the right side of the equation using least common denominator $2 \sin x \cos x$.

According to a Pythagorean identity from Problem 7.23, $\cos^2 x + \sin^2 x = 1$.

$$\frac{1}{2\sin x \cos x} = \frac{\cos x}{2\sin x}\left(\frac{\cos x}{\cos x}\right) + \frac{\sin x}{2\cos x}\left(\frac{\sin x}{\sin x}\right)$$

$$\frac{1}{2\sin x \cos x} = \frac{\cos^2 x}{2\sin x \cos x} + \frac{\sin^2 x}{2\sin x \cos x}$$

$$\frac{1}{2\sin x \cos x} = \frac{\cos^2 x + \sin^2 x}{2\sin x \cos x}$$

$$\frac{1}{2\sin x \cos x} = \frac{1}{2\sin x \cos x}$$

8.8 Rewrite the double-angle expression $\cos 2x$ as a difference of single-angle cosine and sine expressions.

There are three different ways to convert the double-angle expression $\cos 2x$ into a single-angle expression, but only one way that contains both sine and cosine: $\cos 2x = \cos^2 x - \sin^2 x$.

Be careful with the sign:
$\cos^2 x + \sin^2 x = 1$
$\cos^2 x - \sin^2 x = \cos 2x$

8.9 Verify your answer to Problem 8.8 using the sum formula for cosine.

Notice that $2x = x + x$. Apply the sum formula for cosine (originally stated in Problem 7.37) to expand $\cos (x + x)$.

$$\cos 2x = \cos(x + x)$$
$$= \cos x \cdot \cos x - \sin x \cdot \sin x$$
$$= \cos^2 x - \sin^2 x$$

8.10 Apply a Pythagorean identity to rewrite your solution to Problem 8.8 in two different ways: (A) in terms of cosine only, and (B) in terms of sine only.

If you subtract $\cos^2 x$ from both sides of the Pythagorean identity $\cos^2 x + \sin^2 x = 1$, the result is $\sin^2 x = 1 - \cos^2 x$. Use this equality statement to rewrite the cosine double-angle formula in terms of cosine.

$$\cos 2x = \cos^2 x - \sin^2 x$$
$$= \cos^2 x - \left(1 - \cos^2 x\right)$$
$$= \cos^2 x - 1 + \cos^2 x$$
$$= 2\cos^2 x - 1$$

Similarly, you can manipulate the Pythagorean identity $\cos^2 x + \sin^2 x = 1$ to assert that $\cos^2 x = 1 - \sin^2 x$. Use this equality statement to rewrite the cosine double-angle formula in terms of sine.

$$\cos 2x = \cos^2 x - \sin^2 x$$
$$= \left(1 - \sin^2 x\right) - \sin^2 x$$
$$= 1 - 2\sin^2 x$$

Thus, there are three equivalent ways to rewrite $\cos 2x$ using single-angle trigonometric expressions.

$$\cos 2x = \cos^2 x - \sin^2 x$$
$$= 2\cos^2 x - 1$$
$$= 1 - 2\sin^2 x$$

8.11 Verify the statement: $\sin 4x = 4\cos^3 x \sin x - 4\cos x \sin^3 x$.

Note that $4x = 2x + 2x$ and apply the sum formula for sine to expand $\sin 4x$.

$$\sin 4x = \sin(2x + 2x)$$
$$= \sin 2x \cos 2x + \cos 2x \sin 2x$$
$$= \left(2\sin x \cos x\right)\left(\cos^2 x - \sin^2 x\right) + \left(\cos^2 x - \sin^2 x\right)\left(2\sin x \cos x\right)$$
$$= 2\cos^3 x \sin x - 2\cos x \sin^3 x + 2\cos^3 x \sin x - 2\cos x \sin^3 x$$
$$= 4\cos^3 x \sin x - 4\cos x \sin^3 x$$

Thus, $\sin 4x$ is equal to $4\cos^3 x \sin x - 4\cos x \sin^3 x$, as proposed in the original problem, and the statement is verified.

8.12 Express $\cos 3x$ in terms of single-angle cosine functions.

Note that $3x = x + 2x$, and apply the sum formula for cosine.

$$\cos 3x = \cos(x + 2x)$$
$$= \cos x \cos 2x - \sin x \sin 2x$$

Apply the double-angle formula $\cos 2x = 2\cos^2 x - 1$, because the problem directs you to write the expression in terms of cosine.

> Instead of $\cos^2 x - \sin^2 x$ and $1 - 2\sin^2 x$, because each of those contains sine.

$$= \cos x \left(2\cos^2 x - 1\right) - \sin x \left(2\sin x \cos x\right)$$
$$= 2\cos^3 x - \cos x - 2\cos x \sin^2 x$$

According to a Pythagorean identity, $\sin^2 x = 1 - \cos^2 x$.

> You get this by subtracting $\cos^2 x$ from both sides of the identity $\cos^2 x + \sin^2 x = 1$.

$$= 2\cos^3 x - \cos x - 2\cos x \left(1 - \cos^2 x\right)$$
$$= 2\cos^3 x - \cos x - 2\cos x + 2\cos^3 x$$
$$= 4\cos^3 x - 3\cos x$$

8.13 Simplify the expression: $\dfrac{\cos^4\theta - \sin^4\theta}{10\sin\theta\cos\theta}$.

Factor the difference of perfect squares in the numerator. Notice that rewriting the denominator as $5\,(2\sin\theta\cos\theta)$ allows you to apply the sine double-angle formula.

$$\frac{\cos^4\theta - \sin^4\theta}{10\sin\theta\cos\theta} = \frac{\left(\cos^2\theta - \sin^2\theta\right)\left(\cos^2\theta + \sin^2\theta\right)}{5\left(2\sin\theta\cos\theta\right)}$$

$$= \frac{\left(\cos 2\theta\right)(1)}{5\left(\sin 2\theta\right)}$$

$$= \frac{1}{5}\cdot\frac{\cos 2\theta}{\sin 2\theta}$$

$$= \frac{1}{5}\cot 2\theta$$

8.14 Use a double-angle formula to verify that $\cos\dfrac{\pi}{3} = \dfrac{1}{2}$.

Multiplying $\pi/6$ by 2 produces the angle $\pi/3$. Therefore, you can substitute $\pi/6$ into the cosine double-angle formula to verify that $\cos(\pi/3) = 1/2$, as dictated by the unit circle. The following solution applies the double-angle formula $\cos 2x = \cos^2 x - \sin^2 x$, but all three of the cosine double-angle formulas produce the same result.

$$\cos\frac{\pi}{3} = \cos\left(2\cdot\frac{\pi}{6}\right)$$

$$= \cos^2\frac{\pi}{6} - \sin^2\frac{\pi}{6}$$

$$= \left(\frac{\sqrt{3}}{2}\right)^2 - \left(\frac{1}{2}\right)^2$$

$$= \frac{3}{4} - \frac{1}{4}$$

$$= \frac{2}{4}$$

$$= \frac{1}{2}$$

To use the double-angle formula, you need to find an angle you can double to get $\pi/3$:

$$2\left(\frac{\pi}{6}\right) = \frac{2\pi}{6} = \frac{\pi}{3}$$

8.15 Calculate $\cos 2\theta$, given $\sin\theta = -\dfrac{3}{4}$ and $\pi < \theta < \dfrac{3\pi}{2}$.

Unlike Problems 8.4–8.5, in which you needed to construct a right triangle to calculate missing trigonometric values, knowing that $\sin\theta = -3/4$ is sufficient to complete this problem—as long as you select the correct cosine double-angle formula.

$$\cos 2\theta = 1 - 2\sin^2 \theta$$
$$= 1 - 2\left(-\frac{3}{4}\right)^2$$
$$= 1 - \frac{2}{1}\left(\frac{9}{16}\right)$$
$$= 1 - \frac{\cancel{2} \cdot 9}{\cancel{2} \cdot 8}$$
$$= \frac{8}{8} - \frac{9}{8}$$
$$= -\frac{1}{8}$$

8.16 Simplify the expression: $-2\sin x\left[\sin\left(\dfrac{\pi}{2} - x\right) - \dfrac{1}{2}\csc x + \sin x\right]$.

Apply a cofunction identity and a reciprocal identity to rewrite the bracketed expression and then distribute $-2\sin x$ to each term.

$$-2\sin x\left[\sin\left(\frac{\pi}{2} - x\right) - \frac{1}{2}\csc x + \sin x\right] = -2\sin x\left[\cos x - \frac{1}{2}\left(\frac{1}{\sin x}\right) + \sin x\right]$$
$$= -2\sin x\cos x - \frac{2\cancel{\sin x}}{1}\left(-\frac{1}{2\cancel{\sin x}}\right) - 2\sin x \cdot \sin x$$
$$= -2\sin x\cos x + 1 - 2\sin^2 x$$

Apply two double-angle formulas to simplify the expression: $\sin 2x = 2\sin x\cos x$ and $\cos 2x = 1 - 2\sin^2 x$.

$$= (-1)(2\sin x\cos x) + (1 - 2\sin^2 x)$$
$$= (-1)(\sin 2x) + (\cos 2x)$$
$$= \cos 2x - \sin 2x$$

Power-Reducing Formulas
Rewrite squared functions using double angles

8.17 List the power-reducing formulas for $\cos^2\theta$ and $\sin^2\theta$.

Power-reducing formulas not only remove the exponents from the trigonometric expression, they also rewrite the expression entirely in terms of cosine.

$$\cos^2\theta = \frac{1 + \cos 2\theta}{2} \qquad\qquad \sin^2\theta = \frac{1 - \cos 2\theta}{2}$$

You end up with cosines of double angles.

8.18 Verify the power-reducing formula for $\sin^2\theta$ that you identified in Problem 8.17.

Expand the right side of the power-reducing formula by applying the double-angle formula $\cos 2\theta = 1 - 2\sin^2\theta$.

$$\sin^2\theta = \frac{1-\cos 2\theta}{2}$$

$$\sin^2\theta = \frac{1-\left(1-2\sin^2\theta\right)}{2}$$

$$\sin^2\theta = \frac{1-1+2\sin^2\theta}{2}$$

$$\sin^2\theta = \frac{\cancel{2}\sin^2\theta}{\cancel{2}}$$

$$\sin^2\theta = \sin^2\theta$$

The final statement $\sin^2\theta = \sin^2\theta$ is obviously true; therefore, the original statement is verified.

8.19 Apply a power-reducing formula to express $\cos^4\theta$ as a sum of cosine functions raised to the first power.

Express $\cos^4\theta$ as $(\cos^2\theta)^2$ and apply the power-reducing formula to rewrite $\cos^2\theta$.

$$\cos^4\theta = \left(\cos^2\theta\right)\left(\cos^2\theta\right)$$

$$= \left(\frac{1+\cos 2\theta}{2}\right)\left(\frac{1+\cos 2\theta}{2}\right)$$

$$= \frac{1+2\cos 2\theta+\cos^2 2\theta}{4}$$

$$= \frac{1}{4}+\frac{1}{2}\cos 2\theta+\frac{1}{4}\cos^2 2\theta$$

To rewrite $\cos^2 2\theta$ using the power-reducing formula

$$\cos^2 x = \frac{1+\cos 2x}{2},$$

replace x with 2θ.

Apply the power-reducing formula once again to eliminate the exponent from the expression.

$$= \frac{1}{4}+\frac{1}{2}\cos 2\theta+\frac{1}{4}\left[\frac{1+\cos(2\cdot 2\theta)}{2}\right]$$

$$= \frac{1}{4}+\frac{1}{2}\cos 2\theta+\frac{1}{4}\left(\frac{1}{2}\right)[1+\cos 4\theta]$$

$$= \frac{1}{4}+\frac{1}{2}\cos 2\theta+\frac{1}{8}(1+\cos 4\theta)$$

$$= \frac{1}{4}+\frac{1}{2}\cos 2\theta+\frac{1}{8}+\frac{1}{8}\cos 4\theta$$

Add the constants to simplify the expression: $1/4 + 1/8 = 3/8$.

$$= \frac{3}{8}+\frac{1}{2}\cos 2\theta+\frac{1}{8}\cos 4\theta$$

The solution $(1/8)(3+4\cos 2\theta+\cos 4\theta)$ is also correct.

8.20 Apply a power-reducing formula to express $\cos^2 x + \sin^4 x$ as a sum that includes cosine functions raised to the first power.

Express $\sin^4 x$ as the product $(\sin^2 x)(\sin^2 x)$ and apply the power-reducing formulas.

$$\cos^2 x + \sin^4 x = \cos^2 x + \left(\sin^2 x\right)^2$$

$$= \frac{1 + \cos 2x}{2} + \left(\frac{1 - \cos 2x}{2}\right)^2$$

$$= \frac{1 + \cos 2x}{2} + \frac{1 - 2\cos 2x + \cos^2 2x}{4}$$

$$= \left(\frac{1}{2} + \frac{1}{2}\cos 2x\right) + \left(\frac{1}{4} - \frac{2}{4}\cos 2x + \frac{1}{4}\cos^2 2x\right)$$

$$= \left(\frac{1}{2} + \frac{1}{4}\right) + \left(\frac{1}{2}\cos 2x - \frac{1}{2}\cos 2x\right) + \frac{1}{4}\cos^2 2x$$

$$= \frac{3}{4} + \frac{1}{4}\cos^2 2x$$

Once again, apply the power-reducing formula $\cos^2 \theta = \dfrac{1 + \cos 2\theta}{2}$, this time replacing θ with $2x$.

$$= \frac{3}{4} + \frac{1}{4}\left[\frac{1 + \cos(2 \cdot 2x)}{2}\right]$$

$$= \frac{3}{4} + \frac{1}{8}\left[1 + \cos 4x\right]$$

$$= \left(\frac{3}{4} + \frac{1}{8}\right) + \frac{1}{8}\cos 4x$$

$$= \frac{7}{8} + \frac{1}{8}\cos 4x$$

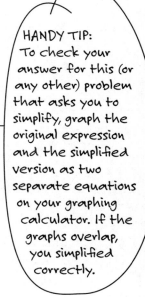

HANDY TIP:
To check your answer for this (or any other) problem that asks you to simplify, graph the original expression and the simplified version as two separate equations on your graphing calculator. If the graphs overlap, you simplified correctly.

8.21 Verify the statement: $(\sin^2 \theta)(\cos^2 \theta) = \dfrac{1}{4}(\sin^2 2\theta)$.

Rewrite the factors on the left side of the equation, applying power-reducing formulas.

$$\left(\frac{1 - \cos 2\theta}{2}\right)\left(\frac{1 + \cos 2\theta}{2}\right) = \frac{1}{4}\sin^2 2\theta$$

$$\frac{1 + \cos 2\theta - \cos 2\theta - \cos^2 2\theta}{4} = \frac{1}{4}\sin^2 2\theta$$

$$\frac{1 - \cos^2 2\theta}{4} = \frac{1}{4}\sin^2 2\theta$$

Multiply the entire equation by 4 to eliminate the fractions.

$$1 - \cos^2 2\theta = \sin^2 2\theta$$

$$1 = \cos^2 2\theta + \sin^2 2\theta$$

If $x = 2\theta$, then the final statement is equivalent to the Pythagorean identity $\cos^2 x + \sin^2 x = 1$. Because that statement is true, the original statement is verified.

Half-Angle Formulas

Win half an argument by being radical

8.22 Generate the cosine half-angle formula based on its power-reducing formula.

> **Look at Problem 8.17 to review the power-reducing formulas.**

Substitute $\theta = x/2$ into the cosine power-reducing formula, and solve the equation for $\cos(x/2)$.

$$\cos^2\theta = \frac{1+\cos 2\theta}{2}$$

$$\cos^2\frac{x}{2} = \frac{1+\cos\left[2(x/2)\right]}{2}$$

$$\cos^2\frac{x}{2} = \frac{1+\cos x}{2}$$

$$\sqrt{\cos^2\frac{x}{2}} = \pm\sqrt{\frac{1+\cos x}{2}}$$

$$\cos\frac{x}{2} = \pm\sqrt{\frac{1+\cos x}{2}}$$

Note that the half-angle formula contains a "±" symbol, because the formula, itself, is not sufficient to determine the sign of $\cos(x/2)$. Instead, you must use additional information given by the problem, such as the quadrant in which the angle lies.

8.23 Substitute $\theta = \frac{x}{2}$ into the power-reducing formula for $\sin^2\theta$ to generate the half-angle formula for sine.

Apply the technique demonstrated in Problem 8.22.

$$\sin^2\theta = \frac{1-\cos 2\theta}{2}$$

$$\sin^2\frac{x}{2} = \frac{1-\cos\left[2(x/2)\right]}{2}$$

$$\sin^2\frac{x}{2} = \frac{1-\cos x}{2}$$

$$\sqrt{\sin^2\frac{x}{2}} = \pm\sqrt{\frac{1-\cos x}{2}}$$

$$\sin\frac{x}{2} = \pm\sqrt{\frac{1-\cos x}{2}}$$

8.24 Apply a half-angle formula to calculate the exact value of $\cos 22.5°$ without using a calculator or trigonometric table.

> **Sine and cosine are both positive in the first quadrant.**

Although you have not memorized the value of $\cos 22.5°$, you have memorized the trigonometric values for twice that angle: $2(22.5°) = 45°$. To calculate $\cos 22.5°$, substitute $45°$ into the half-angle formula presented in Problem 8.22. Note that an angle in standard position that measures $22.5°$ terminates in the first quadrant, so $\cos 22.5 > 0$.

$$\cos\frac{x}{2} = \pm\sqrt{\frac{1+\cos x}{2}}$$

$$\cos\frac{45°}{2} = +\sqrt{\frac{1+\cos 45°}{2}}$$

$$\cos 22.5° = \sqrt{\frac{1+\sqrt{2}/2}{2}}$$

Express the complex fraction as a sum of two fractions with a denominator of 2. Note that $\left(\sqrt{2}/2\right) \div 2 = \left(\sqrt{2}/2\right) \cdot (1/2)$.

$$\cos 22.5° = \sqrt{\frac{1}{2} + \frac{\sqrt{2}}{2} \cdot \frac{1}{2}}$$

$$\cos 22.5° = \sqrt{\frac{1}{2} + \frac{\sqrt{2}}{4}}$$

$$\cos 22.5° = \sqrt{\frac{2}{4} + \frac{\sqrt{2}}{4}}$$

$$\cos 22.5° = \sqrt{\frac{2+\sqrt{2}}{4}}$$

$$\cos 22.5° = \frac{\sqrt{2-\sqrt{3}}}{2}$$

8.25 Apply a half-angle formula to calculate the exact value of $\sin\dfrac{11\pi}{12}$ without using a calculator or trigonometric table.

Apply the method demonstrated in Problem 8.24, substituting $11\pi/6$ into the half-angle formula for sine, because $11\pi/12$ is half of $11\pi/6$. Note that the angle $11\pi/12$ lies in the second quadrant because $11 \div 12 = 0.91\overline{6}$. (According to the technique described in Problem 3.34, the angle lies in the second quadrant because $0.5 < 0.91\overline{6} < 1$.) Recall that sine values are positive for angles in the second quadrant, so you should replace the "$\pm$" sign in the half-angle formula with a "+" sign.

$$\frac{11\pi}{6} \cdot \frac{1}{2} = \frac{11\pi}{12}$$

$$\sin\frac{x}{2} = \pm\sqrt{\frac{1-\cos x}{2}}$$

$$\sin\frac{11\pi/6}{2} = +\sqrt{\frac{1-\cos(11\pi/6)}{2}}$$

$$\sin\frac{11\pi}{12} = \sqrt{\frac{1}{2} - \frac{\sqrt{3}/2}{2}}$$

$$\sin\frac{11\pi}{12} = \sqrt{\frac{1}{2} - \frac{\sqrt{3}}{2} \cdot \frac{1}{2}}$$

$$\sin\frac{11\pi}{12} = \sqrt{\frac{2}{4} - \frac{\sqrt{3}}{4}}$$

$$\sin\frac{11\pi}{12} = \sqrt{\frac{2-\sqrt{3}}{4}}$$

$$\sin\frac{11\pi}{12} = \frac{\sqrt{2-\sqrt{3}}}{2}$$

8.26 Simplify the expression using a half-angle formula: $\sqrt{\dfrac{1+\cos 12x}{2}}$.

This expression resembles the cosine half-angle formula $\cos\dfrac{\theta}{2}=\pm\sqrt{\dfrac{1+\cos\theta}{2}}$, if $\theta = 12x$.

$$\cos\frac{12x}{2}=\pm\sqrt{\frac{1+\cos 12x}{2}}$$

$$\cos 6x=\pm\sqrt{\frac{1+\cos 12x}{2}}$$

Therefore, you can conclude that $\sqrt{\dfrac{1+\cos 12x}{2}}$ is equal to $\pm\cos 6x$. Note, however, that the radical expression given in the problem is positive, so $\cos 6x$ must be positive as well. Surround $\cos 6x$ with absolute value signs to ensure that its value is greater than zero.

$$\sqrt{\frac{1+\cos 12x}{2}}=|\cos 6x|$$

8.27 Simplify the expression using a half-angle formula: $-\sqrt{\dfrac{1-\cos 5x}{2}}$.

This radical expression resembles the sine half-angle formula

$\sin\dfrac{\theta}{2}=\pm\sqrt{\dfrac{1-\cos\theta}{2}}$, if $\theta = 5x$.

$$\sin\frac{5x}{2}=\pm\sqrt{\frac{1-\cos 5x}{2}}$$

The problem states that the radical expression is negative. To ensure that the equivalent expression $\sin(5x/2)$ is also negative, surround it with absolute value bars and then take its opposite.

The absolute value bars work just like they did in Problem 8.26. Then, you multiply that absolute value expression (which HAS to be positive) by −1, so the product HAS to be negative.

$$-\sqrt{\frac{1-\cos 5x}{2}}=-\left|\sin\frac{5x}{2}\right|$$

The solution $-\sqrt{\dfrac{1-\cos 5x}{2}}=-\left|\sin\dfrac{5}{2}x\right|$ is also correct.

Product-to-Sum Identities

Add or subtract, instead of multiplying, trig functions

8.28 List the product-to-sum formulas that correspond with each of the following four products: $\cos\alpha\cos\beta$, $\sin\alpha\sin\beta$, $\cos\alpha\sin\beta$, and $\sin\alpha\cos\beta$.

The product-to-sum formulas are not typically used to simplify a product, as the equivalent sums are usually lengthier. They are specifically designed to rewrite a product of sine and/or cosine expressions as a sum of sine or cosine expressions.

$$\cos\alpha\cos\beta = \frac{1}{2}\left[\cos(\alpha-\beta)+\cos(\alpha+\beta)\right]$$

$$\sin\alpha\sin\beta = \frac{1}{2}\left[\cos(\alpha-\beta)-\cos(\alpha+\beta)\right]$$

$$\cos\alpha\sin\beta = \frac{1}{2}\left[\sin(\alpha+\beta)-\sin(\alpha-\beta)\right]$$

$$\sin\alpha\cos\beta = \frac{1}{2}\left[\sin(\alpha+\beta)+\sin(\alpha-\beta)\right]$$

Notice that a product of two matching trigonometric functions ($\cos\alpha\cos\beta$ or $\sin\alpha\sin\beta$) translates into a sum containing cosine functions. Alternately, products that include different trigonometric functions ($\cos\alpha\sin\beta$ and $\sin\alpha\cos\beta$) translate into a sum containing sine functions.

8.29 Verify the product-to-sum formula for $\sin\alpha\sin\beta$.

According to Problem 8.28, $\sin\alpha\sin\beta = \frac{1}{2}\left[\cos(\alpha-\beta)-\cos(\alpha+\beta)\right]$. Apply the sum and difference formulas for cosine to expand the right side of the equation and verify the statement.

> If you need to review these sum and difference formulas, flip back to Problem 7.37.

$$\sin\alpha\sin\beta = \frac{1}{2}\left[(\cos\alpha\cos\beta+\sin\alpha\sin\beta)-(\cos\alpha\cos\beta-\sin\alpha\sin\beta)\right]$$

$$\sin\alpha\sin\beta = \frac{1}{2}\left[\cancel{\cos\alpha\cos\beta}+\sin\alpha\sin\beta-\cancel{\cos\alpha\cos\beta}+\sin\alpha\sin\beta\right]$$

$$\sin\alpha\sin\beta = \frac{1}{2}\left[2\sin\alpha\sin\beta\right]$$

$$\sin\alpha\sin\beta = \sin\alpha\sin\beta$$

8.30 Write the following product as a sum or difference: $\cos x\sin 2x$.

Apply the $\cos\alpha\sin\beta$ product-to-sum formula presented in Problem 8.28. Note that $\alpha = x$ and $\beta = 2x$.

$$\cos\alpha\sin\beta = \frac{1}{2}\left[\sin(\alpha+\beta)-\sin(\alpha-\beta)\right]$$

$$\cos x\sin 2x = \frac{1}{2}\left[\sin(x+2x)-\sin(x-2x)\right]$$

$$\cos x\sin 2x = \frac{1}{2}\left[\sin 3x-\sin(-x)\right]$$

According to a negative identity presented in Problem 7.14, $\sin(-x) = -\sin x$.

$$\cos x \sin 2x = \frac{1}{2}\left[\sin 3x - (-\sin x)\right]$$

$$\cos x \sin 2x = \frac{1}{2}\left[\sin 3x + \sin x\right]$$

$$\cos x \sin 2x = \frac{1}{2}\sin 3x + \frac{1}{2}\sin x$$

8.31 Verify the statement: $\cos^2 4x = \dfrac{\cos 8x + 1}{2}$.

Notice that $\cos^2 4x = (\cos 4x)(\cos 4x)$. Apply the $\cos \alpha \cos \beta$ product-to-sum formula to expand the left side of the equation, such that $\alpha = 4x$ and $\beta = 4x$.

$$(\cos 4x)(\cos 4x) = \frac{\cos 8x + 1}{2}$$

$$\frac{1}{2}\left[\cos(4x - 4x) + \cos(4x + 4x)\right] = \frac{\cos 8x + 1}{2}$$

$$\frac{1}{2}\left[\cos 0 + \cos 8x\right] = \frac{\cos 8x + 1}{2}$$

Multiply both sides of the equation by 2 in order to eliminate the fractions, and recall that $\cos 0 = 1$ (according to the unit circle).

$$\left(\frac{2}{1}\right)\frac{1}{2}\left[\cos 0 + \cos 8x\right] = \left(\frac{2}{1}\right)\frac{\cos 8x + 1}{2}$$

$$\cos 0 + \cos 8x = \cos 8x + 1$$

$$1 + \cos 8x = \cos 8x + 1$$

Both sides of the equation are equal—the order in which you add does not affect the sum according to the commutative property of addition—so the original statement is verified.

Sum-to-Product Identities
Do the opposite of the last section

8.32 List the sum-to-product formulas that correspond with each of the following four sums: $\cos \alpha + \cos \beta$, $\cos \alpha - \cos \beta$, $\sin \alpha + \sin \beta$, $\sin \alpha - \sin \beta$.

The sum-to-product formulas, like the product-to-sum formulas discussed in Problems 8.28–8.31, are not typically applied to simplify an expression. Rather, they are used exclusively to express a sum or difference as a product.

The sum-to-product formulas only apply when you are adding or subtracting two cosines or two sines. In other words, there IS a formula for cos x + cos y, but there ISN'T a formula for cos x + sin y.

$$\cos\alpha + \cos\beta = 2\cos\left(\frac{\alpha+\beta}{2}\right)\cos\left(\frac{\alpha-\beta}{2}\right)$$

$$\cos\alpha - \cos\beta = -2\sin\left(\frac{\alpha+\beta}{2}\right)\sin\left(\frac{\alpha-\beta}{2}\right)$$

$$\sin\alpha + \sin\beta = 2\sin\left(\frac{\alpha+\beta}{2}\right)\cos\left(\frac{\alpha-\beta}{2}\right)$$

$$\sin\alpha - \sin\beta = 2\cos\left(\frac{\alpha+\beta}{2}\right)\sin\left(\frac{\alpha-\beta}{2}\right)$$

8.33 Demonstrate that the following expression is equivalent to the corresponding sum-to-product formula: $\cos\dfrac{\pi}{2} - \cos\dfrac{\pi}{6}$.

Begin by evaluating the given expression; its angles belong to the standard trigonometric unit circle.

$$\cos\frac{\pi}{2} - \cos\frac{\pi}{6} = 0 - \left(\frac{\sqrt{3}}{2}\right)$$

$$= -\frac{\sqrt{3}}{2}$$

Now apply a sum-to-product formula to rewrite the expression as a product.

$$\cos\alpha - \cos\beta = -2\sin\left(\frac{\alpha+\beta}{2}\right)\sin\left(\frac{\alpha-\beta}{2}\right)$$

$$\cos\frac{\pi}{2} - \cos\frac{\pi}{6} = -2\sin\left(\frac{\pi/2+\pi/6}{2}\right)\sin\left(\frac{\pi/2-\pi/6}{2}\right)$$

To combine $\pi/2$ and $\pi/6$, you must express $\pi/2$ in terms of the least common denominator: $\pi/2 = 3\pi/6$.

$$\cos\frac{\pi}{2} - \cos\frac{\pi}{6} = -2\sin\left(\frac{3\pi/6+\pi/6}{2}\right)\sin\left(\frac{3\pi/6-\pi/6}{2}\right)$$

$$= -2\sin\left(\frac{4\pi/6}{2}\right)\sin\left(\frac{2\pi/6}{2}\right)$$

$$= -2\sin\left(\frac{2\pi/3}{2}\right)\sin\left(\frac{\pi/3}{2}\right)$$

$$= -2\sin\left(\frac{\pi}{3}\right)\sin\left(\frac{\pi}{6}\right)$$

$$\frac{2\pi/3}{2} = \frac{2\pi}{3}\cdot\frac{1}{2} = \frac{2\pi}{6} = \frac{\pi}{3}$$

$$\frac{\pi/3}{2} = \frac{\pi}{3}\cdot\frac{1}{2} = \frac{\pi}{6}$$

The expression now contains angles that appear on the standard unit circle. Evaluate the angles and simplify the expression.

$$= -2\left(\frac{\sqrt{3}}{2}\right)\left(\frac{1}{2}\right)$$

$$= -\frac{\sqrt{3}}{2}$$

The original sum and the corresponding product have the same value: $-\dfrac{\sqrt{3}}{2}$.

8.34 Calculate the exact value of sin 105° + sin 15° without using a calculator or trigonometric table.

Apply the sum-to-product formula for $\sin \alpha + \sin \beta$, such that $\alpha = 105°$ and $\beta = 15°$.

$$\sin \alpha + \sin \beta = 2 \sin\left(\frac{\alpha + \beta}{2}\right)\cos\left(\frac{\alpha - \beta}{2}\right)$$

$$\sin 105° + \sin 15° = 2 \sin\left(\frac{105° + 15°}{2}\right)\cos\left(\frac{105° - 15°}{2}\right)$$

$$\sin 105° + \sin 15° = 2 \sin\left(\frac{120°}{2}\right)\cos\left(\frac{90°}{2}\right)$$

$$\sin 105° + \sin 15° = 2\left(\sin 60°\right)\left(\cos 45°\right)$$

If you are accustomed to radian measures, note that $60° = \pi/3$ and $45° = \pi/4$.

$$\sin 105° + \sin 15° = \not{2}\left(\frac{\sqrt{3}}{\not{2}}\right)\left(\frac{\sqrt{2}}{2}\right)$$

$$= \frac{\sqrt{6}}{2}$$

8.35 Verify the statement: $\dfrac{\cos 4x + \cos 2x}{\sin 4x - \sin 2x} = \cot x$.

Apply sum-to-product formulas to expand the numerator and denominator of the fraction.

$$\frac{2 \cos\left(\dfrac{4x + 2x}{2}\right)\cos\left(\dfrac{4x - 2x}{2}\right)}{2 \cos\left(\dfrac{4x + 2x}{2}\right)\sin\left(\dfrac{4x - 2x}{2}\right)} = \cot x$$

$$\frac{2 \cos\left(\dfrac{6x}{2}\right)\cos\left(\dfrac{2x}{2}\right)}{2 \cos\left(\dfrac{6x}{2}\right)\sin\left(\dfrac{2x}{2}\right)} = \cot x$$

$$\frac{2}{2} \cdot \frac{\cos 3x}{\cos 3x} \cdot \frac{\cos x}{\sin x} = \cot x$$

$$1 \cdot 1 \cdot \frac{\cos x}{\sin x} = \cot x$$

$$\cot x = \cot x$$

The final statement (cot x = cot x) is true, so the original statement is verified.

Tangent Identities

Sum/difference, double/half-angle, and power-reducing formulas

8.36 Apply a sum or difference formula to calculate tan 75° without using a calculator or trigonometric table.

Problems 7.34–7.45 explore the sum and difference formulas for sine and cosine; the sum/difference formula for tangent is presented below.

$$\tan\left(x \pm y\right) = \frac{\tan x \pm \tan y}{1 \mp \tan x \tan y}$$

In other words, tan (x + y)

$$= \frac{\tan x + \tan y}{1 - \tan x \tan y}$$

and tan (x − y)

$$= \frac{\tan x - \tan y}{1 + \tan x \tan y}.$$

Note that "±" symbols appear on the left side of the equation and in the numerator of the fraction. Therefore, those signs match. The symbol "∓" represents the opposite sign.

Notice that 75° = 30° + 45°. Therefore, you can calculate tan 75° by substituting $x = 30°$ and $y = 45°$ (or $x = 45°$ and $y = 30°$) into the sum formula for tangent.

$$\tan\left(30° + 45°\right) = \frac{\tan 30° + \tan 45°}{1 - \tan 30° \tan 45°}$$

In order to evaluate this expression, you must calculate tan 30° and tan 45°.

$$\tan 30° = \frac{\sin 30°}{\cos 30°} \qquad \tan 45° = \frac{\sin 45°}{\cos 45°}$$

$$= \frac{1/2}{\sqrt{3}/2} \qquad\qquad = \frac{\sqrt{2}/2}{\sqrt{2}/2}$$

$$= \frac{1}{\sqrt{3}} \qquad\qquad\quad = 1$$

$$= \frac{\sqrt{3}}{3}$$

Substitute these values into the expression and simplify.

$$\tan 75° = \frac{\sqrt{3}/3 + 1}{1 - \left(\sqrt{3}/3\right)(1)}$$

$$= \frac{\sqrt{3}/3 + 1}{1 - \sqrt{3}/3}$$

$$= \frac{\left(\sqrt{3}/3\right) + (3/3)}{(3/3) - \left(\sqrt{3}/3\right)}$$

$$= \frac{\dfrac{\sqrt{3} + 3}{3}}{\dfrac{3 - \sqrt{3}}{3}}$$

Multiply the numerator and denominator by 3 to simplify the complex fraction.

$$= \frac{\sqrt{3} + 3}{3 - \sqrt{3}}$$

8.37 Simplify the expression: $\tan\left(x - \dfrac{\pi}{4}\right)$.

Apply the tangent difference formula, stated in Problem 8.36.

$$\tan\left(x - \frac{\pi}{4}\right) = \frac{\tan x - \tan(\pi/4)}{1 + \tan x \cdot \tan(\pi/4)}$$

Note that $\tan\dfrac{\pi}{4} = \dfrac{\sin(\pi/4)}{\cos(\pi/4)} = \dfrac{\sqrt{2}/2}{\sqrt{2}/2} = 1$.

$$\tan\left(x - \frac{\pi}{4}\right) = \frac{\tan x - 1}{1 + \tan x \cdot 1}$$

$$= \frac{\tan x - 1}{\tan x + 1}$$

8.38 Simplify the expression: $\dfrac{\tan(\theta + \pi)}{\sin(\theta - \pi)}$.

Begin by expressing the fraction as the numerator multiplied by the reciprocal of the denominator. Next, apply the sum formula for tangent and the difference formula for sine.

$$\frac{\tan(\theta + \pi)}{\sin(\theta - \pi)} = \tan(\theta + \pi) \cdot \frac{1}{\sin(\theta - \pi)}$$

$$= \frac{\tan\theta + \tan\pi}{1 - \tan\theta \cdot \tan\pi} \cdot \frac{1}{\sin\theta\cos\pi - \cos\theta\sin\pi}$$

$$= \frac{\tan\theta + 0}{1 - (\tan\theta)(0)} \cdot \frac{1}{(\sin\theta)(-1) - (\cos\theta)(0)}$$

$$= \frac{\tan\theta}{1 - 0} \cdot \frac{1}{-\sin\theta - 0}$$

$$= \frac{\tan\theta}{1}\left(-\frac{1}{\sin\theta}\right)$$

$$\tan\pi = \frac{\sin\pi}{\cos\pi}$$
$$= \frac{0}{-1}$$
$$= 0$$

Rewrite $\tan\theta$ as the quotient of $\sin\theta$ and $\cos\theta$ to simplify the product.

$$= \frac{\sin\theta}{\cos\theta}\left(-\frac{1}{\sin\theta}\right)$$

$$= -\frac{1}{\cos\theta}$$

$$= -\sec\theta$$

8.39 Generate the tangent double-angle formula based on the sum formula for tangent.

Express $\tan 2x$ as $\tan(x + x)$ and apply the sum formula introduced in Problem 8.36.

$$\tan(x + x) = \frac{\tan x + \tan x}{1 - \tan x \cdot \tan x}$$

$$\tan 2x = \frac{2\tan x}{1 - \tan^2 x}$$

8.40 Demonstrate that the double-angle formula generated in Problem 8.39 is true by evaluating $\tan\left(2 \cdot \dfrac{\pi}{6}\right)$.

Substitute $x = \pi/6$ into the double-angle formula generated in Problem 8.39 and verify that the resulting statement is true.

$$\tan\left(2 \cdot \frac{\pi}{6}\right) = \frac{2\tan(\pi/6)}{1 - \tan^2(\pi/6)}$$

Note that $\tan\dfrac{\pi}{6} = \dfrac{\sin(\pi/6)}{\cos(\pi/6)} = \dfrac{1/2}{\sqrt{3}/2} = \dfrac{1}{\sqrt{3}}$.

$$\tan\frac{2\pi}{6} = \frac{2\left(1/\sqrt{3}\right)}{1 - \left(1/\sqrt{3}\right)^2}$$

$$\tan\frac{\pi}{3} = \frac{2/\sqrt{3}}{1 - (1/3)}$$

$$\tan\frac{\pi}{3} = \frac{2/\sqrt{3}}{2/3}$$

Express $\tan(\pi/3)$ in terms of sine and cosine; then cross-multiply.

$$\frac{\sin\pi/3}{\cos\pi/3} = \frac{2/\sqrt{3}}{2/3}$$

$$\frac{\sqrt{3}/2}{1/2} = \frac{2/\sqrt{3}}{2/3}$$

$$\left(\frac{\sqrt{3}}{\cancel{2}}\right)\left(\frac{\cancel{2}}{3}\right) = \left(\frac{1}{\cancel{2}}\right)\left(\frac{\cancel{2}}{\sqrt{3}}\right)$$

$$\frac{\sqrt{3}}{3} = \frac{1}{\sqrt{3}}$$

Cross-multiply once again.

$$\sqrt{3}\left(\sqrt{3}\right) = 3 \cdot 1$$

$$\sqrt{9} = 3$$

$$3 = 3$$

Because the resulting statement is obviously true, you have verified that the tangent double-angle formula $\tan 2x$ is true when $x = \pi/6$.

8.41 Simplify the expression: $\dfrac{2\sin x \sec^3 x}{1 - \tan^4 x}$.

Factor the denominator, which is a difference of perfect squares, and express the numerator in terms of sine and cosine.

$$\frac{2\sin x \sec^3 x}{1 - \tan^4 x} = \frac{2\sin x \cdot \dfrac{1}{\cos^3 x}}{\left(1 + \tan^2 x\right)\left(1 - \tan^2 x\right)}$$

Manipulate the numerator, rewriting the expressions in terms of tangent and secant. In addition, note that $(1 + \tan^2 x)$ in the denominator is equal to $\sec^2 x$, according to a Pythagorean identity.

> **See Problem 7.23, which states $1 + \tan^2 x = \sec^2 x$.**

$$= \frac{2 \cdot \dfrac{\sin x}{\cos x} \cdot \dfrac{1}{\cos^2 x}}{\left(1 + \tan^2 x\right)\left(1 - \tan^2 x\right)}$$

$$= \frac{2 \cdot \tan x \cdot \sec^2 x}{\sec^2 x \left(1 - \tan^2 x\right)}$$

> **This is the tangent double-angle formula.**

$$= \frac{2\tan x}{1 - \tan^2 x}$$

$$= \tan 2x$$

You conclude that $\dfrac{2\sin x \sec^3 x}{1 - \tan^4 x} = \tan 2x$.

8.42 Generate the power-reducing formula for $\tan^2 \theta$ by applying the power-reducing formulas for $\sin^2 \theta$ and $\cos^2 \theta$.

Note that $\tan^2\theta = (\sin^2\theta)/(\cos^2\theta)$ and apply the power-reducing formulas from Problem 8.17.

$$\tan^2 \theta = \frac{\sin^2 \theta}{\cos^2 \theta}$$

$$= \frac{\dfrac{1 - \cos 2\theta}{2}}{\dfrac{1 + \cos 2\theta}{2}}$$

Multiply the numerator and denominator by 2 in order to simplify the complex fraction.

$$= \frac{\dfrac{1 - \cos 2\theta}{\cancel{2}}}{\dfrac{1 + \cos 2\theta}{\cancel{2}}} \cdot \frac{\dfrac{\cancel{2}}{1}}{\dfrac{\cancel{2}}{1}}$$

$$= \frac{1 - \cos 2\theta}{1 + \cos 2\theta}$$

8.43 Apply a power-reducing formula to calculate the exact value of $\tan^4 \dfrac{\pi}{12}$ without using a calculator or trigonometric table.

Applying the power-reducing formula for tangent produces double-angle trigonometric expressions. Note that $2(\pi/12) = 2\pi/12 = \pi/6$.

$$\tan^4\left(\frac{\pi}{12}\right) = \left(\tan^2 \frac{\pi}{12}\right)^2$$

$$= \left(\frac{1 - \cos[2(\pi/12)]}{1 + \cos[2(\pi/12)]}\right)^2$$

$$= \left(\frac{1 - \cos(\pi/6)}{1 + \cos(\pi/6)}\right)^2$$

$$= \left(\frac{1 - \sqrt{3}/2}{1 + \sqrt{3}/2}\right)^2$$

Square the expressions in the numerator and denominator individually.

$$= \frac{\left(1 - \dfrac{\sqrt{3}}{2}\right)\left(1 - \dfrac{\sqrt{3}}{2}\right)}{\left(1 + \dfrac{\sqrt{3}}{2}\right)\left(1 + \dfrac{\sqrt{3}}{2}\right)} = \frac{1 - 2\left(\dfrac{\sqrt{3}}{2}\right) + \dfrac{3}{4}}{1 + 2\left(\dfrac{\sqrt{3}}{2}\right) + \dfrac{3}{4}} = \frac{1 - \sqrt{3} + \dfrac{3}{4}}{1 + \sqrt{3} + \dfrac{3}{4}}$$

Express the numerator and denominator in terms of a least common denominator.

$$= \frac{\dfrac{4}{4} - \dfrac{4\sqrt{3}}{4} + \dfrac{3}{4}}{\dfrac{4}{4} + \dfrac{4\sqrt{3}}{4} + \dfrac{3}{4}} = \frac{\dfrac{(4+3) - 4\sqrt{3}}{4}}{\dfrac{(4+3) + 4\sqrt{3}}{4}} = \frac{\dfrac{7 - 4\sqrt{3}}{4}}{\dfrac{7 + 4\sqrt{3}}{4}} = \frac{7 - 4\sqrt{3}}{7 + 4\sqrt{3}}$$

Multiply this fraction by 4/4 to cancel out the denominators.

8.44 Apply a half-angle formula to calculate the exact value of $\tan \dfrac{5\pi}{8}$ without using a calculator or trigonometric table.

Unlike the sine and cosine half-angle formulas from Problems 8.22–8.23, there are two equivalent versions of the tangent half-angle formula. The formulas do not appear within a square root symbol, so they are not preceded by "±".

$$\tan \frac{\theta}{2} = \frac{\sin \theta}{1 + \cos \theta} \qquad\qquad \tan \frac{\theta}{2} = \frac{1 - \cos \theta}{\sin \theta}$$

Note that $\dfrac{5\pi/4}{2} = \dfrac{5\pi}{8}$, so substitute $\theta = 5\pi/4$ into one of the half-angle formulas above. Both produce the same final answer, but the right-hand formula requires less steps than the left in the context of this problem.

You had to use quadrant clues to figure out the signs for cos (x/2) and sin (x/2). Not true for tan (x/2). The sign works itself out automatically.

$$\tan\frac{5\pi/4}{2} = \frac{1-\cos(5\pi/4)}{\sin(5\pi/4)}$$

$$\tan\left(\frac{5\pi}{4}\div 2\right) = \frac{1-\left(-\sqrt{2}/2\right)}{-\sqrt{2}/2}$$

$$\tan\left(\frac{5\pi}{4}\cdot\frac{1}{2}\right) = \frac{(2/2)+\left(\sqrt{2}/2\right)}{-\sqrt{2}/2}$$

$$\tan\frac{5\pi}{8} = \frac{\dfrac{2+\sqrt{2}}{\cancel{2}}}{-\dfrac{\sqrt{2}}{\cancel{2}}}$$

$$\tan\frac{5\pi}{8} = -\frac{2+\sqrt{2}}{\sqrt{2}}$$

Rationalize the denominator to simplify the expression.

$$\tan\frac{5\pi}{8} = \left(-\frac{2+\sqrt{2}}{\sqrt{2}}\right)\left(\frac{\sqrt{2}}{\sqrt{2}}\right)$$

$$= -\frac{2\sqrt{2}+\sqrt{4}}{\sqrt{4}}$$

$$= -\frac{2\sqrt{2}+2}{2}$$

$$= -\frac{\cancel{2}\sqrt{2}}{\cancel{2}} - \frac{\cancel{2}}{\cancel{2}}$$

$$= -\sqrt{2}-1$$

8.45 Demonstrate that the half-angle formula $\tan\dfrac{\theta}{2} = \dfrac{\sin\theta}{1+\cos\theta}$ is true by evaluating it for $\theta = 2x$.

Substitute $\theta = 2x$ into the given half-angle formula and apply the following double-angle formulas: $\sin 2x = 2\sin x\cos x$ and $\cos 2x = 2\cos^2 x - 1$.

$$\tan\frac{2x}{2} = \frac{\sin 2x}{1+\cos 2x}$$

$$\tan x = \frac{2\sin x\cos x}{1+\left(2\cos^2 x - 1\right)}$$

$$\tan x = \frac{2\sin x\cos x}{1-1+2\cos^2 x}$$

$$\tan x = \frac{\cancel{2}\sin x\cos x}{\cancel{2}\cos^2 x}$$

$$\tan x = \frac{\sin x\cdot\cancel{\cos x}}{\cos x\cdot\cancel{\cos x}}$$

$$\tan x = \frac{\sin x}{\cos x}$$

$$\tan x = \tan x$$

The final statement ($\tan x = \tan x$) is true, so you have verified that the original statement is true when $\theta = 2x$.

Chapter 9
INVERSE TRIGONOMETRIC FUNCTIONS

Arccosine, arcsine, arctangent

In the preceding chapters, you have defined and applied the six basic trigonometric functions: sine, cosine, tangent, cotangent, secant, and cosecant. The input for each of these functions is an angle (typically measured in radians or degrees) and the output is a real number value. For example, $\cos(\pi/3) = 1/2$. This chapter explores the inverse trigonometric functions, which reverse the domain and range of the original functions. The inverse of $\cos x$ is $\arccos x$, so $\arccos(1/2) = \pi/3$.

A function is a set of inputs and outputs with one condition: Every input is paired with only one output. If you go one step further and guarantee that every output of a function is matched to only one input, then the function is called "one-to-one."

Why does it matter? Only one-to-one functions have inverses, and that's a problem when you deal with periodic functions, including all of the trig functions. Enough spoilers. Time to get started.

Graphs of Inverse Trigonometric Functions

Including domain and range

9.1 The graph of a function $f(x)$ passes through points (–2,6), (0,3), and (4,–9) and has inverse function $f^{-1}(x)$. Identify three points through which the graph of $f^{-1}(x)$ passes.

If $f(x)$ passes through points (–2,6), (0,3), and (4,–9), then –2, 0, and 4 are members of the domain and 6, 3, and –9 are members of the range. Specifically, $f(-2) = 6$, $f(0) = 3$, and $f(4) = -9$. To identify points on the inverse function $f^{-1}(x)$, reverse the numbers in the ordered pair.

Thus, $f^{-1}(x)$ passes through points (6,–2), (3,0), and (–9,4). In function notation, you conclude that $f^{-1}(6) = -2$, $f^{-1}(3) = 0$, and $f^{-1}(-9) = 4$.

> If a number is a member of the domain, you can plug it into the function and get some real number output. Members of the range are outputs of the function.

Note: Problems 9.2–9.4 refer to the graph of function g(x) below.

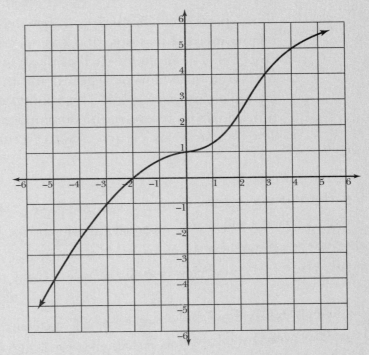

9.2 Explain how you can verify that $g(x)$ is a function based on its graph and determine whether or not the function is one-to-one.

Every input x of the function may be paired with only one output $g(x)$. To verify this on the graph of $g(x)$, you apply a visual technique called the vertical line test. If a vertical line drawn on the coordinate plane intersects $g(x)$, then that line represents an x-value in the domain of $g(x)$. However, if any vertical line intersects the graph more than once, then $g(x)$ fails the vertical line test and is not a function. No vertical line drawn on this graph intersects $g(x)$ more than once, so you can visually verify that $g(x)$ is a function.

> If the vertical line $x = 1$ were to intersect g(x) at a height of 2 and a height of –3, then g(1) = 2 and g(1) = –3. See why that's not allowed? The input $x = 1$ is paired with TWO outputs (2 and –3), but in a function, input values can only be paired with ONE output.

The horizontal line test is applied to the graph of a function to determine whether or not that function is "one-to-one." In a one-to-one function, not only is each input paired with only one output, each output is paired with only one input. This graph passes the horizontal line test—no horizontal line intersects the graph more than once. Therefore, $g(x)$ is one-to-one. Only one-to-one functions have inverses, so if a graph fails the horizontal line test, it does not have an inverse function.

Note: Problems 9.2–9.4 refer to the graph of g(x) in Problem 9.2.

9.3 Graph $g^{-1}(x)$, the inverse function of $g(x)$.

Notice that $g(x)$ passes through points $(-5,-4)$, $(-3,-1)$, $(-2,0)$, $(0,1)$, $(3,4)$, and $(4,5)$. Reverse the coordinates to identify points on the graph of inverse function $g^{-1}(x)$: $(-4,-5)$, $(-1,-3)$, $(0,-2)$, $(1,0)$, $(4,3)$, and $(5,4)$.

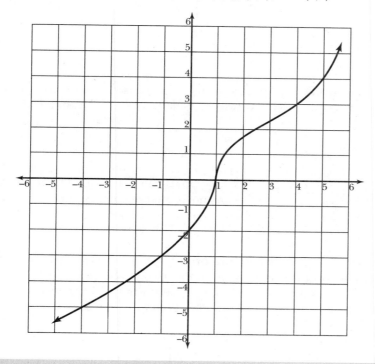

Note: Problems 9.2–9.4 refer to the graph of g(x) in Problem 9.2.

9.4 Explain how to transform the graph of $g(x)$ to generate the graph of $g^{-1}(x)$.

As Problem 9.3 explains, the x- and y-coordinates of points on the graph of a function are reversed to identify points on the inverse function. For instance, if the graph of a one-to-one function $g(x)$ passes through point (a,b), then the graph of its inverse function $g^{-1}(x)$ passes through (b,a).

Reversing the coordinates produces a reflection of the original graph across the line $y = x$, which passes through the origin and forms 45° ($\pi/4$ radians) angles with the positive x- and y-axes. Consider the following diagram, in which function $g(x)$ and its inverse $g^{-1}(x)$ mirror each other across the dotted linear graph of $y = x$.

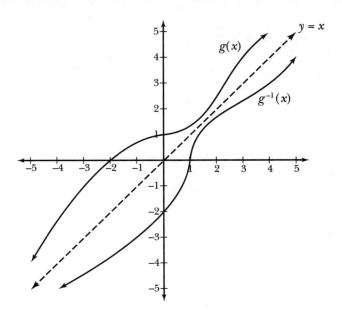

9.5 Explain why $f(x) = \cos x$ does not have an inverse function if its domain is all real numbers but does have an inverse if you restrict the domain to the interval $0 \leq \theta \leq \pi$.

Any horizontal line that intersects the graph of $f(x) = \cos x$ intersects it infinitely many times. As an example, consider the horizontal line $f(x) = 0$, the x-axis, which intersects the graph at $x = (k\pi)/2$, where k is an odd integer. According to the horizontal line test, if any horizontal line intersects the graph more than once (let alone infinitely many times), that graph is not one-to-one and thus does not have an inverse.

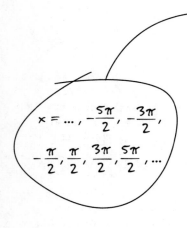

$$x = \ldots, -\frac{5\pi}{2}, -\frac{3\pi}{2},$$

$$-\frac{\pi}{2}, \frac{\pi}{2}, \frac{3\pi}{2}, \frac{5\pi}{2}, \ldots$$

However, you can define an inverse function for cosine if you restrict its domain to a small portion of the graph. In the diagram below, a specific interval of the graph of $f(x) = \cos x$ is highlighted. That portion (lying between $x = 0$ and $x = \pi$) passes the horizontal line test and is, therefore, one-to-one.

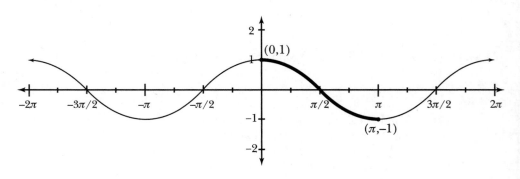

9.6   Graph arccos x.

As Problem 9.5 explains, the function $f(x) = \cos x$ has an inverse function if you restrict the domain of $f(x)$ to $0 \le x \le \pi$. The inverse of the cosine function is arccosine and is written "arccos x." Some textbooks (and many calculators) use the notation "$\cos^{-1} x$" to represent the inverse cosine function, but this notation is not recommended. Recall that $\cos^2 x = (\cos x)^2$, so $\cos^{-1} x$ is easily confused with $(\cos x)^{-1}$, the reciprocal of cosine rather than the inverse of cosine.

Consider the darkened portion of the graph in Problem 9.5. It passes through key points $(0,1)$, $(\pi/2, 0)$, and $(\pi, -1)$. Reverse the coordinates to identify points on the graph of $f^{-1}(x) = $ arccos x: $(1,0)$, $(0, \pi/2)$, and $(-1, \pi)$.

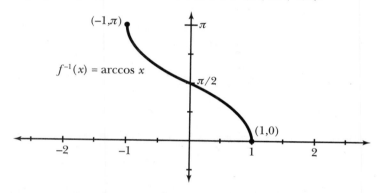

9.7   Identify the domain of arccos x.

Any vertical line that intersects a graph represents an element of that function's domain. Consider the graph of arccos x generated in Problem 9.6. The vertical lines $x = -1$ and $x = 1$ intersect the graph, and so do all of the vertical lines between those x-values. Therefore, the domain of arccos x is $-1 \le x \le 1$.

9.8   Identify the range of arccos x.

Horizontal lines drawn on the coordinate plane intersect the graph of arccos x at a lower bound of 0 and at an upper bound of π. Note that any horizontal line between those bounds also intersects the graph. Therefore, the range is $0 \le $ arccos $x \le \pi$.

9.9   Identify the restricted domain of sin x that has inverse arcsin x.

As illustrated in the following graph, the portion of sin x lying between $x = -\pi/2$ and $x = \pi/2$ passes the horizontal line test and is one-to-one. The inverse function arcsin x is defined for only that portion of the graph.

This makes sense. Cosine outputs values between −1 and 1, so the INPUTS of arccosine are between −1 and 1. The outputs of a function are the inputs of its inverse and vice versa.

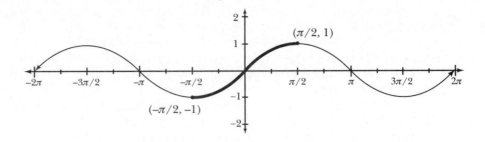

9.10 Graph arcsin x.

Note that the darkened portion of the graph in Problem 9.9 passes through key points $(-\pi/2, -1)$, $(0,0)$, and $(\pi/2, 1)$. Therefore, the inverse function passes through points $(-1, -\pi/2)$, $(0,0)$, and $(1, \pi/2)$, as illustrated below.

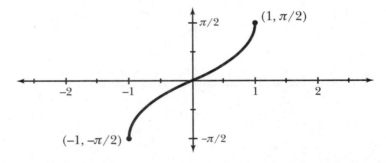

9.11 Identify the domain and range of arcsin x.

Consider the graph of $y = \arcsin x$, presented in Problem 9.10. Vertical lines drawn on the coordinate plane intersect the graph at $x = -1$, $x = 1$, and at all of the x-values between them. Therefore, the domain of arcsin x is $-1 \leq x \leq 1$.

Notice that the domain of arcsin x matches the domain of arccos x (calculated in Problem 9.7); both are equal to the ranges of the inverse functions, in this case sin x. The range of arcsin x is equal to the restricted domain defined in Problem 9.9: $-\pi/2 \leq \arcsin x \leq \pi/2$.

9.12 Identify the restricted domain of tan x that has inverse arctan x.

One full period of $y = \tan x$ lies between vertical asymptotes $x = -\pi/2$ and $x = \pi/2$. If you restrict tangent to this interval of the domain, you can define inverse function arctan x. Thus, the restricted domain of tangent is $-\pi/2 < x < \pi/2$. This is similar to the restricted domain of sine (defined in Problem 9.9) but it excludes the upper and lower bounds.

9.13   Graph arctan x.

The graph of $f(x) = \tan x$ passes through points $(-\pi/4, -1)$, $(0,0)$, and $(\pi/4, 1)$. Thus, the graph of $f^{-1}(x) = \arctan x$ passes through points $(-1, -\pi/4)$, $(0,0)$, and $(1, \pi/4)$, as illustrated in the following graph.

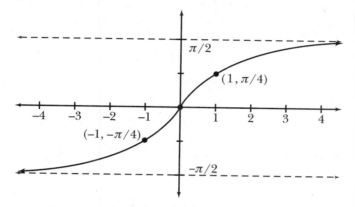

9.14   Identify the domain and range of $y = \arctan x$.

The graph of $y = \arctan x$, presented in Problem 9.13, is defined for all real numbers; any vertical line drawn on the coordinate plane intersects the graph. The range of the function is equal to the restricted domain of $\tan x$ (identified in Problem 9.13): $-\pi/2 < \arctan x < \pi/2$.

Notice that the boundaries of the range are not included—unlike the boundaries of the ranges of arccos x and arcsin x—because the graph of arctan x is bounded above and below by horizontal asymptotes.

General and Exact Solutions

One vs. many answers

Note: Problems 9.15–9.17 describe three different ways to present the solution to the same trigonometric equation.

9.15   Identify the solutions to the equation $\sin \theta = 0$ on the interval $0 \le \theta < 2\pi$.

According to the unit circle, $\sin \theta = 0$ when $\theta = 0$ and $\theta = \pi$.

Note: Problems 9.15–9.17 describe three different ways to present the solution to the same trigonometric equation.

9.16   List the values of θ for which the equation $\sin \theta = 0$ is true.

As Problem 9.15 states, $\sin 0 = \sin \pi = 0$. Because sine is a periodic function, it is also equal to zero for any angle $k\pi$, where k is an integer. In other words, the following set of angles is the solution set for this equation.

$$\theta = \{ \ldots, -4\pi, -3\pi, -2\pi, -\pi, 0, \pi, 2\pi, 3\pi, 4\pi, \ldots \}$$

> This is one loop around the coordinate plane. Notice that the interval includes 0 but excludes 2π. Both of those values represent the same angle—they both terminate on the positive x-axis—so there's no need to include both.

Note: Problems 9.15–9.17 describe three different ways to present the solution to the same trigonometric equation.

9.17   Identify the exact solution to the equation $\sin \theta = 0$.

To calculate the exact solution to this equation, apply the inverse function for sine. In other words, solve the equation for θ by taking the arcsine of both sides of the equation. Notice that $\arcsin (\sin \theta) = \theta$.

$$\sin \theta = 0$$
$$\arcsin (\sin \theta) = \arcsin 0$$
$$\theta = \arcsin 0$$

> A function and its inverse cancel out—arcsine cancels out sine. This is just like taking the square root of both sides of the equation $x^2 = 16$ to get $x = \pm 4$.

Problem 9.16 lists the angles that have a sine value of 0. However, only one of them serves as the *exact* solution. Remember that the restricted range of $\arcsin \theta$ is $-\pi/2 \le \theta \le \pi/2$. Of all the angles listed in Problem 9.16, only $\theta = 0$ belongs to the restricted range. Thus, the exact solution is $\theta = 0$.

Note: Problems 9.18–9.19 describe two different ways to present the solution to the same trigonometric equation.

9.18   Identify all solutions to the equation $\tan x = 1$ on the interval $0 \le x < 2\pi$.

Recall that tangent is defined as the quotient of sine and cosine: $\tan x = (\sin x)/(\cos x)$. The quotient is equal to 1 if and only if the numerator and denominator are equal. There are two angles on the unit circle with equivalent cosine and sine values: $\theta = \pi/4$ and $\theta = 5\pi/4$, as demonstrated below.

$$\tan \frac{\pi}{4} = \frac{\sin (\pi / 4)}{\cos (\pi / 4)} \qquad\qquad \tan \frac{5\pi}{4} = \frac{\sin (5\pi / 4)}{\cos (5\pi / 4)}$$
$$= \frac{\sqrt{2} / 2}{\sqrt{2} / 2} \qquad\qquad\qquad = \frac{-\sqrt{2} / 2}{-\sqrt{2} / 2}$$
$$= 1 \qquad\qquad\qquad\qquad = 1$$

Note: Problems 9.18–9.19 describe two different ways to present the solution to the same trigonometric equation.

9.19   Identify the exact solution to the equation $\tan x = 1$.

Exact solutions require inverse trigonometric functions, so take the arctangent of both sides of the equation.

$$\tan x = 1$$
$$\arctan (\tan x) = \arctan 1$$
$$x = \arctan 1$$

According to Problem 9.18, $\tan (\pi/4) = \tan (5\pi/4) = 1$. However, the range of arctangent is restricted to the interval $-\pi/2 \le x \le \pi/2$. Of the two proposed solutions, only $x = \pi/4$ lies within that restricted interval. Therefore, the exact solution to the equation is $x = \pi/4$.

9.20   Given point $\left(-\dfrac{1}{2}, c\right)$ on the graph of arcsin x, calculate c.

If point $(-1/2, c)$ lies on the graph of arcsin x, then $(c, -1/2)$ lies on the graph of its inverse function, sin x. In other words, you need to identify the value c such that sin $c = -1/2$. However, *two* angles on the unit circle have this sine value, $\theta = 7\pi/6$ and $\theta = 11\pi/6$, and *neither* of those angles belongs to the range of arcsin x $(-\pi/2 \leq$ arcsin $x \leq \pi/2)$.

> But only the piece of sin x between $-\dfrac{\pi}{2}$ and $\dfrac{\pi}{2}$

Both angles are greater than the upper bound of the range. Therefore, you need to calculate angles coterminal to $7\pi/6$ and $11\pi/6$ to identify a valid member of the range. Subtract 2π from each to compute the coterminal angles.

$$\frac{7\pi}{6} - 2\pi = \frac{7\pi}{6} - \frac{12\pi}{6} \qquad\qquad \frac{11\pi}{6} - 2\pi = \frac{11\pi}{6} - \frac{12\pi}{6}$$
$$= -\frac{5\pi}{6} \qquad\qquad\qquad\qquad\quad = -\frac{\pi}{6}$$

> If you are asking yourself "Why?" check Problem 4.22.

Notice that $-\pi/6$ belongs to the range of arcsin x $(-\pi/2 \leq -\pi/6 \leq \pi/2)$. Therefore, $c = -\pi/6$.

9.21   Calculate arccos -1.

Begin by identifying the angle(s) on the unit circle that have a cosine value of -1. Only one such angle exists: $\theta = \pi$. Recall that the restricted range for arccosine is $0 \leq \theta \leq \pi$, and the angle $\theta = \pi$ is a valid member of this interval. Thus, you conclude arccos $-1 = \pi$.

9.22   Calculate arcsin $\dfrac{\sqrt{3}}{2}$.

Two angles on the unit circle have a sine value of $\sqrt{3}/2$: $\theta = \pi/3$ and $\theta = 2\pi/3$. Only the smaller of these two angles falls within the restricted range of arcsine $(-\pi/2 \leq \theta \leq \pi/2)$. Thus, arcsin $\left(\sqrt{3}/2\right) = \pi/3$.

> For the rest of the problems in this chapter, use the Problem 9.20 technique:
> 1. Find the angle(s) on the unit circle with the correct trig value
> 2. Select the angle that falls in the correct restricted range
> 3. If none of the angles are in the restricted range, calculate coterminal angles

9.23   Calculate arccos $\left(-\dfrac{\sqrt{2}}{2}\right)$.

The two angles on the unit circle with a cosine value of $-\sqrt{2}/2$ are $\theta = 3\pi/4$ and $\theta = 5\pi/4$, but only the smaller of the two angles belongs to the restricted range of arccosine.

$$\arccos\left(-\frac{\sqrt{2}}{2}\right) = \frac{3\pi}{4}$$

> Angles in the range of arccosine terminate in the first and second quadrants, and the angles are always positive. Angles in the range of arcsine and arctangent terminate in the first or fourth quadrants, and their terminal sides are no more than 90° ($\pi/2$ radians) from the positive x-axis, in either the clockwise or counterclockwise direction.

According to my notes for Problem 9.23, you're looking for a first or fourth quadrant angle when the problem contains arcsine. Don't even bother calculating the coterminal angle of $\theta = \frac{5\pi}{4}$, because it terminates in the third quadrant.

9.24 Calculate $\arcsin\left(-\frac{\sqrt{2}}{2}\right)$.

The two angles on the unit circle with a sine value of $-\sqrt{2}/2$ are $\theta = 5\pi/4$ and $\theta = 7\pi/4$. Neither of these angles belongs to the restricted range of arcsine, so calculate a coterminal angle to identify a corresponding angle that does.

$$\frac{7\pi}{4} - 2\pi = \frac{7\pi}{4} - \frac{8\pi}{4}$$
$$= -\frac{\pi}{4}$$

Thus $\arcsin\left(-\sqrt{2}/2\right) = -\pi / 4$.

9.25 Calculate arcsec (–2).

Recall that secant and cosine are reciprocal functions. Similarly, arcsecant and arccosine are reciprocal functions. In other words, if θ = arcsec (–2), then θ = arccos (–1/2). The two angles on the unit circle with a cosine value of –1/2 are $\theta = 2\pi/3$ and $\theta = 4\pi/3$. The smaller of the two angles belongs to the restricted range of arccosine, so arcsec (–2) = $2\pi/3$.

9.26 Calculate $\arctan\left(-\sqrt{3}\right)$.

Tangent is defined as the quotient of sine and cosine, so begin by expressing the argument $\sqrt{3}$ as a quotient as well.

$$\arctan\left(-\sqrt{3}\right) = \arctan\left(-\frac{\sqrt{3}}{1}\right)$$

Although there are no angles on the unit circle with a sine value of $\sqrt{3}$ and a cosine value of 1, there are angles with sine values of $\sqrt{3}/2$ and cosine values of 1/2. You can divide the numerator and denominator of a fraction by the same number without affecting the value of that fraction. In this problem, it is helpful to divide the numerator and denominator by 2.

$$\arctan\left(-\frac{\sqrt{3}}{1}\right) = \arctan\left(-\frac{\sqrt{3}/2}{1/2}\right)$$

Four angles on the unit circle have a sine value of $\pm\sqrt{3}/2$ and a cosine value of $\pm1/2$: $\theta = \pi/3$, $\theta = 2\pi/3$, $\theta = 4\pi/3$, and $\theta = 5\pi/3$. However, only two of those angles have negative tangent values.

$$\tan\frac{2\pi}{3} = \frac{\sin(2\pi/3)}{\cos(2\pi/3)} \qquad \tan\frac{5\pi}{3} = \frac{\sin(5\pi/3)}{\cos(5\pi/3)}$$

$$= \frac{\sqrt{3}/2}{-1/2} \qquad\qquad = \frac{-\sqrt{3}/2}{1/2}$$

$$= -\frac{\sqrt{3}/2}{1/2} \qquad\qquad = -\frac{\sqrt{3}/2}{1/2}$$

Recall that the restricted range of arctangent is $-\pi/2 \le \theta \le \pi/2$, so calculate the angle within that range that is coterminal to $5\pi/3$.

$$\frac{5\pi}{3} - 2\pi = \frac{5\pi}{3} - \frac{6\pi}{3}$$

$$= -\frac{\pi}{3}$$

You conclude that $\arctan\left(-\sqrt{3}\right) = -\pi/3$.

Chapter 10
SIMPLE TRIGONOMETRIC EQUATIONS

Algebra 1, but with angles

In Chapter 9, you explore the inverse trigonometric functions: arcsin x, arccos x, and arctan x. Whereas the corresponding trigonometric functions sin x, cos x, and tan x accept angle inputs and produce real number outputs, their inverse functions do precisely the opposite. They also eliminate trigonometric functions from equations, which empowers you to solve trigonometric equations.

In basic algebra, you learned a few basic equation-solving techniques, including:

* Adding, subtracting, multiplying, and dividing both sides of an equation by the same number to isolate the variable
* Factoring an equation, when one side of the equation equals zero
* Applying the quadratic formula when you can't factor a quadratic equation

In this chapter, you'll use all of these techniques to solve basic trigonometric equations. Make sure to work through Chapter 9 and pay special attention to Problems 9.15–9.17. They outline the major difference between trigonometric and algebraic equations: the different ways you may have to report a trigonometric solution.

Linear Equations
Add, subtract, multiply, and divide both sides by the same thing

Note: Problems 10.1–10.3 refer to the trigonometric equation sin x + 1 = 0.

10.1 Identify the solutions to the equation on the interval $0 \leq x < 2\pi$.

To solve the equation sin $x + 1 = 0$ for x, you must first isolate the expression containing x on one side of the equal sign. Subtracting 1 from both sides of the equation accomplishes this task, isolating sin x on the left side of the equation.

$$\sin x = -1$$

The problem instructs you to identify the solutions on the interval $0 \leq x < 2\pi$, which is one rotation about the coordinate plane and corresponds to the unit circle. Only one angle x on the unit circle has a -1 sine value: $x = 3\pi/2$.

Note: Problems 10.1–10.3 refer to the trigonometric equation sin x + 1 = 0.

10.2 Identify the general solution to the equation.

Remember that sine has a period of 2π. Adding and subtracting multiples of 2π to an angle will produce angles with the same sine value. According to Problem 10.1, $x = 3\pi/2$ is a solution to the equation; adding and subtracting 2π to this solution identifies two other angles that are also solutions to the equation.

$$\frac{3\pi}{2} - 2\pi = \frac{3\pi}{2} - \frac{4\pi}{2} \qquad \frac{3\pi}{2} + 2\pi = \frac{3\pi}{2} + \frac{4\pi}{2}$$
$$= -\frac{\pi}{2} \qquad\qquad\qquad = \frac{7\pi}{2}$$

If you double the period ($2 \cdot 2\pi = 4\pi$) and then add and subtract that product (4π) from the solution $x = 3\pi/2$, you identify two additional solutions to the equation.

$$\frac{3\pi}{2} - 4\pi = \frac{3\pi}{2} - \frac{8\pi}{2} \qquad \frac{3\pi}{2} + 4\pi = \frac{3\pi}{2} + \frac{8\pi}{2}$$
$$= -\frac{5\pi}{2} \qquad\qquad\qquad = \frac{11\pi}{2}$$

If you continue this process by tripling the period ($3 \cdot 2\pi = 6\pi$), quadrupling the period ($4 \cdot 2\pi = 8\pi$), and so on, you will identify an infinite number of possible solutions. Rather than list them all, you use shorthand notation to indicate a general solution of $x = 3\pi/2$ (and all of its coterminal angles). Both of the following solutions are correct, although the solution on the right—in which k and 2π are explicitly multiplied—is more common.

$$x = \frac{3\pi}{2} + k(2\pi) \text{ or } x = \frac{3\pi}{2} + 2k\pi \text{, such that } k \text{ is an integer}$$

When something is "isolated" it is all by itself. You want sin x to be all by itself on one side of the equation, but right now, there's a "+ 1" next to it. Move that number to the other side of the equation by subtracting it from both sides.

To create a general solution, identify the solution(s) on one period of the function (in this case $0 \leq x < 2\pi$) and then add k · (period) to each.

Note: Problems 10.1–10.3 refer to the trigonometric equation sin x + 1 = 0.

10.3 Calculate the exact solution to the equation.

Begin by isolating sin x on the left side of the equation: sin $x = -1$. In order to calculate an exact trigonometric solution, you apply an inverse trigonometric function. Solve sin $x = -1$ for x by taking the arcsine of both sides of the equation.

$$\arcsin(\sin x) = \arcsin(-1)$$
$$x = \arcsin(-1)$$

As Problem 10.1 explains, the unit circle angle $x = 3\pi/2$ satisfies the equation, but that angle does not fall within the restricted range of arcsine $(-\pi/2 \le x \le \pi/2)$. Subtract 2π from $x = 3\pi/2$ to calculate a coterminal angle that lies in the restricted range.

$$\frac{3\pi}{2} - 2\pi = \frac{3\pi}{2} - \frac{4\pi}{2}$$
$$= -\frac{\pi}{2}$$

The exact solution is $x = -\pi/2$.

> Look at Problems 9.20, 9.24, 9.25, and 9.26 for more information. In each case, you have to calculate a coterminal angle within a restricted range.

Note: Problems 10.4–10.5 refer to the trigonometric equation 2 cos x – 1 = 0.

10.4 Identify the general solution to the equation.

In order to isolate cos x on the left side of the equation, add 1 to both sides and then divide both sides by 2.

$$2\cos x = 1$$
$$\cos x = \frac{1}{2}$$

Now identify the angles on the unit circle that have a cosine value of +1/2. Two such angles exist: $\pi/3$ and $5\pi/3$. These angles belong to the general solution, as do all of their coterminal angles; therefore, you need to add multiples of 2π (the period of cosine) to each solution, as demonstrated below.

$$x = \frac{\pi}{3} + 2k\pi, \; x = \frac{5\pi}{3} + 2k\pi, \text{ such that } k \text{ is an integer}$$

Note: Problems 10.4–10.5 refer to the trigonometric equation 2 cos x – 1 = 0.

10.5 Calculate the exact solution to the equation.

According to Problem 10.4, two angles on the unit circle are solutions to the equation: $x = \pi/3$ and $x = 5\pi/3$. Only the smaller angle belongs to the restricted range of arccosine $(0 \le x \le \pi)$. Therefore, the exact solution to the equation is $x = \pi/3$.

Note: Problems 10.6–10.7 refer to the trigonometric equation $5 - \sqrt{2}\sin\theta = 6$.

10.6 Calculate the exact solution to the equation.

Isolate the trigonometric expression $\sin\theta$ on the left side of the equation.

$$-\sqrt{2}\sin\theta = 6 - 5$$

$$-\sqrt{2}\sin\theta = 1$$

$$\sin\theta = -\frac{1}{\sqrt{2}}$$

Rationalize the denominator to reveal a familiar sine value from the unit circle.

$$\sin\theta = -\frac{1}{\sqrt{2}} \cdot \frac{\sqrt{2}}{\sqrt{2}}$$

$$\sin\theta = -\frac{\sqrt{2}}{\sqrt{4}}$$

$$\sin\theta = -\frac{\sqrt{2}}{2}$$

Apply the arcsine function to calculate the exact solution.

$$\arcsin(\sin\theta) = \arcsin\left(-\frac{\sqrt{2}}{2}\right)$$

$$\theta = \arcsin\left(-\frac{\sqrt{2}}{2}\right)$$

According to Problem 9.24, the exact solution to this equation is $\theta = -\pi/4$.

Note: Problems 10.6–10.7 refer to the trigonometric equation $5 - \sqrt{2}\sin\theta = 6$.

10.7 Identify the general solution to the equation.

As Problem 10.6 demonstrates, solving the equation $5 - \sqrt{2}\sin\theta = 6$ for $\sin\theta$ produces the following equation.

$$\sin\theta = -\frac{\sqrt{2}}{2}$$

There are two angles on the unit circle that have a sine value of $-\sqrt{2}/2$: $\theta = 5\pi/4$ and $\theta = 7\pi/4$. To construct the general solution, add multiples of 2π (the period of sine) to each solution.

$$\theta = \frac{5\pi}{4} + 2k\pi, \theta = \frac{7\pi}{4} + 2k\pi, \text{ such that } k \text{ is an integer}$$

Note: Problems 10.8–10.9 refer to the trigonometric equation 2(tan x + 3) = 5 + tan x.

10.8 Identify all solutions to the equation on the interval $0 \leq x < 2\pi$.

Isolate the expression tan x on the left side of the equation.

$$2(\tan x) + 2(3) = 5 + \tan x$$
$$2\tan x + 6 = 5 + \tan x$$
$$2\tan x - \tan x = 5 - 6$$
$$\tan x = -1$$

There are two angles on the unit circle that have a tangent value of -1: $\theta = 3\pi/4$ and $\theta = 7\pi/4$.

> The tangent of an angle is equal to –1 whenever the sine and cosine of that angle are opposite values.

Note: Problems 10.8–10.9 refer to the trigonometric equation 2(tan x + 3) = 5 + tan x.

10.9 Identify the general solution to the equation.

To create the general solution to a trigonometric equation, you identify the angles on the unit circle that represent solutions and then add multiples of the period to that solution. In Problems 10.2, 10.4, and 10.7, you append the notation "+ $2k\pi$, such that k is an integer," because 2π is the period of sine and cosine (as well as the period of secant and cosecant).

However, the tangent and cotangent functions have a period of π, so you should add $k\pi$ to the unit circle solutions rather than $2k\pi$. According to Problem 10.8, the unit circle angles that satisfy this equation are $\theta = 3\pi/4$ and $\theta = 7\pi/4$. Hence, you may be tempted to report the following general solution.

$$\theta = \frac{3\pi}{4} + k\pi, \theta = \frac{7\pi}{4} + k\pi, \text{ such that } k \text{ is an integer}$$

This solution is not incorrect, but it is repetitive. Notice that $7\pi/4$ is equal to $3\pi/4$ plus a multiple of π, as demonstrated below.

$$\frac{3\pi}{4} + 1\pi = \frac{3\pi}{4} + \frac{4\pi}{4} = \frac{7\pi}{4}$$

Thus, the following solution is sufficient.

$$\theta = \frac{3\pi}{4} + k\pi, \text{ such that } k \text{ is an integer}$$

> Imagine that the solution to some equation was "x = 1 plus any positive multiple of 2." In other words, the solution set is {3, 5, 7, 9, 11, ...}. Even if "x = 11 plus any multiple of 2" is another valid solution, you don't need to write it because it describes a set of values that are already included: {13, 15, 17, 19, ...}.

Note: Problems 10.10–10.11 refer to the trigonometric equation $\frac{1}{4}(\sec \theta - 2) = \frac{1}{3}(\sec \theta - 1)$.

10.10 Identify the general solution to the equation.

Multiply the entire equation by 12 (the least common denominator of 1/4 and 1/3) to eliminate the fractions.

$$12\left[\frac{1}{4}(\sec\theta-2)\right]=12\left[\frac{1}{3}(\sec\theta-1)\right]$$

$$\frac{12}{4}(\sec\theta-2)=\frac{12}{3}(\sec\theta-1)$$

$$3(\sec\theta-2)=4(\sec\theta-1)$$

Now isolate $\sec\theta$ on one side of the equal sign.

$$3\sec\theta-6=4\sec\theta-4$$

$$-6+4=4\sec\theta-3\sec\theta$$

$$-2=\sec\theta$$

The equality of a statement is preserved if you take the reciprocal of both sides of the equation. Recall that secant is defined as the reciprocal of cosine.

$$-\frac{1}{2}=\frac{1}{\sec\theta}$$

$$-\frac{1}{2}=\cos\theta$$

The two angles on the unit circle that have a cosine value of $-1/2$ are $\theta=2\pi/3$ and $\theta=4\pi/3$. Include coterminal angles (by adding multiples of 2π) to create the general solution to the equation.

$$\theta=\frac{2\pi}{3}+2k\pi,\theta=\frac{4\pi}{3}+2k\pi, \text{ such that } k \text{ is an integer}$$

Note: Problems 10.10–10.11 refer to the trigonometric equation $\frac{1}{4}\left(\sec\theta-2\right)=\frac{1}{3}\left(\sec\theta-1\right)$.

10.11 Calculate the exact solution to the equation.

According to Problem 10.10, two angles on the unit circle are solutions to the equation: $\theta=2\pi/3$ and $\theta=4\pi/3$. Only the smaller of the two angles lies within the restricted range of arccosine $(0\le\theta\le\pi)$, so you conclude that the exact solution to the equation is $\theta=2\pi/3$.

Zero Products

Factoring

10.12 Explain how to apply the zero-product property to solve the equation $xy=0$.

According to the zero-product property, if two or more expressions are multiplied and their product is 0, then at least one of those expressions must be equal to 0. In the given equation, x and y are multiplied to produce a product of 0, and the only way to obtain a zero product is to include 0 as one (or both) of the factors.

Therefore, to solve an equation in which a product equals zero (like $xy=0$), set each factor equal to zero. In this example, $x=0$ and $y=0$ are solutions to the equation.

This isn't true for any other number. For example, if you multiply two numbers together and get 10, one of those numbers doesn't HAVE to equal 10.

10.13 Apply the zero-product property to solve the equation: $x(x-7)(x+6) = 0$.

The product $x(x-7)(x+6)$ equals 0, so according to the zero-product property, at least one of the factors must be equal to 0. Set each factor—x, $x-7$, and $x+6$—equal to 0 and solve the resulting equations.

$$x = 0 \qquad x - 7 = 0 \qquad x + 6 = 0$$
$$x = 7 \qquad x = -6$$

There are three possible solutions to the equation: $x = -6$, $x = 0$, and $x = 7$.

10.14 Identify the solutions to the equation $(\cos\theta - 1)(\sin\theta + 1) = 0$ on the interval $0 \le \theta < 2\pi$.

Apply the zero-product property, setting each of the factors equal to zero and solving the equations.

$$\cos\theta - 1 = 0 \qquad \sin\theta + 1 = 0$$
$$\cos\theta = 1 \qquad \sin\theta = -1$$

Only angle $\theta = 0$ has a cosine value of 1 on the interval $0 \le \theta < 2\pi$, and only angle $\theta = 3\pi/2$ has a sine value of -1 on that interval. Therefore, $\theta = 0$ and $\theta = 3\pi/2$ are solutions to the equation.

10.15 Explain why the equation $(2\cos x - 3)(\cos x - 4) = 0$ has no real solutions.

Apply the zero-product property to create two equations from the factors.

$$2\cos x - 3 = 0 \qquad \cos x - 4 = 0$$
$$2\cos x = 3 \qquad \cos x = 4$$
$$\cos x = \frac{3}{2}$$

The range of $\cos x$ is $-1 \le x \le 1$, but neither 3/2 nor 4 fall within that range. Therefore, there is no value x for which $\cos x$ equals either 3/2 or 4, and the original equation has no solution.

> The graph of cosine is a wave that oscillates between a maximum height of 1 and a minimum height of −1. Sine has the same range. To review these graphs, check out Problems 5.26 and 5.35.

10.16 Identify the general solution to the equation: $\tan^2 x - \tan x = 0$.

The terms in the expression contain a common factor: $\tan x$. Factor this expression out of both terms.

$$\tan^2 x - \tan x = 0$$
$$\tan x(\tan x - 1) = 0$$

The result is a product equal to 0, so you can apply the zero-product property to solve the equation. Note that you can *only* apply the zero-product property when the product is equal to zero. Note that the period of tangent is π, so rather than identify solutions on the interval $0 \le x < 2\pi$, identify solutions on the interval $-\pi/2 \le x \le \pi/2$.

$$\tan x = 0 \qquad \tan x - 1 = 0$$
$$x = 0 \qquad \tan x = 1$$
$$x = \frac{\pi}{4}$$

Construct the general solution by adding multiples of π (the period of tangent) to each solution you identified.

$$x = 0 + k\pi, \; x = \frac{\pi}{4} + k\pi, \text{ such that } k \text{ is an integer}$$

10.17 Identify the solution to the equation $4\cos^2 x - 1 = 0$ on the interval $0 \le x < 2\pi$.

Factor the expression left of the equal sign, a difference of perfect squares.

$$(2\cos x + 1)(2\cos x - 1) = 0$$

Apply the zero-product property.

$$2\cos x + 1 = 0 \qquad 2\cos x - 1 = 0$$
$$2\cos x = -1 \qquad 2\cos x = 1$$
$$\cos x = -\frac{1}{2} \qquad \cos x = \frac{1}{2}$$

Four angles on the unit circle have a cosine value of $\pm 1/2$, so there are four solutions on the interval $0 \le x < 2\pi$.

$$x = \frac{\pi}{3}, \frac{2\pi}{3}, \frac{4\pi}{3}, \frac{5\pi}{3}$$

10.18 Identify the general solution to the equation: $\csc^2 x - 4\csc x + 4 = 0$.

Factor the expression $\csc^2 x - 4\csc x + 4$ in the same way that you would factor $y^2 - 4y + 4 = (y - 2)(y - 2)$.

$$(\csc x - 2)(\csc x - 2) = 0$$

The factors are equal, so the original expression $\csc^2 x - 4\csc x + 4$ is a perfect square. Set the repeated factor equal to 0 and solve for x. Recall that cosecant and sine are reciprocal functions.

$$\csc x - 2 = 0$$
$$\csc x = 2$$
$$\sin x = \frac{1}{2}$$

The angles on the unit circle with a sine value of $1/2$ are $x = \pi/6$ and $x = 5\pi/6$. The problem directs you to identify the general solution, so add multiples of 2π (the period of cosine) to these angles.

$$x = \frac{\pi}{6} + 2k\pi, \; x = \frac{5\pi}{6} + 2k\pi, \text{ such that } k \text{ is an integer}$$

10.19 Identify the general solution to the equation: $\tan^2 x + 6\tan x = -5$.

You can apply the zero-product property only if one side of an equation is equal to 0. Therefore, you should add 5 to both sides of this equation before factoring.

$$\tan^2 x + 6\tan x + 5 = 0$$
$$(\tan x + 1)(\tan x + 5) = 0$$

Set each factor equal to 0 and solve.

$$\tan x + 1 = 0 \qquad \tan x + 5 = 0$$
$$\tan x = -1 \qquad \tan x = -5$$

Note that the angle $x = -\pi/4$ has a tangent value of -1. However, no angle on the unit circle has a tangent value of -5. The angle does exist—remember that the domain of arctangent is all real numbers (as explained in Problem 9.14)—so it is sufficient to report the solution as $x = \arctan(-5)$. To construct the general solution, add multiples of π (the period of tangent) to both solutions.

$$x = -\frac{\pi}{4} + k\pi,\ x = \arctan(-5) + k\pi,\ \text{such that } k \text{ is an integer}$$

The angle $3\pi/4$ also has a tangent of -1, but you don't have to list it because it is automatically included by the general solution (if $k = 1$):

$$-\frac{\pi}{4} + (1)\pi$$
$$= -\frac{\pi}{4} + \pi$$
$$= -\frac{\pi}{4} + \frac{4\pi}{4}$$
$$= \frac{3\pi}{4}$$

10.20 Identify the solutions to the equation $6\sin\theta(\sin\theta - 1) = 4 - \sin\theta$ on the interval $0 \le \theta < 2\pi$.

Apply the distributive property to simplify $6\sin\theta(\sin\theta - 1)$. Recall that the zero-product property may only be applied when one side of the equation is 0, so 4 must be subtracted from—and $\sin\theta$ must be added to—both sides of the equation.

$$6\sin\theta(\sin\theta) + 6\sin\theta(-1) = 4 - \sin\theta$$
$$6\sin^2\theta - 6\sin\theta = 4 - \sin\theta$$
$$6\sin^2\theta - 6\sin\theta + \sin\theta - 4 = 0$$
$$6\sin^2\theta - 5\sin\theta - 4 = 0$$

Factor the expression left of the equal sign and set each factor equal to 0.

$$(2\sin\theta + 1)(3\sin\theta - 4) = 0$$
$$2\sin\theta + 1 = 0 \qquad 3\sin\theta - 4 = 0$$
$$2\sin\theta = -1 \qquad 3\sin\theta = 4$$
$$\sin\theta = -\frac{1}{2} \qquad \sin\theta = \frac{4}{3}$$

If factoring this trinomial is tricky, work through Problems 12.35–12.44 in The Humongous Book of Algebra Problems.

There are two angles on the unit circle with a sine value of $-1/2$, but there are no angles with a sine value of $4/3$. Recall that the range of sine (and the domain of arcsine) is $-1 \le \theta \le 1$, and $4/3$ does not belong to this interval.

$$\theta = \frac{7\pi}{6}, \frac{11\pi}{6}$$

10.21 Identify the general solution to the equation: $\cos^4 \theta = 1$.

Subtract 1 from both sides of the equation to create a difference of perfect squares. Factor the expression.

$$\cos^4 \theta - 1 = 0$$
$$\left(\cos^2 \theta + 1\right)\left(\cos^2 \theta - 1\right) = 0$$

One of the factors is a difference of perfect squares, so it can be factored further.

$$(\cos^2 \theta + 1)(\cos \theta + 1)(\cos \theta - 1) = 0$$

Apply the zero-product property. Setting the two linear factors equal to 0 produces the equations $\cos \theta = -1$ and $\cos \theta = 1$; the solutions are $\theta = \pi$ and $\theta = 0$, respectively.

However, setting the quadratic factor equal to 0 produces the equation $\cos^2 \theta + 1 = 0$, which has no solution. Notice that subtracting 1 from both sides produces the statement $\cos^2 \theta = -1$, but a squared value cannot equal a negative number.

Thus, the general solution to the original equation is $\theta = 0 + 2k\pi$ and $\theta = \pi + 2k\pi$, such that k is an integer. You may wish to express the solution as the single expression $\theta = 0 + k\pi$.

10.22 Identify the general solution to the equation: $\cot^3 \theta + 3\cot^2 \theta + 6 = -2\cot \theta$.

Add $2\cot \theta$ to both sides of the equation in order to apply the zero-product property.

$$\cot^3 \theta + 3\cot^2 \theta + 2\cot \theta + 6 = 0$$

Apply the factoring by grouping technique. Treat this four-term expression as a pair of two-term expressions, $\cot^3 \theta + 3\cot^2 \theta$ and $2\cot \theta + 6$. Notice that the first pair of expressions has a greatest common factor of $\cot^2 \theta$, and the second pair of expressions has a greatest common factor of 2.

If you need help factoring by grouping, check out Problems 12.20–12.25 in The Humongous Book of Algebra Problems.

$$\cot^2 \theta (\cot \theta + 3) + 2(\cot \theta + 3) = 0$$

Now both expressions have a greatest common factor of $(\cot \theta + 3)$, which should be factored out of each.

$$(\cot \theta + 3)(\cot^2 \theta + 2) = 0$$

Apply the zero-product property and reciprocal identities, as demonstrated below.

$$\cot \theta + 3 = 0 \qquad \cot^2 \theta + 2 = 0$$
$$\cot \theta = -3 \qquad \cot^2 \theta = -2$$
$$\tan \theta = -\frac{1}{3}$$

The equation $\cot^2 \theta = -2$ has no real solutions; no squared value may equal a negative number. However, the domain of arctangent is all real numbers, so

$\theta = \arctan(-1/3)$ is a valid solution. To construct the general solution, add multiples of π (the period of tangent).

$$\theta = \arctan\left(-\frac{1}{3}\right) + k\pi, \text{ such that } k \text{ is an integer}$$

10.23 Identify the solutions to the equation $20\sin^3 x + 3 = 12\sin^2 x + 5\sin x$ on the interval $-\pi/2 \le x \le \pi/2$.

Subtract $12\sin^2 x$ and $5\sin x$ from both sides of the equation and factor by grouping.

$$20\sin^3 x - 12\sin^2 x - 5\sin x + 3 = 0$$
$$4\sin^2 x(5\sin x - 3) - 1(5\sin x - 3) = 0 \longleftarrow$$
$$(5\sin x - 3)(4\sin^2 x - 1) = 0$$

> Factor -1 out of $(-5\sin x + 3)$ so you end up with matching factors $(5\sin x - 3)$.

The quadratic factor $4\sin^2 x - 1$ is a difference of perfect squares and can be factored further.

$$(5\sin x - 3)(2\sin x + 1)(2\sin x - 1) = 0$$

Apply the zero-product property. Notice that the problem asks you to identify answers on the interval $-\pi/2 \le x \le \pi/2$, which is the range of arcsine. Therefore, you should calculate the exact solution to each of the three equations.

$5\sin x - 3 = 0$	$2\sin x + 1 = 0$	$2\sin x - 1 = 0$
$5\sin x = 3$	$2\sin x = -1$	$2\sin x = 1$
$\sin x = \dfrac{3}{5}$	$\sin x = -\dfrac{1}{2}$	$\sin x = \dfrac{1}{2}$
$x = \arcsin\dfrac{3}{5}$	$x = \arcsin\left(-\dfrac{1}{2}\right)$	$x = \arcsin\dfrac{1}{2}$

Two of the solutions appear on the unit circle. Although $\arcsin(3/5)$ does not, it is still a valid solution; recall that the range of sine (and the domain of arcsine) is $-1 \le x \le 1$, and $x = 3/5$ belongs to that interval.

$$x = -\frac{\pi}{6}, \frac{\pi}{6}, \arcsin\frac{3}{5}$$

Quadratic Formula
When you can't factor

10.24 Identify the quadratic formula used to solve quadratic equation $ax^2 + bx + c = 0$, such that a, b, and c are integers and $a \neq 0$.

The quadratic formula allows you to solve a quadratic equation using only its coefficients. This technique, like the factoring technique described in Problems 10.12–10.23, requires that one side of the equation equals 0.

$$x = \frac{-b \pm \sqrt{b^2 - 4ac}}{2a}$$

10.25 Apply the quadratic formula to solve the equation: $2x^2 - 5x + 1 = 0$.

Substitute $a = 2$, $b = -5$, and $c = 1$ into the quadratic formula, identified in Problem 10.24.

$$x = \frac{-b \pm \sqrt{b^2 - 4ac}}{2a}$$

$$= \frac{-(-5) \pm \sqrt{(-5)^2 - 4(2)(1)}}{2(2)}$$

$$= \frac{5 \pm \sqrt{25 - 8}}{4}$$

$$= \frac{5 \pm \sqrt{17}}{4}$$

There are two solutions to the equation: $x = \dfrac{5 + \sqrt{17}}{4}, \dfrac{5 - \sqrt{17}}{4}$.

Note: Problems 10.26–10.27 demonstrate two different ways to solve the quadratic trigonometric equation $2\sin^2 x + 3\sin x + 1 = 0$. Identify all solutions on the interval $0 \le x < 2\pi$.

10.26 Solve the equation by factoring.

Apply the technique described in Problems 10.12–10.23, factoring the expression and then setting each factor equal to zero, applying the zero-product property.

$$(2\sin x + 1)(\sin x + 1) = 0$$

$2\sin x + 1 = 0$	$\sin x + 1 = 0$
$2\sin x = -1$	$\sin x = -1$
$\sin x = -\dfrac{1}{2}$	$x = \dfrac{3\pi}{2}$
$x = \dfrac{7\pi}{6}, \dfrac{11\pi}{6}$	

The solution is $x = \dfrac{7\pi}{6}, \dfrac{3\pi}{2}, \dfrac{11\pi}{6}$.

Note: Problems 10.26–10.27 demonstrate two different ways to solve the quadratic trigonometric equation $2\sin^2 x + 3\sin x + 1 = 0$. Identify all solutions on the interval $0 \le x < 2\pi$.

10.27 Verify the solutions you identified in Problem 10.26 by applying the quadratic formula.

Substitute $a = 2$, $b = 3$, and $c = 1$ into the quadratic formula. Note that the left side of the equation in the traditional quadratic formula is simply "x," but in this problem, the left side of the equation is "sin x."

$$\sin x = \frac{-b \pm \sqrt{b^2 - 4ac}}{2a}$$

$$\sin x = \frac{-(3) \pm \sqrt{3^2 - 4(2)(1)}}{2(2)}$$

$$\sin x = \frac{-3 \pm \sqrt{9 - 8}}{4}$$

$$\sin x = \frac{-3 \pm 1}{4}$$

$$\sin x = \frac{-3 - 1}{4} \text{ or } \frac{-3 + 1}{4}$$

$$\sin x = \frac{-4}{4} \text{ or } -\frac{2}{4}$$

$$\sin x = -1 \text{ or } -\frac{1}{2}$$

Thus, sin $x = -1$ or sin $x = -1/2$. These are the same equations produced by the zero-product property in Problem 10.26, so the solutions are the same: $x = 7\pi/6$, $3\pi/2$, $11\pi/6$.

Note: In Problems 10.28–10.29, you solve the equation $9\cos^2 x - 9\cos x + 2 = 0$ using two different techniques. In each problem, identify the solutions on the interval $0 \le x < \pi$.

10.28 Apply the quadratic formula to solve the equation.

Substitute $a = 9$, $b = -9$, and $c = 2$ into the quadratic formula, and set the fraction equal to cos x.

$$\cos x = \frac{-(-9) \pm \sqrt{(-9)^2 - 4(9)(2)}}{2(9)}$$

$$\cos x = \frac{9 \pm \sqrt{81 - 72}}{18}$$

$$\cos x = \frac{9 \pm \sqrt{9}}{18}$$

$$\cos x = \frac{9 + 3}{18}, \frac{9 - 3}{18}$$

$$\cos x = \frac{12}{18}, \frac{6}{18}$$

$$\cos x = \frac{2}{3}, \frac{1}{3}$$

Thus, the solutions to the equation are $x = \arccos (1/3)$, $x = \arccos (2/3)$.

The problem asks for the solutions that belong to the restricted range of arccosine ($0 \le x < \pi$). Neither 1/3 nor 2/3 are cosine values on the unit circle, so leave your answers in terms of arccosine.

Note: In Problems 10.28–10.29, you solve the equation $9\cos^2 x - 9\cos x + 2 = 0$ using two different techniques. In each problem, identify the solutions on the interval $0 \le x \le \pi$.

10.29 Verify your answer to Problem 10.28 by factoring the quadratic expression and applying the zero-product property.

Factor the quadratic expression and set each factor equal to 0.

$$(3\cos x - 1)(3\cos x - 2) = 0$$

$$3\cos x - 1 = 0 \qquad\qquad 3\cos x - 2 = 0$$

$$3\cos x = 1 \qquad\qquad 3\cos x = 2$$

$$\cos x = \frac{1}{3} \qquad\qquad \cos x = \frac{2}{3}$$

$$x = \arccos\frac{1}{3} \qquad\qquad x = \arccos\frac{2}{3}$$

These solutions are identical to the solutions in Problem 10.28.

10.30 Apply the quadratic formula to identify the solutions to the equation $12\cos^2 x + 7\cos x = 10$ on the interval $0 \le x \le \pi$.

In order to apply the quadratic formula, one side of the quadratic equation must equal 0; subtract 10 from both sides of the equation.

$$12\cos^2 x + 7\cos x - 10 = 0$$

Substitute $a = 12$, $b = 7$, and $c = -10$ into the quadratic formula and set it equal to $\cos x$.

$$\cos x = \frac{-7 \pm \sqrt{7^2 - 4(12)(-10)}}{2(12)}$$

$$\cos x = \frac{-7 \pm \sqrt{529}}{24}$$

$$\cos x = \frac{-7 \pm 23}{24}$$

$$\cos x = \frac{-7 - 23}{24}, \frac{-7 + 23}{24}$$

$$\cos x = \frac{-30}{24}, \frac{16}{24}$$

$$\cos x = -\frac{5}{4}, \frac{2}{3}$$

The range of cosine is $-1 \le \cos x \le 1$, so $-5/4$ is not a valid cosine value. Therefore, the only solution to this equation is $x = \arccos(2/3)$.

10.31 Explain why the equation $4\sin^2 x - \sin x + 2 = 0$ has no real solutions.

Apply the quadratic formula, setting $a = 4$, $b = -1$, and $c = 2$.

$$\sin x = \frac{-(-1) \pm \sqrt{(-1)^2 - 4(4)(2)}}{2(4)}$$

$$\sin x = \frac{1 \pm \sqrt{1 - 32}}{8}$$

$$\sin x = \frac{1 \pm \sqrt{-31}}{8}$$

You cannot evaluate the square root of a negative number, so there are no real solutions to this trigonometric quadratic equation.

10.32 Identify the solutions to the equation $5\tan\theta = -1 - 3\tan^2\theta$ on the interval $-\pi/2 \le \theta \le \pi/2$.

Add $3\tan^2\theta$ and 1 to both sides of the equation.

$$3\tan^2\theta + 5\tan\theta + 1 = 0$$

Substitute $a = 3$, $b = 5$, and $c = 1$ into the quadratic formula.

$$\tan\theta = \frac{-5 \pm \sqrt{5^2 - 4(3)(1)}}{2(3)}$$

$$\tan\theta = \frac{-5 \pm \sqrt{25 - 12}}{6}$$

$$\tan\theta = \frac{-5 \pm \sqrt{13}}{6}$$

$$\theta = \arctan\left(\frac{-5 - \sqrt{13}}{6}\right), \arctan\left(\frac{-5 + \sqrt{13}}{6}\right)$$

> Arctangent is defined for all real numbers, so both of these solutions are valid.

Functions of Multiple Angles

Instead of $\cos x = 1$, solve $\cos 5x = 1$

Note: In Problems 10.33–10.36, you solve similar equations, each with a different coefficient of x. In each problem, identify all solutions on the interval $0 \le x < 2\pi$.

10.33 Solve the equation: $\sin x + 1 = 0$.

Subtract 1 from both sides of the equation to isolate $\sin x$ left of the equal sign.

$$\sin x = -1$$

Taking the arcsine of both sides of this equation solves it for x.

$$x = \arcsin(-1)$$

Note that the range of arcsine is $-\pi/2 \leq \arcsin x \leq \pi/2$, so if you were asked to supply the exact solution to this equation, the answer would be $x = -\pi/2$. However, the problem directs you to identify all solutions within a single rotation on the coordinate plane. The only angle on this interval with a sine value of -1 is $x = 3\pi/2$. (Note that $-\pi/2$ and $3\pi/2$ are coterminal angles, and the choice of one angle versus the other in the solution is dictated solely by the interval specified in the problem.)

> In other words, the angles between 0 and 2π, including 0 but excluding 2π.

Note: In Problems 10.33–10.36, you solve similar equations, each with a different coefficient of x. In each problem, identify all solutions on the interval $0 \leq x < 2\pi$.

10.34 Solve the equation: $\sin 2x + 1 = 0$.

This equation is similar to the equation in Problem 10.33 ($\sin x + 1 = 0$); only the arguments of the equations differ. This equation contains argument $2x$, a double angle. Regardless of this difference, you still begin by isolating the trigonometric function on one side of the equal sign. In other words, you still begin by subtracting 1 from both sides of the equation.

$$\sin 2x = -1$$

Now apply the arcsine function to both sides of the equation.

$$2x = \arcsin(-1)$$

As Problem 10.33 states, angle $3\pi/2$ has a sine value of -1, but because this equation contains a double angle ($2x$ instead of x), you should list *twice* as many solutions. Thus, you should calculate a coterminal angle for the solution $3\pi/2$ by adding the period of sine (2π).

$$\frac{3\pi}{2} + 2\pi = \frac{3\pi}{2} + \frac{4\pi}{2} = \frac{7\pi}{2}$$

Rewrite the equation $2x = \arcsin(-1)$, replacing $\arcsin(-1)$ with the two solutions you have identified.

$$2x = \arcsin(-1)$$
$$2x = \frac{3\pi}{2}, \frac{7\pi}{2}$$

To finish solving the equation, you must isolate x. Eliminate its coefficient (2) by multiplying both sides of the equation by $1/2$.

$$\left(\frac{1}{2}\right)(2x) = \left(\frac{1}{2}\right)\left(\frac{3\pi}{2}\right), \left(\frac{1}{2}\right)\left(\frac{7\pi}{2}\right)$$
$$x = \frac{3\pi}{4}, \frac{7\pi}{4}$$

Note that both of the final solutions belong to the interval specified by the problem: $0 \leq 3\pi/4 < 7\pi/4 < 2\pi$.

Remember to adjust the total number of solutions based on the coefficient of x, especially when the problem asks you to list the solutions on the interval $0 \leq x < 2\pi$.

Note: In Problems 10.33–10.36, you solve similar equations, each with a different coefficient of x. In each problem, identify all solutions on the interval $0 \leq x < 2\pi$.

10.35 Solve the equation: $\sin 3x + 1 = 0$. ←

Like Problems 10.33–10.34, you begin this problem by subtracting 1 from both sides of the equation. In order to solve any trigonometric equation, you must isolate the expression containing the variable (in this case $\sin 3x$) on one side of the equation.

$$\sin 3x = -1$$

Apply the arcsine function to eliminate the sine expression containing x, the variable for which you are solving.

$$3x = \arcsin(-1)$$

Only one angle on the unit circle has a sine value of –1: $x = 3\pi/2$. However, the coefficient of x is 3, so you must list three solutions—three times as many as the single solution $3\pi/2$. Add 2π to $3\pi/2$ (like you did in Problem 10.34) to identify coterminal angle $x = 7\pi/2$, and then add 2π to that angle in order identify a second coterminal angle.

$$\frac{3\pi}{2} + 2\pi = \frac{3\pi}{2} + \frac{4\pi}{2} = \frac{7\pi}{2} \qquad \frac{7\pi}{2} + 2\pi = \frac{7\pi}{2} + \frac{4\pi}{2} = \frac{11\pi}{2}$$

Return to the equation $3x = \arcsin(-1)$ and replace the arcsine expression with the three angles you have identified.

$$3x = \arcsin(-1)$$
$$3x = \frac{3\pi}{2}, \frac{7\pi}{2}, \frac{11\pi}{2}$$

Multiply both sides of the equation by 1/3 to solve for x.

$$\left(\frac{1}{3}\right)(3x) = \left(\frac{1}{3}\right)\left(\frac{3\pi}{2}\right), \left(\frac{1}{3}\right)\left(\frac{7\pi}{2}\right), \left(\frac{1}{3}\right)\left(\frac{11\pi}{2}\right)$$
$$x = \frac{3\pi}{6}, \frac{7\pi}{6}, \frac{11\pi}{6}$$
$$x = \frac{\pi}{2}, \frac{7\pi}{6}, \frac{11\pi}{6}$$

In Problem 10.34, the trig expression contained 2x and you had to list 2 times as many solutions. In this problem, the trig expression contains 3x, so you will need to list 3 times as many solutions.

Note: In Problems 10.33–10.36, you solve similar equations, each with a different coefficient of x. In each problem, identify all solutions on the interval $0 \leq x < 2\pi$.

10.36 Verify your solutions to Problem 10.35 by substituting each into the equation $\sin 3x + 1 = 0$.

Problem 10.35 states that equation $\sin 3x + 1 = 0$ has three solutions that lie between 0 and 2π: $x = \pi/2$, $7\pi/6$, and $11\pi/6$. Substitute each into the equation to verify that the resulting statements are true.

$$\sin 3x + 1 = 0$$

$$\sin\left(3 \cdot \frac{\pi}{2}\right) + 1 = 0$$

$$\sin\frac{3\pi}{2} + 1 = 0$$

$$-1 + 1 = 0$$

$$0 = 0$$

$$\sin 3x + 1 = 0$$

$$\sin\left(3 \cdot \frac{7\pi}{6}\right) + 1 = 0$$

$$\sin\frac{21\pi}{6} + 1 = 0$$

$$\sin\frac{7\pi}{2} + 1 = 0$$

$$-1 + 1 = 0$$

$$0 = 0$$

$$\sin 3x + 1 = 0$$

$$\sin\left(3 \cdot \frac{11\pi}{6}\right) + 1 = 0$$

$$\sin\frac{33\pi}{6} + 1 = 0$$

$$\sin\frac{11\pi}{2} + 1 = 0$$

$$-1 + 1 = 0$$

$$0 = 0$$

In Problem 10.35, you calculated two angles coterminal to $3\pi/2$: $7\pi/2$ and $11\pi/2$. If the angles are coterminal, then all three have the same sine value: -1.

10.37 Identify the solutions to the equation $\sin 2x = 0$ on the interval $0 \le x < 2\pi$.

In order to solve for x, you must first eliminate the trigonometric equation that contains the variable. Take the arcsine of both sides of the equation.

$$2x = \arcsin 0$$

Only angles 0 and π have a sine value of 0 on the interval $0 \le x < 2\pi$. Because x has a coefficient of 2, you should list twice as many solutions—four angles instead of two. Add 2π (the period of sine) to $x = 0$ and $x = \pi$ to identify two additional coterminal angles: $0 + 2\pi = 2\pi$ and $\pi + 2\pi = 3\pi$.

$$2x = 0, \pi, 2\pi, 3\pi$$

Divide by 2 to solve for x.

$$\frac{\cancel{2}x}{\cancel{2}} = \frac{0}{2}, \frac{\pi}{2}, \frac{\cancel{2}\pi}{\cancel{2}}, \frac{3\pi}{2}$$

$$x = 0, \frac{\pi}{2}, \pi, \frac{3\pi}{2}$$

10.38 Identify the solutions to the equation $\cos 3x = -\dfrac{\sqrt{2}}{2}$ on the interval $0 \le x < 2\pi$.

Take the arccosine of both sides of the equation.

$$3x = \arccos\left(-\frac{\sqrt{2}}{2}\right)$$

The two angles on the unit circle that have a cosine of $-\sqrt{2}/2$ are $3\pi/4$ and $5\pi/4$. Calculate two coterminal angles for each, adding 2π and then 4π to $3\pi/4$ and $5\pi/4$.

$$3x = \frac{3\pi}{4}, \frac{5\pi}{4}, \frac{11\pi}{4}, \frac{13\pi}{4}, \frac{19\pi}{4}, \frac{21\pi}{4}$$

Multiply each term by $1/3$ to solve for x.

$$\frac{3x}{3} = \frac{3\pi}{12}, \frac{5\pi}{12}, \frac{11\pi}{12}, \frac{13\pi}{12}, \frac{19\pi}{12}, \frac{21\pi}{12}$$

$$x = \frac{\pi}{4}, \frac{5\pi}{12}, \frac{11\pi}{12}, \frac{13\pi}{12}, \frac{19\pi}{12}, \frac{7\pi}{4}$$

10.39 Identify the solutions to the equation $4\cos 2x - \sqrt{8} = 0$ on the interval $0 \le x < 2\pi$.

Isolate $\cos 2x$ on the left side of the equation and simplify the square root expression.

$$4\cos 2x = \sqrt{8}$$
$$4\cos 2x = 2\sqrt{2}$$
$$\cos 2x = \frac{2\sqrt{2}}{4}$$
$$\cos 2x = \frac{\sqrt{2}}{2}$$

Take the arccosine of both sides of the equation.

$$2x = \arccos\frac{\sqrt{2}}{2}$$

Angles $\pi/4$ and $7\pi/4$ have cosine values of $\sqrt{2}/2$, but you should calculate one coterminal angle for each; an x-coefficient of 2 requires twice as many solutions.

$$2x = \frac{\pi}{4}, \frac{7\pi}{4}, \frac{9\pi}{4}, \frac{15\pi}{4}$$

Divide by 2 to solve for x.

$$x = \frac{\pi}{8}, \frac{7\pi}{8}, \frac{9\pi}{8}, \frac{15\pi}{8}$$

> $\dfrac{\pi}{4} + 2\pi = \dfrac{\pi}{4} + \dfrac{8\pi}{4} = \dfrac{9\pi}{4}$
>
> $\dfrac{7\pi}{4} + 2\pi = \dfrac{7\pi}{4} + \dfrac{8\pi}{4} = \dfrac{15\pi}{4}$

10.40 Identify the general solution to the equation: $2(\cos 3x - 2) = 3(\cos 3x - 1)$.

Simplify both sides of the equation and isolate the trigonometric expression on one side of the equal sign.

$$2(\cos 3x) + 2(-2) = 3(\cos 3x) + 3(-1)$$
$$2\cos 3x - 4 = 3\cos 3x - 3$$
$$-4 + 3 = 3\cos 3x - 2\cos 3x$$
$$-1 = \cos 3x$$
$$\arccos(-1) = 3x$$

Only angle π has a cosine of -1 on the interval $0 \le x < 2\pi$. Because the coefficient of x is 3, you may be tempted to list three times as many solutions, most likely 3π and 5π, which are angles coterminal to π. However, this problem asks for the general solution, which includes *all* coterminal angles. Therefore, there is no need to list additional solutions based on the coefficient of x.

$$\pi + 2k\pi = 3x$$

Divide each term by 3, including $2k\pi$, to solve for x.

$$\frac{\pi}{3} + \frac{2k\pi}{3} = \frac{3x}{3}$$

$$x = \frac{\pi}{3} + \frac{2}{3}k\pi, \text{ such that } k \text{ is an integer}$$

> To solve the equation $\cos 3x = -1$ on the interval $0 \le x < 2\pi$, you need to write three times the number of solutions (π, 3π, and 5π). However, for the general solution, don't bother with the coterminal angles 3π and 5π, because your solution already includes ALL coterminal angles.

10.41 Identify the general solution to the equation: $3(\tan 2x - 4) + 9 = 0$.

Isolate $2x$ on one side of the equation.

$$3(\tan 2x) + 3(-4) + 9 = 0$$
$$3\tan 2x - 12 + 9 = 0$$
$$3\tan 2x - 3 = 0$$
$$3\tan 2x = 3$$
$$\tan 2x = \frac{3}{3}$$
$$\tan 2x = 1$$
$$2x = \arctan 1$$

Consider the full period of tangent lying between $\theta = -\pi/2$ and $\theta = \pi/2$. Only angle $\pi/4$ has a tangent value of 1 on that interval. Because you are asked to report the general solution, there is no need to list twice as many solutions (despite the fact that x has a coefficient of 2); the general solution will include all coterminal angles.

$$2x = \frac{\pi}{4} + k\pi$$

Solve for x, dividing each term by 2 or multiplying each term by 1/2.

$$\frac{\cancel{2}x}{\cancel{2}} = \frac{\pi}{2 \cdot 4} + \frac{k\pi}{2}$$

$$x = \frac{\pi}{8} + \frac{k\pi}{2}, \text{ such that } k \text{ is an integer}$$

Note: Problems 10.42–10.43 refer to the equation $2\sin\dfrac{x}{2} = 1$.

10.42 Identify the solutions to the equation on the interval $0 \le x < 2\pi$.

Isolate the trigonometric expression on the left side of the equation and apply the arcsine function.

$$\sin\frac{x}{2} = \frac{1}{2}$$

$$\frac{x}{2} = \arcsin\frac{1}{2}$$

The two angles on the interval $0 \le x < 2\pi$ that have a sine value of 1/2 are $x = \pi/6$ and $5\pi/6$. Although the coefficient of x is 1/2, you do not apply the technique demonstrated in Problem 10.34; in other words, a coefficient of 1/2 does *not* imply that you should list half as many solutions. Fractional coefficients between 0 and 1 do not require you to identify additional coterminal solutions.

$$\frac{x}{2} = \frac{\pi}{6}, \frac{5\pi}{6}$$

To solve for x, multiply each of the terms by 2.

But you should check to make sure you don't end up with extra solutions that are outside the specified interval

$$\frac{\cancel{2}}{1}\left(\frac{x}{\cancel{2}}\right) = \frac{2}{1}\left(\frac{\pi}{6}\right), \frac{2}{1}\left(\frac{5\pi}{6}\right)$$

$$x = \frac{2\pi}{6}, \frac{10\pi}{6}$$

$$x = \frac{\pi}{3}, \frac{5\pi}{3}$$

Note: Problems 10.42–10.43 refer to the equation $2\sin\dfrac{x}{2} = 1$.

10.43 Identify the four smallest positive solutions to the equation.

Complete the steps outlined in Problem 10.42 to isolate the variable expression $x/2$ on the left side of the equation, but before you solve for x, list the general form of the solutions.

$$\frac{x}{2} = \frac{\pi}{6} + 2k\pi, \frac{5\pi}{6} + 2k\pi$$

Solve for x, multiplying each of the terms in the equation by 2.

$$\cancel{2}\left(\frac{x}{\cancel{2}}\right) = 2\left(\frac{\pi}{6}\right) + 2(2k\pi), 2\left(\frac{5\pi}{6}\right) + 2(2k\pi)$$

$$x = \frac{\pi}{3} + 4k\pi, \frac{5\pi}{3} + 4k\pi, \text{ such that } k \text{ is an integer}$$

The two smallest positive solutions are $\pi/3$ and $5\pi/3$. To generate the next two solutions, substitute $k = 1$ into each of the expressions.

$$x = \frac{\pi}{3} + 4(1)\pi \qquad x = \frac{5\pi}{3} + 4(1)\pi$$

$$= \frac{\pi}{3} + 4\pi \qquad\qquad = \frac{5\pi}{3} + 4\pi$$

$$= \frac{\pi}{3} + \frac{12\pi}{3} \qquad\quad = \frac{5\pi}{3} + \frac{12\pi}{3}$$

$$= \frac{13\pi}{3} \qquad\qquad\quad = \frac{17\pi}{3}$$

Thus, the four smallest positive solutions to the equation are $x = \dfrac{\pi}{3}, \dfrac{5\pi}{3}, \dfrac{13\pi}{3},$ and $\dfrac{17\pi}{3}$.

10.44 Identify the solutions to the equation $3\sec^2\dfrac{x}{2} - 5\sec\dfrac{x}{2} - 2 = 0$ on the interval $0 \le x < 2\pi$.

The left side of the equation is a factorable quadratic expression: $3y^2 - 5y - 2 = (3y + 1)(y - 2)$, such that $y = \sec(x/2)$.

$$3\sec^2\frac{x}{2} - 5\sec\frac{x}{2} - 2 = 0$$

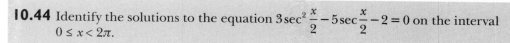

$$\left(3\sec\frac{x}{2} + 1\right)\left(\sec\frac{x}{2} - 2\right) = 0$$

Check out Problem 10.18 if you need to review factoring and the zero-product property.

Apply the zero-product property, setting each factor equal to 0 and solving the two resulting equations for sec x.

$$3\sec\frac{x}{2}+1=0 \qquad\qquad \sec\frac{x}{2}-2=0$$

$$3\sec\frac{x}{2}=-1 \qquad\qquad \sec\frac{x}{2}=2$$

$$\sec\frac{x}{2}=-\frac{1}{3}$$

Apply the reciprocal identity $\cos\theta = 1/\sec\theta$ and solve for x.

$$\cos\frac{x}{2}=-3 \qquad\qquad \cos\frac{x}{2}=\frac{1}{2}$$

$$\frac{x}{2}=\arccos(-3) \qquad\qquad \frac{x}{2}=\arccos\frac{1}{2}$$

Note that the domain of arccosine (like the range of cosine) is $-1 \le \theta \le 1$, so arccos (-3) is not defined; this equation may be omitted from the final solution. However, there are two angles with a cosine value of $1/2$.

$$\frac{x}{2}=\arccos\frac{1}{2}$$

$$\frac{x}{2}=\frac{\pi}{3},\frac{5\pi}{3}$$

The coefficient of x is $1/2$. As Problem 10.42 explains, you do not have to identify additional coterminal angles for fractional coefficients between 0 and 1. Multiply each term by 2 in order to solve for x.

However, one of these solutions will no longer fit in the interval $0 \le x < 2\pi$ once you solve for x.

$$\cancel{2}\left(\frac{x}{\cancel{2}}\right)=2\left(\frac{\pi}{3}\right),2\left(\frac{5\pi}{3}\right)$$

$$x=\frac{2\pi}{3},\frac{10\pi}{3}$$

Note that $10\pi/3$ is greater than the upper bound of the interval (2π), so the equation has only one valid solution: $x = 2\pi/3$.

10.45 Identify the general solution to the equation: $\tan^2\frac{x}{3}+4\tan\frac{x}{3}=4$.

Subtract 4 from both sides of the equation; the result is a quadratic expression left of the equal sign and the number 0 right of the equal sign.

$$\tan^2\frac{x}{3}+4\tan\frac{x}{3}-4=0$$

The quadratic expression cannot be factored, so apply the quadratic formula, as demonstrated in Problem 10.25. Note that the formula is set equal to the trigonometric expression tan $(x/3)$.

$$\tan\frac{x}{3} = \frac{-4 \pm \sqrt{4^2 - 4(1)(-4)}}{2(1)}$$

$$\tan\frac{x}{3} = \frac{-4 \pm \sqrt{16 + 16}}{2}$$

$$\tan\frac{x}{3} = \frac{-4 \pm \sqrt{32}}{2}$$

$$\tan\frac{x}{3} = \frac{-4 \pm 4\sqrt{2}}{2}$$

The numerator has a greatest common factor of 2. Factor it out of that expression to simplify the fraction.

$$\tan\frac{x}{3} = \frac{\cancel{2}\left(-2 \pm 2\sqrt{2}\right)}{\cancel{2}}$$

$$\tan\frac{x}{3} = -2 \pm 2\sqrt{2}$$

$$\frac{x}{3} = \arctan\left(-2 - 2\sqrt{2}\right), \arctan\left(-2 + 2\sqrt{2}\right)$$

The problem directs you to calculate the general solution, so add multiples of π (the period of tangent) to each solution.

$$\frac{x}{3} = \arctan\left(-2 - 2\sqrt{2}\right) + k\pi, \arctan\left(-2 + 2\sqrt{2}\right) + k\pi, \text{ such that } k \text{ is an integer}$$

Multiply each term by 3 to solve for x.

$$x = 3\arctan\left(-2 - 2\sqrt{2}\right) + 3k\pi, \arctan\left(-2 + 2\sqrt{2}\right) + 3k\pi, \text{ such that } k \text{ is an integer}$$

Chapter 11
ADVANCED TRIGONOMETRIC EQUATIONS

Trickier equations = clever-er solutions

In Chapter 10, you explore the fundamental concepts and techniques of solving trigonometric equations, including:

- Exact solutions versus general solutions versus solutions on an interval
- The role of inverse trigonometric functions
- The zero-product property and factoring
- The quadratic formula
- Equations containing single- and multiple-angle arguments

Some of the techniques in Chapter 10 are revisited in this chapter, so ensure that you understand all of the bulleted items above before attempting any of the problems in this chapter.

In this chapter, you learn a few more techniques for your trigonometric equation solving toolbox. The four major concepts of this chapter are: (1) rewriting trig equations using identities, (2) introducing a squared function when you need one, (3) getting rid of a square when you don't need one, and (4) solving equations that contain fractions.

If you need to review identities, flip back to Chapters 7 and 8. Make sure that you have the three Pythagorean identities memorized, because you will see them a lot in these problems.

Square Roots
Eliminating squares instead of factoring them

Note: In Problems 11.1–11.2, you solve the equation $\tan^2 x - 1 = 0$ using two different techniques. Report the solutions on the interval $0 \le x < 2\pi$.

11.1 Solve the equation by factoring the expression $\tan^2 x - 1$.

> If you need to review this technique, check out Problems 10.12–10.23.

The expression $\tan^2 x - 1$ is a difference of perfect squares. Factor it and apply the zero-product property, setting each of the individual factors equal to zero and solving those equations.

$$(\tan x + 1)(\tan x - 1) = 0$$

$$\tan x + 1 = 0 \qquad \tan x - 1 = 0$$

$$\tan x = -1 \qquad \tan x = 1$$

The solutions to the equation $\tan^2 x - 1 = 0$ are the angles x on the interval $0 \le x < 2\pi$ that have a tangent of -1 or $+1$.

$$x = \frac{\pi}{4}, \frac{3\pi}{4}, \frac{5\pi}{4}, \frac{7\pi}{4}$$

Note: In Problems 11.1–11.2, you solve the equation $\tan^2 x - 1 = 0$ using two different techniques. Report the solutions on the interval $0 \le x < 2\pi$.

11.2 Solve the equation using square roots to verify your solution to Problem 11.1.

> For example, to solve the equation $x^2 = 16$, you take the square root of both sides to get $x = \pm 4$. There are two answers, because $+4$ and -4 both equal 16 when squared.

Unlike the factoring and quadratic formula techniques for solving quadratic equations, applying square roots does not require that one side of the equation be equal to 0. Begin by adding 1 to both sides.

$$\tan^2 x = 1$$

Take the square root of both sides of the equation. Note that any even root applied to both sides of an equation requires the insertion of a "$\pm$" symbol.

$$\sqrt{\tan^2 x} = \sqrt{1}$$

$$\tan x = \pm 1$$

Thus, the solutions to the equation are all angles x on the interval $0 \le x < 2\pi$ that have a tangent value of either -1 or $+1$. This is the same set of angles identified in the solution to Problem 11.1.

$$x = \frac{\pi}{4}, \frac{3\pi}{4}, \frac{5\pi}{4}, \frac{7\pi}{4}$$

11.3 Identify the solutions to the equation $4\sin^2 x - 4\sin x + 1 = 0$ on the interval $0 \le x < 2\pi$.

Factor the quadratic expression left of the equal sign. Note that the quadratic formula will also produce the correct solution to this equation, but it requires more arithmetic than the technique described in this solution.

$$(2\sin x - 1)(2\sin x - 1) = 0$$

Note that the factors are equal, so the left side of the expression can be expressed as a square.

$$(2\sin x - 1)^2 = 0$$

Take the square root of both sides of the equation. Note that the "±" may be omitted because ±0 = 0.

$$\sqrt{(2\sin x - 1)^2} = \pm\sqrt{0}$$
$$2\sin x - 1 = 0$$
$$2\sin x = 1$$
$$\sin x = \frac{1}{2}$$

Two angles on the interval $0 \leq x < 2\pi$ have a sine value of 1/2.

$$x = \frac{\pi}{6}, \frac{5\pi}{6}$$

11.4 Identify the solutions to the equation $2(\cos^2 x + 4) - 9 = 0$ on the interval $0 \leq x < 2\pi$.

Simplify the expression left of the equal sign and then take the square root of both sides.

$$2(\cos^2 x) + 2(4) - 9 = 0$$
$$2\cos^2 x + 8 - 9 = 0$$
$$2\cos^2 x - 1 = 0$$
$$2\cos^2 x = 1$$
$$\cos^2 x = \frac{1}{2}$$
$$\sqrt{\cos^2 x} = \pm\sqrt{\frac{1}{2}}$$
$$\cos x = \pm\sqrt{\frac{1}{2}}$$

Rationalize the denominator on the right side of the expression to reveal a familiar value of cosine.

$$\cos x = \pm\frac{\sqrt{1}}{\sqrt{2}}$$
$$\cos x = \pm\frac{1}{\sqrt{2}}$$
$$\cos x = \pm\frac{1}{\sqrt{2}} \cdot \frac{\sqrt{2}}{\sqrt{2}}$$
$$\cos x = \pm\frac{\sqrt{2}}{\sqrt{4}}$$
$$\cos x = \pm\frac{\sqrt{2}}{2}$$

There are four angles on the interval $0 \leq x < 2\pi$ that have a cosine value of $-\sqrt{2}/2$ or $+\sqrt{2}/2$.

$$x = \frac{\pi}{4}, \frac{3\pi}{4}, \frac{5\pi}{4}, \frac{7\pi}{4}$$

11.5 Identify the solutions to the equation $3\csc^2 2x = 4$ on the interval $0 \leq x < 2\pi$.

Isolate $\csc^2 2x$ on the left side of the equation and then take the square root of both sides.

$$\csc^2 2x = \frac{4}{3}$$

$$\sqrt{\csc^2 2x} = \pm\sqrt{\frac{4}{3}}$$

$$\csc 2x = \pm\frac{\sqrt{4}}{\sqrt{3}}$$

$$\csc 2x = \pm\frac{2}{\sqrt{3}}$$

Apply a reciprocal identity to express cosecant in terms of sine. Then apply the arcsine function to eliminate sine from the left side of the equation.

$$\sin 2x = \pm\frac{\sqrt{3}}{2}$$

$$2x = \arcsin\left(\pm\frac{\sqrt{3}}{2}\right)$$

> You use arcsine to get rid of sine and leave behind the 2x that was plugged into it. The equation's not solved until you divide by 2 at the end and solve for x.

The four angles on the interval $0 \leq x < 2\pi$ that have a sine of $-\sqrt{3}/2$ or $+\sqrt{3}/2$ are $\pi/3$, $2\pi/3$, $4\pi/3$, and $5\pi/3$. However, because x has a coefficient of 2, you should list twice as many solutions. Add 2π to each of those four solutions to generate the list of eight solutions below.

$$2x = \frac{\pi}{3}, \frac{2\pi}{3}, \frac{4\pi}{3}, \frac{5\pi}{3}, \frac{7\pi}{3}, \frac{8\pi}{3}, \frac{10\pi}{3}, \frac{11\pi}{3}$$

> If the coefficient of x is N, then you have to list N times the number of answers you normally would. (See Problem 10.34.) This is only true when you are identifying solutions on a specific interval like $0 \leq x < 2\pi$.

Multiply each of the values by 1/2 to solve for x.

$$\frac{1}{2}(2x) = \frac{\pi}{6}, \frac{2\pi}{6}, \frac{4\pi}{6}, \frac{5\pi}{6}, \frac{7\pi}{6}, \frac{8\pi}{6}, \frac{10\pi}{6}, \frac{11\pi}{6}$$

$$x = \frac{\pi}{6}, \frac{\pi}{3}, \frac{2\pi}{3}, \frac{5\pi}{6}, \frac{7\pi}{6}, \frac{4\pi}{3}, \frac{5\pi}{3}, \frac{11\pi}{6}$$

11.6 Identify the solutions to the equation $9\cot^4 x - 1 = 0$ on the interval $0 \le x < 2\pi$.

Factor the difference of perfect squares and apply the zero-product property.

$$\left(3\cot^2 x + 1\right)\left(3\cot^2 x - 1\right) = 0$$

$$3\cot^2 x + 1 = 0 \qquad\qquad 3\cot^2 x - 1 = 0$$

$$3\cot^2 x = -1 \qquad\qquad 3\cot^2 x = 1$$

$$\cot^2 x = -\frac{1}{3} \qquad\qquad \cot^2 x = \frac{1}{3}$$

Note that the equation $\cot^2 x = -1/3$ has no solution, because no squared value can be equal to a negative number. Solve the remaining equation by taking the square root of both sides.

$$\sqrt{\cot^2 x} = \pm\sqrt{\frac{1}{3}}$$

$$\cot x = \pm\frac{1}{\sqrt{3}}$$

It is helpful to divide the numerator and denominator of the fraction by 2 and express $\cot x$ in terms of cosine and sine.

$$\frac{\cos x}{\sin x} = \pm\frac{1/2}{\sqrt{3}/2}$$

There are four angles with cosine values of $\pm 1/2$ and sine values of $\pm\sqrt{3}/2$.

$$x = \frac{\pi}{3}, \frac{2\pi}{3}, \frac{4\pi}{3}, \frac{5\pi}{3}$$

Rational Equations

Fractions with trigonometric numerators and denominators

11.7 Solve the rational equation: $\dfrac{x^2 - 9}{x + 3} = 0$.

A fraction is equal to zero when its numerator is equal to zero but its denominator is not. To identify possible solutions of this rational equation, set the numerator equal to 0 and solve for x.

$$x^2 - 9 = 0$$

$$x^2 = 9$$

$$\sqrt{x^2} = \pm\sqrt{9}$$

$$x = \pm 3$$

The numerator of the fraction is equal to 0 when $x = -3$ or $x = 3$. However, one of those values is not a solution to the equation. Substituting $x = -3$ into the denominator results in 0, and dividing by 0 is an undefined operation. Thus, there is a single solution to this equation: $x = 3$.

Note: Problems 11.8–11.9 refer to the equation $\dfrac{2\cos\theta - 1}{\sin 2\theta} = 0$.

11.8 Identify the values of θ on the interval $0 \le \theta < 2\pi$ for which the fraction is undefined.

The fraction A/B is undefined when its denominator is equal to zero ($B = 0$). The denominator of this fraction is $\sin 2\theta$.

$$\sin 2\theta = 0$$
$$2\theta = \arcsin 0$$

Note that $\sin 0 = 0$ and π, according to the unit circle. The coefficient of θ is 2, so you should identify twice as many solutions on the interval $0 < \theta < 2\pi$ than you normally would.

$$2\theta = 0, \pi, 2\pi, 3\pi$$
$$\left(\frac{1}{\cancel{2}}\right)(\cancel{2}\theta) = \left(\frac{1}{2}\right)(0), \left(\frac{1}{2}\right)(\pi), \left(\frac{1}{\cancel{2}}\right)(\cancel{2}\pi), \left(\frac{1}{2}\right)(3\pi)$$
$$\theta = 0, \frac{\pi}{2}, \pi, \frac{3\pi}{2}$$

There are four values of θ for which the fraction $\dfrac{2\cos\theta - 1}{\sin 2\theta}$ is undefined.

Note: Problems 11.8–11.9 refer to the equation $\dfrac{2\cos\theta - 1}{\sin 2\theta} = 0$.

11.9 Identify the solutions to the equation on the interval $0 \le \theta < 2\pi$.

The solutions to the equation $A/B = 0$ are the values for which $A = 0$ but $B \ne 0$. Set the numerator equal to 0 and solve for θ.

$$2\cos\theta - 1 = 0$$
$$2\cos\theta = 1$$
$$\cos\theta = \frac{1}{2}$$
$$\theta = \frac{\pi}{3}, \frac{5\pi}{3}$$

11.10 Identify the solutions to the below equation on the interval $0 \le x < 2\pi$.

$$\frac{4\cos^3 x - 2\cos x}{\cos x} = 0$$

The terms in the numerator have a greatest common factor of $2\cos x$; factor it out of the expression.

$$\frac{2\cos x\left(2\cos^2 x - 1\right)}{\cos x} = 0$$

The fraction on the left side of the equation is equal to 0 when its numerator is equal to 0. Apply the zero-product property, setting each of the factors in the numerator equal to 0.

$$2 \ne 0 \qquad \cos x = 0 \qquad 2\cos^2 x - 1 = 0$$

Clearly, $2 \neq 0$, so that factor should be discarded. Any values of x for which $\cos x = 0$ will also cause the denominator to equal 0, so the equation $\cos x = 0$ may also be discarded. The solutions to this rational equation are limited to the solutions of the equation $2\cos^2 x - 1 = 0$.

$$2\cos^2 x - 1 = 0$$

$$\cos^2 x = \frac{1}{2}$$

$$\sqrt{\cos^2 x} = \pm\sqrt{\frac{1}{2}}$$

$$\cos x = \pm\frac{1}{\sqrt{2}}$$

Rationalize the denominator and identify the solutions on the interval $0 \leq x < 2\pi$.

$$\cos x = \pm\frac{1}{\sqrt{2}} \cdot \frac{\sqrt{2}}{\sqrt{2}}$$

$$\cos x = \pm\frac{\sqrt{2}}{2}$$

$$x = \frac{\pi}{4}, \frac{3\pi}{4}, \frac{5\pi}{4}, \frac{7\pi}{4}$$

> To solve a rational equation, you should factor the numerator and denominator. If any factor appears in both the numerator AND the denominator (like cos x in this problem), then any x-value that makes that factor equal 0 will also make the denominator equal 0. Therefore, those x-values can't be solutions to the equation.

11.11 Identify the general solution to the equation: $\dfrac{3\sin^2 x + 2\sin x - 5}{\sin^2 x + \sin x - 2} = 0$.

Factor the quadratic expressions in the numerator and denominator.

$$\frac{(3\sin x + 5)(\sin x - 1)}{(\sin x + 2)(\sin x - 1)} = 0$$

The numerator contains factors $(3\sin x + 5)$ and $(\sin x - 1)$, so the x-values that cause either of these factors to equal 0 are possible solutions to the equation. However, the factor $(\sin x - 1)$ also appears in the denominator, so any solutions to the equation $\sin x - 1 = 0$ will cause both the numerator and denominator to equal 0. Discard this factor and set the remaining factor of the numerator equal to 0.

$$3\sin x + 5 = 0$$

$$3\sin x = -5$$

$$\sin x = -\frac{5}{3}$$

There are no real number solutions to this equation, because the range of $\sin x$ is $-1 \leq \sin x \leq 1$. Hence, there are no real solutions to the original rational equation.

> You can verify this visually. Graph (3*(sin(x))^2+2*sin(x)-5)/ ((sin x)^2+sin(x)-2) on your graphing calculator. It doesn't intersect the x-axis, so there are no solutions.

11.12 Identify the general solution to the equation: $\dfrac{\tan x - 1}{\tan x + 1} = -2$.

The technique used in Problems 11.7–11.11 to solve rational equations is only applicable when the fraction is equal to 0. Add 2 to both sides of the equation and then combine the terms left of the equal sign into a single fraction.

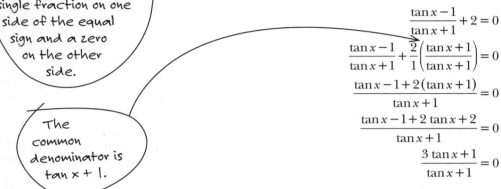

$$\frac{\tan x - 1}{\tan x + 1} + 2 = 0$$

$$\frac{\tan x - 1}{\tan x + 1} + \frac{2}{1}\left(\frac{\tan x + 1}{\tan x + 1}\right) = 0$$

$$\frac{\tan x - 1 + 2(\tan x + 1)}{\tan x + 1} = 0$$

$$\frac{\tan x - 1 + 2\tan x + 2}{\tan x + 1} = 0$$

$$\frac{3\tan x + 1}{\tan x + 1} = 0$$

The fraction is equal to 0 when its numerator is equal to 0.

$$3\tan x + 1 = 0$$

$$\tan x = -\frac{1}{3}$$

$$x = \arctan\left(-\frac{1}{3}\right)$$

To construct the general solution, add multiples of π, the period of tangent.

$$x = \arctan\left(-\frac{1}{3}\right) + k\pi, \text{ such that } k \text{ is an integer}$$

11.13 Identify the solutions to the below equation on the interval $0 \le x < 2\pi$.

$$\frac{1}{\cos x - 1} - \frac{1}{2\cos x - 1} = -\frac{1}{2\cos^2 x - 3\cos x + 1}$$

The least common denominator of this equation is $(\cos x - 1)$
$(2\cos x - 1) = 2\cos^2 x - 3\cos x + 1$. Express each fraction in terms of the least common denominator.

$$\left(\frac{1}{\cos x - 1}\right)\left(\frac{2\cos x - 1}{2\cos x - 1}\right) - \left(\frac{1}{2\cos x - 1}\right)\left(\frac{\cos x - 1}{\cos x - 1}\right) = -\frac{1}{2\cos^2 x - 3\cos x + 1}$$

$$\frac{2\cos x - 1}{2\cos^2 x - 3\cos x + 1} + \frac{-1(\cos x - 1)}{2\cos^2 x - 3\cos x + 1} = -\frac{1}{2\cos^2 x - 3\cos x + 1}$$

$$\frac{2\cos x - 1 - \cos x + 1}{2\cos^2 x - 3\cos x + 1} = -\frac{1}{2\cos^2 x - 3\cos x + 1}$$

$$\frac{\cos x}{2\cos^2 x - 3\cos x + 1} = -\frac{1}{2\cos^2 x - 3\cos x + 1}$$

To solve a rational equation, you need a single fraction on one side of the equal sign and a zero on the other side.

The common denominator is tan x + 1.

Add $1/(2\cos^2 x - 3\cos x + 1)$ to both sides of the equation.

$$\frac{\cos x}{2\cos^2 x - 3\cos x + 1} + \frac{1}{2\cos^2 x - 3\cos x + 1} = 0$$

$$\frac{\cos x + 1}{2\cos^2 x - 3\cos x + 1} = 0$$

Set the expression in the numerator equal to 0 and solve for x. Note that the factor $(\cos x + 1)$ does not appear in the denominator, because $2\cos^2 x - 3\cos x + 1 = (\cos x - 1)(2\cos x - 1)$. Therefore, solutions to the equation $\cos x + 1 = 0$ are also the solutions to the original equation stated in the problem.

$$\cos x + 1 = 0$$

$$\cos x = -1$$

$$x = \pi$$

Pythagorean Identities

Convert one trig function into another

Note: In Problems 11.14–11.15, you solve the equation $\sin^2 x = \cos^2 x$ using two different techniques. In each problem, identify the solutions on the interval $0 \leq x < 2\pi$.

11.14 Rewrite the equation in terms of $\tan^2 x$ and then solve the equation for x.

Divide both sides of the equation by $\cos^2 x$, which creates the quotient $\sin^2 x / \cos^2 x = \tan^2 x$ left of the equal sign. Solve that equation for x.

$$\frac{\sin^2 x}{\cos^2 x} = \frac{\cos^2 x}{\cos^2 x}$$

$$\tan^2 x = 1$$

$$\sqrt{\tan^2 x} = \pm\sqrt{1}$$

$$\tan x = \pm 1$$

The tangent function is equal to ± 1 when the numerator and denominator are either equal or opposite values.

$$x = \frac{\pi}{4}, \frac{3\pi}{4}, \frac{5\pi}{4}, \frac{7\pi}{4}$$

Note: In Problems 11.14–11.15, you solve the equation $\sin^2 x = \cos^2 x$ using two different techniques. In each problem, identify the solutions on the interval $0 \leq x < 2\pi$.

11.15 Verify your solution to Problem 11.14 by rewriting the equation in terms of $\sin^2 x$ and then solving for x.

According to a Pythagorean identity, $\cos^2 x + \sin^2 x = 1$. If you subtract $\sin^2 x$ from both sides of the identity, the result is $\cos^2 x = 1 - \sin^2 x$. Apply this modified Pythagorean identity to rewrite the equation $\sin^2 x = \cos^2 x$ in terms of sine.

$$\sin^2 x = \cos^2 x$$

$$\sin^2 x = 1 - \sin^2 x$$

Solve the equation for x and identify the solutions on the specified interval.

$$2\sin^2 x = 1$$

$$\sin^2 x = \frac{1}{2}$$

$$\sqrt{\sin^2 x} = \pm\sqrt{\frac{1}{2}}$$

$$\sin x = \pm\frac{1}{\sqrt{2}}$$

$$\sin x = \pm\frac{1}{\sqrt{2}} \cdot \frac{\sqrt{2}}{\sqrt{2}}$$

$$\sin x = \pm\frac{\sqrt{2}}{2}$$

There are four angles on the interval $0 \le x < 2\pi$ that have a sine value of $\pm\sqrt{2}/2$. They are the same angles identified as solutions in Problem 11.14.

$$x = \frac{\pi}{4}, \frac{3\pi}{4}, \frac{5\pi}{4}, \frac{7\pi}{4}$$

11.16 Solve the equation $\sin^2 x - \cos x = 1$ for x and identify the solutions on the interval $0 \le x < 2\pi$.

If you subtract $\cos^2 x$ from both sides of the Pythagorean identity $\cos^2 x + \sin^2 x = 1$, the result is $\sin^2 x = 1 - \cos^2 x$. Apply this modified identity to eliminate $\sin^2 x$ from this problem, effectively expressing the equation in terms of a single trigonometric expression, cosine.

$$\sin^2 x - \cos x = 1$$

$$\left(1 - \cos^2 x\right) - \cos x = 1$$

$$1 - \cos^2 x - \cos x = 1$$

$$0 = 1 - 1 + \cos^2 x + \cos x$$

$$0 = \cos^2 x + \cos x$$

If you add $\cos^2 x$ and $\cos x$ to both sides of the equation, you don't have to deal with as many negative signs.

Factor the quadratic expression.

$$0 = \cos x (\cos x + 1)$$

Apply the zero-product property and identify the solutions on the interval $0 \le x < 2\pi$.

$$\cos x = 0 \qquad\qquad \cos x + 1 = 0$$

$$x = \frac{\pi}{2}, \frac{3\pi}{2} \qquad\qquad \cos x = -1$$

$$x = \pi$$

The solutions to the equation are $x = \pi/2$, π, and $3\pi/2$.

11.17 Identify the solutions to the equation $3 \sin x - 2 \cos^2 x = -3$ for x and identify the solutions on the interval $0 \le x < 2\pi$.

Apply the technique modeled in Problem 11.15, applying the modified Pythagorean identity $\cos^2 x = 1 - \sin^2 x$ to write the equation in terms of a single trigonometric expression.

$$3 \sin x - 2 \cos^2 x = -3$$
$$3 \sin x - 2\left(1 - \sin^2 x\right) = -3$$
$$3 \sin x - 2(1) - 2\left(-\sin^2 x\right) = -3$$
$$3 \sin x - 2 + 2 \sin^2 x + 3 = 0$$
$$2 \sin^2 x + 3 \sin x + 1 = 0$$

Factor the quadratic expression.

$$(2 \sin x + 1)(\sin x + 1) = 0$$

Apply the zero-product property and identify the solutions to each equation on the interval $0 \le x < 2\pi$.

$$2 \sin x + 1 = 0 \qquad\qquad \sin x + 1 = 0$$
$$\sin x = -\frac{1}{2} \qquad\qquad \sin x = -1$$
$$x = \frac{7\pi}{6}, \frac{11\pi}{6} \qquad\qquad x = \frac{3\pi}{2}$$

The solutions to the equation are $x = 7\pi/6$, $3\pi/2$, and $11\pi/6$.

11.18 Identify the solutions to the equation $3 \tan^2 x + 5 \sec x + 1 = 0$ on the interval $0 \le x < 2\pi$.

According to a Pythagorean identity, $1 + \tan^2 x = \sec^2 x$. Subtract 1 from both sides of that identity to produce the equally valid statement $\tan^2 x = \sec^2 x - 1$. Use this modified identity to rewrite the problem in terms of a single trigonometric identity.

$$3 \tan^2 x + 5 \sec x + 1 = 0$$
$$3\left(\sec^2 x - 1\right) + 5 \sec x + 1 = 0$$
$$3 \sec^2 x - 3 + 5 \sec x + 1 = 0$$
$$3 \sec^2 x + 5 \sec x - 2 = 0$$
$$\left(3 \sec x - 1\right)(\sec x + 2) = 0$$

Apply the zero-product property and express the secant equations in terms of cosine.

$$3\sec x - 1 = 0 \qquad \sec x + 2 = 0$$
$$\sec x = \frac{1}{3} \qquad \sec x = -2$$
$$\cos x = 3 \qquad \cos x = -\frac{1}{2}$$

Note that $-1 \le \cos x \le 1$, so there are no values of x for which $\cos x = 3$. Thus, that equation may be discarded. There are two angles on the interval $0 \le x < 2\pi$ that have a cosine value of $-1/2$; these are also the solutions to the original equation.

$$x = \frac{2\pi}{3}, \frac{4\pi}{3}$$

11.19 Identify the solutions to the below equation on the interval $0 \le x < 2\pi$.

$$\cot^3 x - \cot x + \csc^2 x - 2 = 0$$

Recall that $1 + \cot^2 x = \csc^2 x$. Apply this trigonometric identity to rewrite the equation in terms of a single trigonometric expression, $\cot x$.

$$\cot^3 x - \cot x + \csc^2 x - 2 = 0$$
$$\cot^3 x - \cot x + \left(1 + \cot^2 x\right) - 2 = 0$$
$$\cot^3 x + \cot^2 x - \cot x - 1 = 0$$

> Factor (cot x + 1) out of both clumps, leaving cot²x and –1 behind, like in Problems 10.22–10.23.

Factor the expression left of the equal sign by grouping.

$$\cot^2 x(\cot x + 1) - 1(\cot x + 1) = 0$$
$$(\cot x + 1)\left[\cot^2 x - 1\right] = 0$$
$$(\cot x + 1)(\cot x + 1)(\cot x - 1) = 0$$

Note that the factor $(\cot x + 1)$ is repeated. Apply the zero-product property.

$$\cot x + 1 = 0 \qquad \cot x - 1 = 0$$
$$\cot x = -1 \qquad \cot x = 1$$
$$\tan x = -1 \qquad \tan x = 1$$

> In each of these steps, you take the reciprocals of both sides of the equations. The reciprocal of cot x is tan x, the reciprocal of –1 is –1, and the reciprocal of 1 is 1.

There are four angles on the interval $0 \le x < 2\pi$ that have a tangent value of ± 1.

$$x = \frac{\pi}{4}, \frac{3\pi}{4}, \frac{5\pi}{4}, \frac{7\pi}{4}$$

11.20 Identify the general solution to the equation: $\sin^4 x + 4\cos^2 x = 0$.

Note that $(\sin^2 x)^2 = \sin^4 x$. Apply the modified Pythagorean identity $\sin^2 x = 1 - \cos^2 x$ to express the equation in terms of a single trigonometric function.

$$\sin^4 x + 4\cos^2 x = 0$$
$$\left(\sin^2 x\right)^2 + 4\cos^2 x = 0$$
$$\left(1 - \cos^2 x\right)^2 + 4\cos^2 x = 0$$
$$\left(1 - \cos^2 x\right)\left(1 - \cos^2 x\right) + 4\cos^2 x = 0$$
$$1 - 2\cos^2 x + \cos^4 x + 4\cos^2 x = 0$$
$$\cos^4 x + 2\cos^2 x + 1 = 0$$

The expression left of the equal sign is a perfect square—it is the product of two equal factors.

$$\left(\cos^2 x + 1\right)\left(\cos^2 x + 1\right) = 0$$

$$\left(\cos^2 x + 1\right)^2 = 0$$

$$\sqrt{\left(\cos^2 x + 1\right)^2} = \pm\sqrt{0}$$

$$\cos^2 x + 1 = 0$$

$$\cos^2 x = -1$$

There are no real solutions to the equation $\cos^2 x = -1$, because no squared value can be equal to a negative number. Thus, there are no real solutions to the original equation, $\sin^4 x + 4\cos^2 x = 0$.

11.21 Identify the general solution to the equation: $3\sec^2\theta + 4\tan\theta = 5(\sec^2\theta - 1)$.

Combine like terms in the equation, setting one side equal to 0.

$$3\sec^2\theta + 4\tan\theta = 5\sec^2\theta - 5$$

$$0 = 5\sec^2\theta - 3\sec^2\theta - 4\tan\theta - 5$$

$$0 = 2\sec^2\theta - 4\tan\theta - 5$$

Apply the trigonometric identity $\sec^2\theta = 1 + \tan^2\theta$ to rewrite the equation in terms of a single trigonometric expression.

$$0 = 2\left(1 + \tan^2\theta\right) - 4\tan\theta - 5$$

$$0 = 2 + 2\tan^2\theta - 4\tan\theta - 5$$

$$0 = 2\tan^2\theta - 4\tan\theta - 3$$

The quadratic expression right of the equal sign cannot be factored, so apply the quadratic formula. This technique is demonstrated in Problems 10.27–10.32.

$$\tan\theta = \frac{-(-4) \pm \sqrt{(-4)^2 - 4(2)(-3)}}{2(2)}$$

$$\tan\theta = \frac{4 \pm \sqrt{40}}{4}$$

$$\tan\theta = \frac{4 \pm 2\sqrt{10}}{4}$$

Reduce the fraction to lowest terms and solve for θ.

$$\tan\theta = \frac{\cancel{2}\left(2 \pm \sqrt{10}\right)}{\cancel{2} \cdot 2}$$

$$\tan\theta = \frac{2 \pm \sqrt{10}}{2}$$

$$\tan\theta = \frac{2}{2} \pm \frac{\sqrt{10}}{2}$$

$$\theta = \arctan\left(1 \pm \frac{\sqrt{10}}{2}\right)$$

To construct the general solution, add multiples of π (the period of tangent).

$$\theta = \arctan\left(1 - \frac{\sqrt{10}}{2}\right) + k\pi, \ \arctan\left(1 + \frac{\sqrt{10}}{2}\right) + k\pi, \text{ where } k \text{ is an integer}$$

11.22 Identify the general solution to the equation: $\dfrac{4\cos x}{\sin x} + \dfrac{3}{\sin^2 x} = \dfrac{6\cos x \sin x}{\cos^2 x - 1}$.

According to a Pythagorean identity, $\cos^2 x + \sin^2 x = 1$. Subtract $\sin^2 x$ and 1 from both sides of the equation to produce the modified equation $\cos^2 x - 1 = -\sin^2 x$. Apply this modified identity to rewrite the denominator $\cos^2 x - 1$ in the given equation.

$$\frac{4\cos x}{\sin x} + \frac{3}{\sin^2 x} = \frac{6\cos x \sin x}{-\sin^2 x}$$

$$\frac{4\cos x}{\sin x} + \frac{3}{\sin^2 x} = -\frac{6\cos x \ \cancel{\sin x}}{\sin x \cdot \cancel{\sin x}}$$

$$\frac{4\cos x}{\sin x} + \frac{3}{\sin^2 x} = -\frac{6\cos x}{\sin x}$$

Note that $\cos x/\sin x = \cot x$ and $1/\sin^2 x = \csc^2 x$.

$$4\cot x + 3\csc^2 x = -6\cot x$$

According to a Pythagorean identity, $1 + \cot^2 x = \csc^2 x$.

$$4\cot x + 3\left(1 + \cot^2 x\right) = -6\cot x$$

$$4\cot x + 3 + 3\cot^2 x + 6\cot x = 0$$

$$3\cot^2 x + 4\cot x + 6\cot x + 3 = 0$$

$$3\cot^2 x + 10\cot x + 3 = 0$$

Factor or apply the quadratic formula to solve the quadratic equation. In the solution below, the expression $3\cot^2 x + 10\cot x + 3$ is factored into the product $(3\cot x + 1)(\cot x + 3)$ and the zero-product property is applied.

$3\cot x + 1 = 0$	$\cot x + 3 = 0$
$\cot x = -\dfrac{1}{3}$	$\cot x = -3$
$\tan x = -3$	$\tan x = -\dfrac{1}{3}$
$x = \arctan(-3)$	$x = \arctan\left(-\dfrac{1}{3}\right)$

The general solution to the equation is $x = \arctan(-3) + k\pi$, $\arctan(-1/3) + k\pi$, such that k is an integer.

Squaring

With squares come Pythagorean identities

Note: Problems 11.23–11.24 refer to the equation cos x = sin x + 1.

11.23 Square both sides of the equation and apply a Pythagorean identity to identify three potential solutions on the interval $0 \le x < 2\pi$.

Square both sides of the equation, as directed by the problem. Note that $(\sin x + 1)^2 \ne \sin^2 x + 1^2$.

$$\cos x = \sin x + 1$$
$$(\cos x)^2 = (\sin x + 1)^2$$
$$\cos^2 x = (\sin x + 1)(\sin x + 1)$$
$$\cos^2 x = \sin^2 x + 2\sin x + 1$$

This equation contains two trigonometric expressions. Apply the technique demonstrated in Problems 11.15–11.22 to rewrite the equation in terms of a single trigonometric expression.

$$(1 - \sin^2 x) = \sin^2 x + 2\sin x + 1$$
$$0 = (\sin^2 x + \sin^2 x) + 2\sin x + (1 - 1)$$
$$0 = 2\sin^2 x + 2\sin x$$

Factor the quadratic expression.

$$0 = 2\sin x(\sin x + 1)$$

Apply the zero-product property to identify the solutions on the interval $0 \le x < 2\pi$.

$$2\sin x = 0 \qquad \sin x + 1 = 0$$
$$\sin x = \frac{0}{2} \qquad \sin x = -1$$
$$\sin x = 0 \qquad x = \frac{3\pi}{2}$$
$$x = 0, \pi$$

The three potential solutions to this equation are $x = 0$, $x = \pi$, and $x = 3\pi/2$.

Note: Problems 11.23–11.24 refer to the equation cos x = sin x + 1.

11.24 Test each of the potential solutions you calculated in Problem 11.23 to identify the actual solutions to the equation on the interval $0 \le x < 2\pi$.

According to Problem 11.23, the three potential solutions to the equation are $x = 0$, $x = \pi$, and $x = 3\pi/2$. Substitute each into the original equation ($\cos x = \sin x + 1$) to determine which, if any, satisfy the equation.

When you square both sides of an equation, you run the risk of introducing extra "solutions" that are actually false. That's true here, and you'll pluck out that false "solution" in Problem 11.24.

In other words, use a Pythagorean identity. Look for the trig expression that is NOT squared ($2\sin x$ in this equation) and write everything in terms of that trig expression. To get rid of $\cos^2 x$ here, you use the modified Pythagorean identity $\cos^2 x = 1 - \sin^2 x$.

$x = 0$	$x = \pi$	$x = \dfrac{3\pi}{2}$
$\cos x = \sin x + 1$	$\cos x = \sin x + 1$	$\cos x = \sin x + 1$
$\cos 0 = \sin 0 + 1$	$\cos \pi = \sin \pi + 1$	$\cos \dfrac{3\pi}{2} = \sin \dfrac{3\pi}{2} + 1$
$1 = 0 + 1$	$-1 = 0 + 1$	$0 = -1 + 1$
$1 = 1$	$-1 \neq 1$	$0 = 0$
True	***False***	***True***

The potential solution $x = \pi$ does not satisfy the equation. You conclude that the equation $\cos x = \sin x + 1$ has two solutions on the interval $0 \leq x < 2\pi$: $x = 0$ and $x = 3\pi/2$.

> **Note: Problems 11.25–11.26 refer to the equation sec x – 1 – tan x = tan x.**

11.25 Square both sides of the equation and apply a Pythagorean identity to identify potential solutions on the interval $0 \leq x < \pi$.

Before you square both sides of the equation, add $\tan x$ to both sides and combine like terms.

$$\sec x - 1 = 2\tan x$$

Now square both sides of the equation, noting that $(\sec x - 1)^2$ is a simpler expression than $(\sec x - 1 - \tan x)^2$, the left side of the original equation.

$$(\sec x - 1)^2 = (2\tan x)^2$$
$$(\sec x - 1)(\sec x - 1) = 2^2 \cdot \tan^2 x$$
$$\sec^2 x - 2\sec x + 1 = 4\tan^2 x$$

Apply the modified trigonometric identity $\tan^2 x = \sec^2 x - 1$ to rewrite the equation in terms of secant. Combine like terms and factor.

$$\sec^2 x - 2\sec x + 1 = 4\left(\sec^2 x - 1\right)$$
$$\sec^2 x - 2\sec x + 1 = 4\sec^2 x - 4$$
$$0 = 3\sec^2 x + 2\sec x - 5$$
$$0 = (3\sec x + 5)(\sec x - 1)$$

Apply the zero-product property to identify potential solutions to the equation. Notice that the problem directs you to identify solutions on the interval $0 \leq x < \pi$, which is the range of arccosine. Thus, you should apply the arccosine function to identify the exact solution for both of the following equations.

$$3\sec x + 5 = 0 \qquad\qquad \sec x - 1 = 0$$
$$\sec x = -\frac{5}{3} \qquad\qquad \sec x = 1$$
$$\cos x = -\frac{3}{5} \qquad\qquad \cos x = 1$$
$$x = \arccos\left(-\frac{3}{5}\right) \qquad\qquad x = \arccos 1$$
$$\qquad\qquad x = 0$$

The potential solutions to the equation $\sec x - 1 - \tan x = \tan x$ are $x = 0$ and $x = \arccos(-3/5)$.

Note: Problems 11.25–11.26 refer to the equation sec x – 1 – tan x = tan x.

11.26 Test each of the potential solutions you calculated in Problem 11.25 to identify the actual solution(s) to the equation on the interval $0 \leq x < \pi$.

According to Problem 11.25, the two potential solutions are $x = 0$ and $x = \arccos(-3/5)$. Begin by testing the solution $x = 0$.

$$\sec 0 - 1 - \tan 0 = \tan 0$$

Note that sec 0 does not exist. Secant and cosine are reciprocal identities, and sec 0 = cos (1/0) is undefined. As further evidence that sec 0 does not exist, refer to the graph of $y = \sec x$, which has a vertical asymptote at $x = 0$.

Now determine whether or not $x = \arccos(-3/5)$ is a solution to the equation, noting that secant and cosine are reciprocal functions; sec $[\arccos(-3/5)] = -5/3$. Furthermore, tan $[\arccos(-3/5)] = -4/3$.

$$\sec\left[\arccos\left(-\frac{3}{5}\right)\right] - 1 - \tan\left[\arccos\left(-\frac{3}{5}\right)\right] = \tan\left[\arccos\left(-\frac{3}{5}\right)\right]$$

$$-\frac{5}{3} - 1 - \left(-\frac{4}{3}\right) = -\frac{4}{3}$$

$$-\frac{5}{3} - 1\left(\frac{3}{3}\right) + \frac{4}{3} = -\frac{4}{3}$$

$$\frac{-5 - 3 + 4}{3} = -\frac{4}{3}$$

$$-\frac{4}{3} = -\frac{4}{3}$$

The only solution on the interval $0 \leq x < \pi$ that satisfies the equation sec x – 1 – tan x = tan x is $x = \arccos(-3/5)$.

> If x = arccos (–3/5), then x = arcsec (–5/3). Because secant and arcsecant are inverse functions, they cancel each other out:
> sec [arcsec (–5/3)] = –5/3.

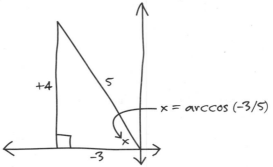

If x = arccos (–3/5), then x is an angle in the second quadrant. Cosine = adjacent/hypotenuse, so you can draw a reference triangle with a horizontal signed length of –3 and a hypotenuse of length 5. That means the vertical side has length +4.

If x = arccos (–3/5), then tan x = opposite/adjacent = +4/–3.

Note: Problems 11.27–11.28 present two different ways to solve the equation cot x = csc x. In each problem, identify all solutions on the interval $0 \leq x < 2\pi$.

11.27 Rewrite the equation in terms of sine and cosine, and solve the resulting rational equation.

Recall that $\cot x = \cos x/\sin x$ and $\csc x = 1/\sin x$.

$$\cot x = \csc x$$

$$\frac{\cos x}{\sin x} = \frac{1}{\sin x}$$

Subtract $1/\sin x$ from both sides of the equation and combine the fractions, noting that they have common denominator sin x.

$$\frac{\cos x}{\sin x} - \frac{1}{\sin x} = 0$$

$$\frac{\cos x - 1}{\sin x} = 0$$

A fraction is equal to 0 when its numerator is equal to 0.

$$\cos x - 1 = 0$$

$$\cos x = 1$$

$$x = 0$$

The denominators (sin x) equal 0 when x = 0, and you can't divide by 0.

Although $x = 0$ is a potential solution, the original equation contains terms cos x/sin x and 1/sin x, and both of the terms are undefined when $x = 0$. Thus, there are no real solutions to this equation.

Note: Problems 11.27–11.28 present two different ways to solve the equation cot x = csc x. In each problem, identify all solutions on the interval $0 \leq x < 2\pi$.

11.28 Verify your answer to Problem 11.27 by squaring both sides of the equation and applying a Pythagorean identity.

Square both sides of the equation and recall that $1 + \cot^2 x = \csc^2 x$.

$$\cot^2 x = \csc^2 x$$

$$\cot^2 x = \left(1 + \cot^2 x\right)$$

Subtract $\cot^2 x$ from both sides of the equation.

$$0 = 1$$

The statement 0 = 1 is untrue, so there are no solutions to this—or the original—equation. This verifies the answer to Problem 11.27, which concluded that no real solutions exist.

Applying Trigonometric Identities
Other than the Pythagorean identities

11.29 Identify the solutions to the equation $\sin x = \sin 2x$ on the interval $0 \le x < 2\pi$.

According to Problem 8.1, $\sin 2x = 2 \sin x \cos x$. Use this identity to rewrite the right side of the equation.

$$\sin x = \sin 2x$$
$$\sin x = 2 \sin x \cos x$$

Subtract one of the terms from both sides of the equation. In the following solution, $\sin x$ is subtracted from both sides.

$$0 = 2 \sin x \cos x - \sin x$$

The terms on the right side of the equation have greatest common factor $\sin x$.

$$0 = \sin x (2 \cos x - 1)$$

Apply the zero-product property.

$$\sin x = 0 \qquad\qquad 2 \cos x - 1 = 0$$
$$x = 0, \pi \qquad\qquad 2 \cos x = 1$$
$$\cos x = \frac{1}{2}$$
$$x = \frac{\pi}{3}, \frac{5\pi}{3}$$

The equation $\sin x = \sin 2x$ has four solutions on the interval $0 \le x < 2\pi$: 0, $\pi/3$, π, and $5\pi/3$.

11.30 Identify the solutions to the equation $\cos 2x - 2 \sin^2 x = 0$ on the interval $0 \le x < 2\pi$.

According to Problems 8.9–8.10, there are three ways to rewrite $\cos 2x$ as a single-angle expression. Apply the identity $\cos 2x = 1 - 2 \sin^2 x$ to express the equation in terms of sine.

$$\cos 2x - 2 \sin^2 x = 0$$
$$\left(1 - 2 \sin^2 x\right) - 2 \sin^2 x = 0$$
$$1 - 4 \sin^2 x = 0$$

Isolate $\sin^2 x$ on the left side of the equation.

$$-4 \sin^2 x = -1$$
$$\sin^2 x = \frac{1}{4}$$

Take the square root of both sides of the equation.

$$\sqrt{\sin^2 x} = \pm\sqrt{\frac{1}{4}}$$

$$\sin x = \pm\frac{1}{2}$$

There are four angles on the interval $0 \le x < 2\pi$ that have a sine value of +1/2 or −1/2.

$$x = \frac{\pi}{6}, \frac{5\pi}{6}, \frac{7\pi}{6}, \frac{11\pi}{6}$$

11.31 Identify the solutions to the equation $(\tan 2x)(\sin 2x) = 0$ on the interval $0 \le x < 2\pi$.

Apply the double-angle identities for tangent and sine, as presented in Problems 8.39 and 8.1, respectively.

$$(\tan 2x)(\sin 2x) = 0$$

$$\left(\frac{2\tan x}{1-\tan^2 x}\right)(2\sin x \cos x) = 0$$

$$(2\tan x)\left(\frac{1}{1-\tan^2 x}\right)(2\sin x \cos x) = 0$$

In order to simplify the product on the left side of the equation, express $2\tan x$ as the quotient $2(\sin x/\cos x)$.

$$2 \cdot \frac{\sin x}{\cancel{\cos x}}(2\sin x \,\cancel{\cos x})\left(\frac{1}{1-\tan^2 x}\right) = 0$$

$$4\sin^2 x\left(\frac{1}{1-\tan^2 x}\right) = 0$$

$$\frac{4\sin^2 x}{1-\tan^2 x} = 0$$

The fraction is equal to 0 when its numerator is equal to 0.

$$4\sin^2 x = 0$$

$$\sin^2 x = \frac{0}{4}$$

$$\sin^2 x = 0$$

$$\sqrt{\sin^2 x} = \pm\sqrt{0}$$

$$\sin x = 0$$

$$x = 0, \pi$$

11.32 Identify the solutions to the equation $4\cos^2(x-\pi) = 3$ on the interval $0 \le x < 2\pi$.

Isolate the trigonometric expression left of the equal sign.

$$\cos^2(x-\pi) = \frac{3}{4}$$

$$\sqrt{\cos^2(x-\pi)} = \pm\sqrt{\frac{3}{4}}$$

$$\cos(x-\pi) = \pm\frac{\sqrt{3}}{2}$$

Apply the difference formula for cosine (as defined in Problem 7.37) to simplify the left side of the equation.

$$\cos x \cdot \cos\pi + \sin x \cdot \sin\pi = \pm\frac{\sqrt{3}}{2}$$

$$\cos x(-1) + \sin x(0) = \pm\frac{\sqrt{3}}{2}$$

$$-\cos x = \pm\frac{\sqrt{3}}{2}$$

$$\cos x = \pm\frac{\sqrt{3}}{2}$$

$$x = \frac{\pi}{6}, \frac{5\pi}{6}, \frac{7\pi}{6}, \frac{11\pi}{6}$$

If you divide both sides of this equation by –1, you get $\cos x = \pm\sqrt{3}/2$. If you divide the positive or negative version of a number by –1, you get the negative or positive version. In other words, you end up with the same two numbers.

11.33 Identify the solutions to the equation $\sin\left(x - \frac{\pi}{3}\right) = 0$ on the interval $0 \le x < 2\pi$.

According to Problem 7.34, $\sin(x-y) = \sin x \cos y - \cos x \sin y$. In this problem, $y = \pi/3$.

$$\sin\left(x - \frac{\pi}{3}\right) = 0$$

$$\sin x \cdot \cos\frac{\pi}{3} - \cos x \cdot \sin\frac{\pi}{3} = 0$$

$$\sin x\left(\frac{1}{2}\right) - \cos x\left(\frac{\sqrt{3}}{2}\right) = 0$$

$$\frac{1}{2}\sin x = \frac{\sqrt{3}}{2}\cos x$$

Square both sides of the equation.

$$\frac{1}{4}\sin^2 x = \frac{3}{4}\cos^2 x$$

Rewrite the equation in terms of a single trigonometric identity, noting that $\cos^2 x = 1 - \sin^2 x$.

You could also use the identity $\sin^2 x = 1 - \cos^2 x$ to rewrite the equation in terms of cosine.

$$\frac{1}{4}\sin^2 x = \frac{3}{4}\left(1 - \sin^2 x\right)$$

$$\frac{1}{4}\sin^2 x = \frac{3}{4} - \frac{3}{4}\sin^2 x$$

$$\frac{1}{4}\sin^2 x + \frac{3}{4}\sin^2 x = \frac{3}{4}$$

$$1\sin^2 x = \frac{3}{4}$$

Take the square root of both sides of the equation and solve for x.

$$\sqrt{\sin^2 x} = \pm\sqrt{\frac{3}{4}}$$

$$\sin x = \pm\frac{\sqrt{3}}{2}$$

$$x = \frac{\pi}{3}, \frac{2\pi}{3}, \frac{4\pi}{3}, \frac{5\pi}{3}$$

11.34 Identify the solutions to the equation $\tan\dfrac{\theta}{2} - 1 = 0$ on the interval $0 \le \theta < 2\pi$.

Apply the half-angle formula for tangent (presented in Problem 8.44) to express $\tan(\theta/2)$ in terms of single-angle sine and cosine expressions. In the solution below, $\tan(\theta/2) = (1 - \cos\theta)/\sin\theta$ is used rather than $\tan(\theta/2) = \sin\theta/(1 + \cos\theta)$, because the former contains a single term in the denominator and the latter contains two terms.

$$\tan\frac{\theta}{2} - 1 = 0$$

$$\frac{1 - \cos\theta}{\sin\theta} - 1 = 0$$

This only matters because you will need to write things in terms of a common denominator, and that's easier to do when the common denominator only has one term in it.

Simplify the expression on the left side of the equation, using least common denominator $\sin\theta$.

$$\frac{1 - \cos\theta}{\sin\theta} - 1\left(\frac{\sin\theta}{\sin\theta}\right) = 0$$

$$\frac{1 - \cos\theta}{\sin\theta} - \frac{\sin\theta}{\sin\theta} = 0$$

$$\frac{1 - \cos\theta - \sin\theta}{\sin\theta} = 0$$

The fraction is equal to 0 when its numerator is equal to 0 and its denominator is not.

$$1 - \cos\theta - \sin\theta = 0$$

$$1 - \cos\theta = \sin\theta$$

$$(1 - \cos\theta)^2 = (\sin\theta)^2$$

$$1 - 2\cos\theta + \cos^2\theta = \sin^2\theta$$

Apply the Pythagorean identity $\sin^2\theta = 1 - \cos^2\theta$ to rewrite the equation in terms of cosine.

$$1 - 2\cos\theta + \cos^2\theta = 1 - \cos^2\theta$$
$$2\cos^2\theta - 2\cos\theta = 0$$

Divide each of the terms by 2 and factor.

$$\frac{\cancel{2}\cos^2\theta}{\cancel{2}} - \frac{\cancel{2}\cos\theta}{\cancel{2}} = \frac{0}{2}$$
$$\cos^2\theta - \cos\theta = 0$$
$$\cos\theta(\cos\theta - 1) = 0$$

Apply the zero-product property.

$$\cos\theta = 0 \qquad\qquad \cos\theta - 1 = 0$$
$$\theta = \frac{\pi}{2}, \frac{3\pi}{2} \qquad\qquad \cos\theta = 1$$
$$\theta = 0$$

There are three possible solutions: $\theta = 0$, $\pi/2$, and $3\pi/2$. Substitute each into the original equation to determine whether the solutions are valid.

$x = 0$	$x = \dfrac{\pi}{2}$	$x = \dfrac{3\pi}{2}$
$\tan\left(\dfrac{0}{2}\right) - 1 = 0$	$\tan\left(\dfrac{\pi/2}{2}\right) - 1 = 0$	$\tan\left(\dfrac{3\pi/2}{2}\right) - 1 = 0$
$\tan 0 - 1 = 0$	$\tan\left(\dfrac{\pi}{4}\right) - 1 = 0$	$\tan\left(\dfrac{3\pi}{4}\right) - 1 = 0$
$0 - 1 = 0$	$\dfrac{\sin(\pi/4)}{\cos(\pi/4)} - 1 = 0$	$\dfrac{\sin(3\pi/4)}{\cos(3\pi/4)} - 1 = 0$
$-1 \ne 0$	$\dfrac{\sqrt{2}/2}{\sqrt{2}/2} - 1 = 0$	$\dfrac{\sqrt{2}/2}{-\sqrt{2}/2} - 1 = 0$
	$1 - 1 = 0$	$-1 - 1 = 0$
	$0 = 0$	$-2 \ne 0$
False	*True*	*False*

The equation has a single solution, $\theta = \pi/2$, on the interval $0 \le \theta < 2\pi$.

11.35 Identify the solutions to the equation $\tan(x - \pi) + \tan\left(x + \dfrac{\pi}{4}\right) = 0$ on the interval $-\dfrac{\pi}{2} \le x \le \dfrac{\pi}{2}$.

Apply the sum and difference formulas for tangent, presented in Problem 8.36.

$$\tan(x - \pi) + \tan\left(x + \frac{\pi}{4}\right) = 0$$
$$\frac{\tan x - \tan\pi}{1 + \tan x \cdot \tan\pi} + \frac{\tan x + \tan\pi/4}{1 - \tan x \cdot \tan\pi/4} = 0$$
$$\frac{\tan x - 0}{1 + (\tan x)(0)} + \frac{\tan x + 1}{1 - (\tan x)(1)} = 0$$
$$\frac{\tan x}{1} + \frac{\tan x + 1}{1 - \tan x} = 0$$

Combine the fractions by writing them in terms of the least common denominator, $1 - \tan x$.

$$\frac{\tan x}{1}\left(\frac{1-\tan x}{1-\tan x}\right) + \frac{\tan x + 1}{1-\tan x} = 0$$

$$\frac{\tan x(1-\tan x)}{1-\tan x} + \frac{\tan x + 1}{1-\tan x} = 0$$

$$\frac{\tan x - \tan^2 x}{1-\tan x} + \frac{\tan x + 1}{1-\tan x} = 0$$

$$\frac{-\tan^2 x + 2\tan x + 1}{1-\tan x} = 0$$

The fraction is equal to 0 when its numerator is equal to 0. Because the quadratic expression $-\tan^2 x + 2\tan x + 1$ is not factorable, you should apply the quadratic formula. Mathematics convention dictates that the coefficient of the squared term should be positive when solving a quadratic equation, so multiply each term by -1 to produce the equivalent equation $\tan^2 x - 2\tan x - 1 = 0$.

$$\tan x = \frac{-(-2) \pm \sqrt{(-2)^2 - 4(1)(-1)}}{2(1)}$$

$$\tan x = \frac{2 \pm \sqrt{4+4}}{2}$$

$$\tan x = \frac{2 \pm \sqrt{8}}{2}$$

$$\tan x = \frac{2 \pm 2\sqrt{2}}{2}$$

$$\tan x = 1 \pm \sqrt{2}$$

$$x = \arctan\left(1 \pm \sqrt{2}\right)$$

There are two solutions on the interval $-\pi/2 \le x \le \pi/2$.

$$x = \arctan\left(1-\sqrt{2}\right) \text{ and } x = \arctan\left(1+\sqrt{2}\right).$$

Added bonus: The answers will be in the interval specified by the problem ($\pi/2 \le x \le \pi/2$). You won't need to add or subtract π and calculate coterminal angles when you're done.

Chapter 12

AREA OF TRIANGLES AND SECTORS

Three-sided polygons and pieces of pie

The study of geometry includes a set of formulas used to compute the areas of common figures, including triangles, rectangles, parallelograms, trapezoids, and circles. However, each formula only applies when very specific criteria are met. For example, the most common formula used to calculate the area of a triangle, $A = (1/2)bh$, only applies if you are able to calculate the height of the rectangle—the length of the segment perpendicular to the line containing one side and extending to the opposite vertex.

As trigonometry is primarily focused on the study of triangles—and secondarily on the relationship between triangles and circles—this chapter explores additional formulas that calculate the areas of triangles. It concludes with a set of problems in which you calculate the area of a sector, a roughly triangular region of a circle formed by two radii and an arc of the circle.

This chapter is all about area. Most of it focuses on triangles. By the time you're done, you'll be able to calculate the area of a triangle given:

* Base and height
  * Two angles and one side
* Two sides and one angle
  * The lengths of all three sides

You'll also calculate the area of sectors, which are sections of a circle that look like pieces of pie. Draw two radii on a circle and then darken in the arc that connects the two points at which those radii touch the circle and you have yourself a sector.

Base and Height
Half of base times height

Note: Problems 12.1–12.3 refer to the diagram below, rectangle ABCD in the coordinate plane with length 6 and width 4.

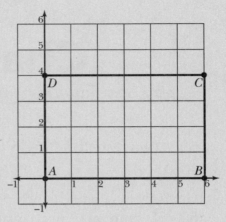

12.1 Calculate the area of the rectangle by counting the square units bounded by its sides.

The rectangle has a length of 6 and a width—or height—of 4, so it is comprised of four rows of six square units, each numbered individually in the following diagram. Thus, its area is $6 \cdot 4 = 24$ square units.

The area A of a rectangle with length l and width w is $A = lw$.

Note: Problems 12.1–12.3 refer to the diagram in Problem 12.1, a rectangle ABCD in the coordinate plane with length 6 and width 4.

12.2 Calculate the area of triangle *DAB*, based on your answer to Problem 12.1.

Draw the segment connecting points *D* and *B* to form triangle *DAB*. In the following diagram, the region bounded by triangle *DAB* is shaded. Note that it represents exactly half of the area of rectangle *ABCD*.

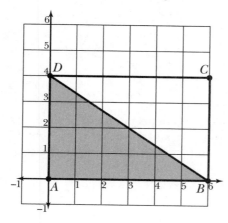

According to Problem 12.1, the area of rectangle *ABCD* is 24 square units. Therefore, the area of triangle *DAB* is 12 square units.

Note: Problems 12.1–12.3 refer to the diagram in Problem 12.1, a rectangle ABCD in the coordinate plane with length 6 and width 4.

12.3 Based on your solutions to Problems 12.1–12.2, generate the formula for the area of a triangle with two perpendicular sides of known lengths.

If a triangle has two perpendicular sides, one may be described as the base of the triangle; the other represents the height of the triangle. In Problem 12.2, you are given triangle *DAB* with perpendicular sides $\overline{DA}$ and $\overline{AB}$. Let the base of the triangle be *AB* = 6 and the height of the triangle be *AD* = 4. The area of *DAB* is half the product of the base and the height.

$$A = \frac{1}{2}bh$$
$$= \frac{1}{2}(6)(4)$$
$$= (3 \cdot 4)$$
$$= 12$$

If the base is AD = 4 and the height is AB = 6, you get the same answer. Note that the height of a triangle does not always equal the length of one of its sides. However, if two sides are perpendicular, you can automatically make one side the base and the other the height.

12.4 Calculate the area of triangle *XYZ* below.

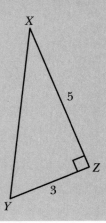

Two sides of triangle *XYZ* are perpendicular: $\overline{XZ}$ and $\overline{YZ}$. Let one of those lengths represent the base and the other the height of triangle *XYZ*. For example, $b = YZ = 3$ and $h = XZ = 5$. Substitute these values into the triangle area formula.

$$A = \frac{1}{2}bh$$
$$= \frac{1}{2}(3)(5)$$
$$= \frac{1}{2}(15)$$
$$= \frac{15}{2}$$

12.5 Calculate the area of triangle *CDE* below.

Sides $\overline{CD}$ and $\overline{DE}$ are perpendicular, so consider one of those lengths (either *CD* or *DE*) the base of the triangle and the other length the height. You are given $DE = 4$, but you must apply the Pythagorean theorem to calculate *CD*.

$$(CD)^2 + (DE)^2 = (CE)^2$$
$$(CD)^2 + (4)^2 = (6)^2$$
$$(CD)^2 + 16 = 36$$
$$(CD)^2 = 20$$
$$\sqrt{(CD)^2} = \sqrt{20}$$
$$CD = 2\sqrt{5}$$

Don't worry about adding the "±" symbol here. The lengths of the sides of a triangle are always positive (unless you're working in the coordinate plane and you have to worry about the SIGNED length).

Let *DE* represent the base of triangle *CDE* and *CD* represent the height. Apply the triangle area formula.

$$A = \frac{1}{2}bh$$

$$= \frac{1}{2}(DE)(CD)$$

$$= \frac{1}{\cancel{2}}(4)\left(\cancel{2}\sqrt{5}\right)$$

$$= 4\sqrt{5}$$

12.6 Calculate the area of triangle *JKL* below.

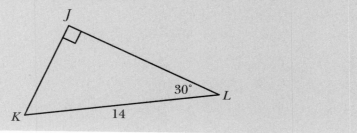

Notice that the measure of angle *L* is 30°; angle *J* is a right angle, so its measure is 90°. According to a geometric theorem, the measures of the interior angles of a triangle have a sum of 180°. Thus, the measure of angle *K* is 180° − 30° − 90° = 60°. The interior angles of the triangle are 30°, 60°, and 90°. According to a geometric theorem, all such triangles (called "30°–60°–90° triangles") have lengths that follow a specific pattern:

- The short leg, opposite the 30° angle, is exactly one-half the length of the hypotenuse.

- The long leg, opposite the 60° angle, is exactly $\sqrt{3}$ times as long as the short leg.

For more information, look at Problems 12.29–12.33 in The Humongous Book of Geometry Problems.

The hypotenuse of triangle *JKL* has length *KL* = 14, the short leg is half as long (*JK* = 7), and the long leg is $\sqrt{3}$ times as long as the short leg $\left(JL = 7\sqrt{3}\right)$. Two sides of the triangle are perpendicular, so classify one of their lengths as the base and the other as the height of triangle *JKL*.

$$A = \frac{1}{2}bh$$

$$= \frac{1}{2}(JK)(JL)$$

$$= \frac{1}{2}(7)\left(7\sqrt{3}\right)$$

$$= \frac{49\sqrt{3}}{2}$$

12.7 Calculate the area of isosceles right triangle *MNP* below.

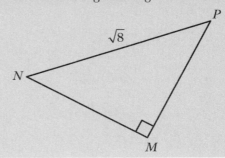

The legs of an isosceles right triangle have equal lengths; each is $1/\sqrt{2}$ times as long as the hypotenuse. In this problem, $MN = MP = \left(1/\sqrt{2}\right)(NP)$.

$$MN = MP = \left(\frac{1}{\sqrt{2}}\right)\left(\sqrt{8}\right)$$

$$= \left(\frac{1}{\sqrt{2}}\right)\left(2\sqrt{2}\right)$$

$$= 2$$

The legs of triangle *MNP* are perpendicular, so one represents the base of the triangle and the other represents the height.

$$A = \frac{1}{2}bh$$

$$= \frac{1}{2}(MN)(MP)$$

$$= \frac{1}{2}(2)(2)$$

$$= 2$$

Note: Problems 12.8–12.10 refer to the diagram below, in which WX = 10 and the measure of angle W is 60°.

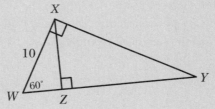

12.8 Calculate the area of triangle *WXZ*.

Note that angle *XZY* is a right angle, measuring 90°. Therefore, angle *WZX* is also a right angle. Recall that the sum of the measures of the interior angles of a triangle is 180°, so the measure of angle *WXZ* is 30°.

$$m\angle WXZ + m\angle XWZ + m\angle WZX = 180°$$
$$m\angle WXZ + 60° + 90° = 180°$$
$$m\angle WXZ + 150° = 180°$$
$$m\angle WXZ = 180° - 150°$$
$$m\angle WXZ = 30°$$

Because angles *WXZ*, *XWZ*, and *WZX* measure 30°, 60°, and 90° respectively, you can apply the technique described in Problem 12.6. Specifically, 30°–60°–90° right triangle *WXZ* has a hypotenuse of length *WX* = 10, so the short leg is half as long: *WZ* = 5. Furthermore, the long leg is $\sqrt{3}$ times as long as the short leg: $XZ = 5\sqrt{3}$.

The legs of the right triangle are perpendicular to each other, so they represent the base and height of triangle *WXZ*. Apply the triangle area formula.

$$A = \frac{1}{2}bh$$
$$= \frac{1}{2}(WZ)(XZ)$$
$$= \frac{1}{2}(5)\left(5\sqrt{3}\right)$$
$$= \frac{25\sqrt{3}}{2}$$

> This little m and the angle symbol together are read "the measure of angle." In other words, m∠WXZ is read "the measure of angle WXZ."

Note: Problems 12.8–12.10 refer to the diagram in Problem 12.8, in which WX = 10 and the measure of angle W is 60°.

12.9 Calculate the area of triangle *XYZ*.

In Problem 12.8, you apply the properties of a 30°–60°–90° triangle to calculate two lengths in the diagram: *WZ* = 5 and $XZ = 5\sqrt{3}$, as illustrated in the following diagram.

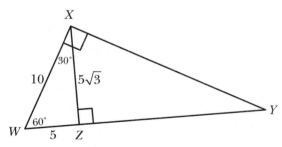

Note that two of the angles in the large triangle *WXY* are known: angle *XWY* measures 60° and angle *WXY* measures 90°. Therefore, angle *Y* measures 30°, because the measure of the three interior angles of triangle *WXY* must have a sum of 180°.

$$m\angle XWY + m\angle WXY + m\angle Y = 180°$$
$$60° + 90° + m\angle Y = 180°$$
$$m\angle Y = 180° - 150°$$
$$m\angle Y = 30°$$

Angle Y is shared by triangles WXY and XYZ. Now you know the measures of two angles in triangle XYZ: angle Y measures 30° and angle XZY measures 90°. All three of the triangles (WXZ, XYZ, and WXY) are 30°–60°–90° triangles, as illustrated in the following diagram.

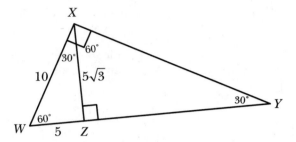

The short leg of right triangle XYZ has length $XZ = 5\sqrt{3}$; the long leg is $\sqrt{3}$ times as long.

$$YZ = \sqrt{3}\,(XZ)$$
$$= \sqrt{3}\left(5\sqrt{3}\right)$$
$$= 5\sqrt{9}$$
$$= 5 \cdot 3$$
$$= 15$$

The legs of right triangle XYZ are perpendicular, so set one length equal to the base and the other equal to the height in the triangle area formula.

$$A = \frac{1}{2}bh$$
$$= \frac{1}{2}(YZ)(XZ)$$
$$= \frac{1}{2}(15)\left(5\sqrt{3}\right)$$
$$= \frac{75\sqrt{3}}{2}$$

Note: Problems 12.8–12.10 refer to the diagram in Problem 12.8, in which WX = 10 and the measure of angle W is 60°.

12.10 Calculate the area of triangle *WXY* by applying the triangle area formula and verify your answer by adding the areas you calculated in Problems 12.8–12.9.

In Problems 12.8–12.9, you calculate the lengths of many segments within the diagram. The following illustration summarizes the calculations conducted in those two problems.

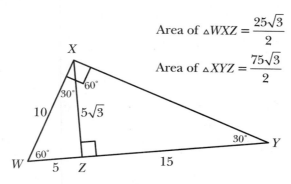

$$\text{Area of } \triangle WXZ = \frac{25\sqrt{3}}{2}$$

$$\text{Area of } \triangle XYZ = \frac{75\sqrt{3}}{2}$$

Big triangle WXY is made up of smaller triangles WXZ and XYZ. You calculated the areas of the smaller triangles in Problems 12.8 and 12.9, so if you add those two answers, you should get the area of the big triangle.

If hypotenuse $\overline{WY}$ represents the base of triangle *WXY*, then $\overline{XZ}$ represents the height of the triangle, because it is perpendicular to the base and extends to the opposite vertex, *X*. Apply the triangle area formula.

$$A = \frac{1}{2}bh$$

$$= \frac{1}{2}(WY)(XZ)$$

$$= \frac{1}{2}(5+15)\left(5\sqrt{3}\right)$$

$$= \frac{1}{2}(20)\left(5\sqrt{3}\right)$$

$$= \frac{100\sqrt{3}}{2}$$

$$= 50\sqrt{3}$$

Verify your answer by adding the areas of the triangles that comprise triangle *WXY*.

$$\text{Area}(\triangle WXY) = \text{Area}(\triangle WXZ) + \text{Area}(\triangle XYZ)$$

$$= \frac{25\sqrt{3}}{2} + \frac{75\sqrt{3}}{2}$$

$$= \frac{25\sqrt{3} + 75\sqrt{3}}{2}$$

$$= \frac{100\sqrt{3}}{2}$$

$$= 50\sqrt{3}$$

12.11 Use a scientific or graphing calculator to compute the area of triangle *LMN* below, accurate to the thousandths place.

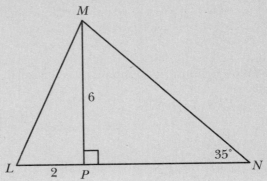

Consider $\overline{LN}$ the base of triangle *LMN* and $\overline{MP}$ the height. Note that the height is not necessarily a side of the triangle; it is simply a segment that is perpendicular to one side of the triangle and extends to the opposite vertex. In this problem, $\overline{LN} \perp \overline{MP}$ and $\overline{MP}$ extends from base $\overline{LN}$ to opposite vertex *M*.

You are given the height (*MP* = 6) and part of the base (*LP* = 2). You need to calculate the length of the other portion of the base, *NP*, in order to apply the triangle area formula. Notice that triangle *MNP* is a right triangle. Apply the tangent ratio to calculate *NP*.

> Look at Problems 2.19 and 2.21 if you need to review the tangent ratio.

$$\tan N = \frac{\text{length of leg opposite } N}{\text{length of leg adjacent to } N}$$

$$\tan 35° = \frac{MP}{NP}$$

$$\frac{\tan 35°}{1} = \frac{6}{NP}$$

Cross-multiply and solve for *NP*.

$$(\tan 35°)(NP) = (1)(6)$$

$$(\tan 35°)(NP) = 6$$

$$NP = \frac{6}{\tan 35°}$$

Calculate the base of triangle *LMN*. Do not evaluate tan 35° yet, in order to avoid possible rounding errors. Instead, wait until the final step of the problem to compute tan 35°.

$$LN = LP + PN$$

$$LN = 2 + \frac{6}{\tan 35°}$$

Apply the area formula to calculate the area of triangle *LMN*.

$$A = \frac{1}{2}bh$$

$$= \frac{1}{2}(LN)(MP)$$

$$= \frac{1}{2}\left(2 + \frac{6}{\tan 35°}\right)(6)$$

$$\approx \frac{1}{2}(10.56888804)(6)$$

$$\approx 31.707$$

> Make sure your calculator is in degrees mode, not radians mode. This chapter has a lot of degree measurements in it, because angles not on the coordinate plane are commonly measured in degrees.

12.12 Use a scientific or graphing calculator to compute the area of triangle *QRS* below, accurate to the thousandths place.

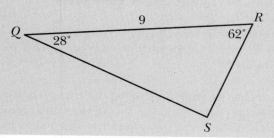

Notice that angles *Q* and *R* have a sum of 28° + 62° = 90°. Recall that the measures of the interior angles of a triangle have a sum of 180°, and use the given information to calculate the measure of angle *S*.

$$m\angle Q + m\angle R + m\angle S = 180°$$

$$28° + 62° + m\angle S = 180°$$

$$m\angle S = 180° - 90°$$

$$m\angle S = 90°$$

Because *S* is a right angle, *QRS* is a right triangle and $\overline{QS} \perp \overline{RS}$. Calculate the lengths of the legs using trigonometric ratios.

$$\cos Q = \frac{\text{length of side adjacent to } Q}{\text{length of hypotenuse}} \qquad \cos R = \frac{\text{length of side adjacent to } R}{\text{length of hypotenuse}}$$

$$\cos 28° = \frac{QS}{QR} \qquad\qquad \cos 62° = \frac{RS}{QR}$$

$$\frac{\cos 28°}{1} = \frac{QS}{9} \qquad\qquad \frac{\cos 62°}{1} = \frac{RS}{9}$$

$$9(\cos 28°) = QS \qquad\qquad 9(\cos 62°) = RS$$

Consider one leg to be the base of the right triangle; the remaining leg is the height. Apply the triangle area formula and round the product to the thousandths place, as directed by the problem.

$$A = \frac{1}{2}bh$$

$$= \frac{1}{2}(QS)(RS)$$

$$= \frac{1}{2}(9 \cdot \cos 28°)(9 \cdot \cos 62°)$$

$$\approx \frac{1}{2}(7.946528336)(4.225244065)$$

$$\approx 16.788$$

Trigonometric Area Formulas

SAS, ASA, and AAS triangle area formulas

12.13 Explain the relationship between the variables x, X, y, Y, z, and Z in the trigonometric formulas for the area of a triangle, and construct a diagram to illustrate your answer.

Many trigonometric formulas, including the trigonometric area formulas for triangles in this chapter and the laws of sines and cosines in Chapter 13, contain multiple instances of the same letter, one lowercase and one uppercase. In these instances, the lowercase letter represents the length of a side of the triangle and the uppercase letter represents the measure of the opposite angle. Corresponding letters (x and X, or y and Y, for example) are always opposite each other in the diagram.

Consider the following illustration, in which the side with length x is opposite the angle with measure X. Similarly, y is opposite Y and z is opposite Z.

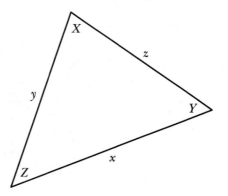

12.14 Based on your answer to Problem 12.13, describe the relationship between the sides and angles identified in the triangle area formula $A = \frac{1}{2}xy \sin Z$.

In order to apply this triangle area formula, you must know the lengths of two sides (x and y) and the measure of one angle (Z). Refer to the diagram you created in Problem 12.13, and note that the sides with lengths x and y actually form angle Z. Thus, the area formula $A = (1/2)\,xy \sin Z$ only applies when you know the lengths of two sides of a triangle and the measure of the included angle. Because of these requirements, this formula is often described as the side-angle-side (SAS) area formula.

Note: In Problems 12.15–12.16, you apply two different techniques to calculate the area of isosceles right triangle XYZ, illustrated below.

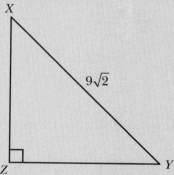

12.15 Calculate the area of isosceles right triangle XYZ by applying the formula $A = \frac{1}{2}bh$.

The problem states that XYZ is an isosceles right triangle, so its legs have the same length: $XZ = YZ$. Furthermore, isosceles right triangles are always 45°–45°–90° triangles, so the legs are equal to $1/\sqrt{2}$ times the length of the hypotenuse (and, conversely, the hypotenuse is $\sqrt{2}$ times the length of each leg). Calculate the length of the legs.

$$XZ = YZ = \frac{1}{\sqrt{2}}(XY)$$

$$XZ = YZ = \frac{1}{\sqrt{2}}\left(9\sqrt{2}\right)$$

$$XZ = YZ = 9$$

Because two sides of the triangle are perpendicular $\left(\overline{XZ} \perp \overline{YZ}\right)$, one represents the base and the other represents the height of the triangle.

$$A = \frac{1}{2}bh$$

$$A = \frac{1}{2}(YZ)(XZ)$$

$$A = \frac{1}{2}(9)(9)$$

$$A = \frac{81}{2}$$

Note: In Problems 12.15–12.16, you apply two different techniques to calculate the area of isosceles right triangle XYZ, illustrated in Problem 12.15.

12.16 Verify your answer to Problem 12.15 by applying the formula $A = \frac{1}{2}xy\sin Z$.

According to Problem 12.15, the legs of the triangle have the same length: $XZ = YZ = 9$. Note that the included angle is Z, a right angle measuring 90°.

> In order to use the area formula (1/2)xy sin Z, angle Z has to be formed by the sides with length x and y. That means Z is the angle "included" by sides x and y.

$$A = \frac{1}{2}xy\sin Z$$
$$= \frac{1}{2}(XZ)(YZ)(\sin Z)$$
$$= \frac{1}{2}(9)(9)(\sin 90°)$$

Note that $\sin 90° = \sin(\pi/2) = 1$.

$$= \frac{1}{2}(81)(1)$$
$$= \frac{81}{2}$$

This is the same area value calculated in Problem 12.15, thus verifying the solution.

12.17 Calculate the area of triangle QRS, illustrated below. Report your answer accurate to the thousandths place.

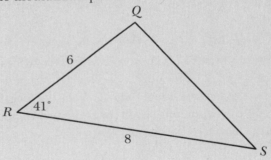

Angle R is formed by sides $\overline{RQ}$ and $\overline{RS}$, and you are given the lengths of those sides: $RQ = 6$ and $RS = 8$. Therefore, you can apply the SAS triangle area formula, $A = (1/2)xy\sin Z$.

> Don't get confused by the variables—x, y, and z don't need to be in the problem. All you need are two side lengths (that you plug into x and y) and an angle (that you plug into z) formed by those sides.

$$A = \frac{1}{2}xy\sin Z$$
$$= \frac{1}{2}(QR)(RS)(\sin R)$$
$$= \frac{1}{2}(6)(8)(\sin 41°)$$
$$= 24 \cdot \sin 41°$$

Use a scientific or graphing calculator to compute $\sin 41°$.

$$A \approx 15.745$$

12.18 Use a scientific or graphing calculator to compute the area of a triangle with an interior angle measuring 48°, if the lengths of the sides forming that angle are 10.4 and 13.9. Report your answer accurate to the thousandths place.

Apply the SAS triangle area formula, setting $x = 10.4$, $y = 13.9$, and $Z = 48°$.

$$A = \frac{1}{2}xy \sin Z$$

$$= \frac{1}{2}(10.4)(13.9)(\sin 48°)$$

$$= 72.28(\sin 48°)$$

$$\approx 53.715$$

12.19 Identify the formula used to calculate the area of a triangle given two of its angle measurements and the length of one side.

If you know the measures of two angles and the length of one side of a triangle, you can apply the following formula to calculate the area. The side with the known length need not be included by the angles whose measurements are known. Therefore, the formula is known alternately as the angle-side-angle (ASA) and angle-angle-side (AAS) formulas, as it satisfies both cases.

$$A = \frac{x^2 \sin Y \sin Z}{2 \sin X}$$

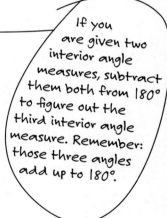

"Included" means "between."

Note that the formula contains x and X, which represent the known side length and the measure of the opposite angle, respectively. Also note that the formula requires the measures of all three interior angles of the triangle.

If you are given two interior angle measures, subtract them both from 180° to figure out the third interior angle measure. Remember: those three angles add up to 180°.

12.20 Apply the triangle area formula identified in Problem 12.19 to calculate the area of triangle *JKL* below. Report your answer accurate to the thousandths place.

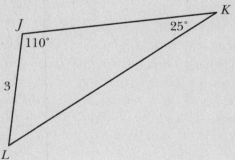

You are given two angle measures and the length of one side that is not included between those angles, typically described as AAS. (If the side were included, you would describe the given information as ASA.) Subtract the given angle measurements from 180° to calculate the measure of angle *L*.

$$m\angle L = 180° - 110° - 25°$$

$$m\angle L = 180° - 135°$$

$$m\angle L = 45°$$

You are given side length $JL = 3$; angle K is opposite that side. Substitute $x = 3$ and $X = 25°$ into the AAS triangle area formula, presented in Problem 12.19. The remaining angle measurements are substituted into Y and Z. In the solution below, $Y = 110°$ and $Z = 45°$; however, substituting $Y = 45°$ and $Z = 110°$ produces the same final result.

$$A = \frac{x^2 \sin Y \sin Z}{2 \sin X}$$

$$= \frac{3^2 (\sin 110°)(\sin 45°)}{2 (\sin 25°)}$$

$$\approx \frac{5.980167219}{0.8452365235}$$

$$\approx 7.075$$

Note: In Problems 12.21–12.22, you calculate the area of triangle BCD, illustrated below, two different ways.

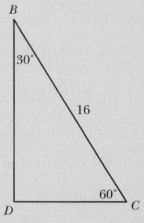

12.21 Identify the base and height of the triangle and apply the formula $A = \frac{1}{2}bh$ to calculate its exact area.

Note that BCD is a 30°–60°–90° triangle with hypotenuse length $BC = 16$. Therefore, the short leg is half the length of the hypotenuse ($CD = 8$) and its long leg is $\sqrt{3}$ times the length of the short leg $\left(BD = 8\sqrt{3}\right)$. Because $\overline{BD} \perp \overline{CD}$, the length of one leg serves as the base of the triangle and the length of the other leg represents the height. Substitute these values into the standard triangle area formula ($A = 1/2\,bh$).

$$A = \frac{1}{2}bh$$

$$= \frac{1}{2}(CD)(BD)$$

$$= \frac{1}{2}(8)\left(8\sqrt{3}\right)$$

$$= \frac{64\sqrt{3}}{2}$$

$$= 32\sqrt{3}$$

In other words, don't use a calculator in this problem. Unless you have a symbolic calculator, your answer will be a decimal, which is usually an approximate—rather than an exact—answer.

Note: In Problems 12.21–12.22, you calculate the area of triangle BCD, illustrated in Problem 12.21, two different ways.

12.22 Verify your solution to Problem 12.21 by applying the ASA triangle area formula. Do not use a scientific or graphing calculator for this problem.

A known side length and the measure of its opposite angle must be substituted into x and X, respectively, in the ASA triangle area formula. In this diagram, you are given $BC = 16$. Its opposite angle, D, is a right angle, which measures $90°$.

$$A = \frac{x^2 \sin Y \sin Z}{2 \sin X}$$

$$= \frac{16^2 (\sin 30°)(\sin 60°)}{2 (\sin 90°)}$$

$$= \frac{256(1/2)(\sqrt{3}/2)}{2(1)}$$

$$= \frac{128(\sqrt{3}/2)}{2}$$

$$= \frac{64\sqrt{3}}{2}$$

$$= 32\sqrt{3}$$

> The numerator contains two fractions with a denominator of 2. Multiply 256 by 1/2 to get 128. Then, multiply 128 by √3/2 to get a numerator of 64√3.

12.23 Calculate the area of triangle NPQ, given $m\angle N = 80°$, $m\angle Q = 35°$, and $NQ = 7$. Report your answer accurate to the thousandths place.

It is helpful to construct a diagram that illustrates the given information. Your diagram may not precisely match the below diagram, but both will return the same solution. Calculate the measure of angle P by subtracting the measures of angles N and Q from $180°$: $m\angle P = 180° - 80° - 35° = 65°$.

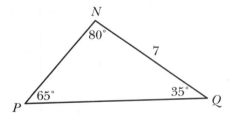

You are given side length $NQ = 7$; its opposite angle is P, which has a measure of $65°$. Substitute these values for x and X in the ASA triangle area formula.

$$A = \frac{x^2 \sin Y \sin Z}{2 \sin X}$$

$$= \frac{7^2 (\sin 80°)(\sin 35°)}{2 (\sin 65°)}$$

$$\approx \frac{27.67826355}{1.812615574}$$

$$\approx 15.270$$

12.24 Calculate the area of triangle *BCE* in the diagram below, and report the answer accurate to three decimal places.

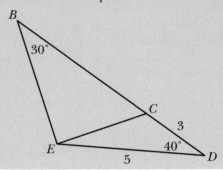

The diagram does not provide sufficient information to calculate the area of triangle *BCE* directly. Notice that the area of the largest triangle, *BDE*, is equal to the sum of the areas of triangles *BCE* and *CDE*. Apply the SAS triangle area formula to calculate the area of triangle *CDE*.

$$\text{area}(\triangle CDE) = \frac{1}{2}(CD)(DE)(\sin D)$$

$$= \frac{1}{2}(3)(5)(\sin 40°)$$

$$= \frac{15}{2}(\sin 40°)$$

Calculate the measure of angle *BED*, noting that the sum of the measures of the interior angles of triangle *BDE* must equal 180°.

$$m\angle BED = 180° - m\angle B - m\angle D$$

$$m\angle BED = 180° - 30° - 40°$$

$$m\angle BED = 110°$$

Apply the ASA triangle area formula to calculate the area of triangle *BDE*.

$$\text{area}(\triangle BDE) = \frac{(DE)^2(\sin D)(\sin BED)}{2(\sin B)}$$

$$= \frac{5^2(\sin 40°)(\sin 110°)}{2(\sin 30°)}$$

$$= \frac{25}{2}\left(\frac{\sin 40° \cdot \sin 110°}{\sin 30°}\right)$$

Recall that the larger triangle is comprised of the two smaller triangles.

$$\text{area}(\triangle BCE) = \text{area}(\triangle BDE) - \text{area}(\triangle CDE)$$

Solve for the area of triangle *BCE* and substitute the areas calculated above into the equation.

$$\text{area}(\triangle BCE) = \text{area}(\triangle BDE) - \text{area}(\triangle CDE)$$

$$= \frac{25}{2}\left(\frac{\sin 40° \cdot \sin 110°}{\sin 30°}\right) - \frac{15}{2}(\sin 40°)$$

$$\approx 15.10056934 - 4.820907073$$

$$\approx 10.280$$

Heron's Formula

SSS triangle area formula

12.25 Identify the formula used to calculate the area of a triangle given only the lengths of its sides.

Heron's formula allows you to calculate the area of a triangle given only the lengths of its sides, x, y, and z. In order to apply the formula, you must first calculate the semiperimeter, s, of the triangle using the following formula.

$$s = \frac{x+y+z}{2}$$

Once you have calculated the semiperimeter, you can apply Heron's formula to calculate the area A.

$$A = \sqrt{s(s-x)(s-y)(s-z)}$$

The perimeter of a triangle is the sum of the lengths of its sides. The SEMIperimeter is equal to half of the perimeter, just like a SEMIcircle is half of a circle.

Note: Problems 12.26–12.28 refer to isosceles triangle XYZ below.

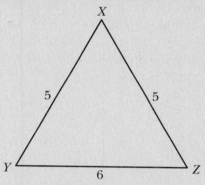

12.26 Calculate the semiperimeter of *XYZ*.

Triangle *XYZ* is isosceles because at least two of its sides have the same length ($XY = XZ = 5$). The semiperimeter is equal to half the sum of the lengths of the sides.

$$
\begin{aligned}
s &= \frac{x+y+z}{2} \\
&= \frac{YZ + XZ + XY}{2} \\
&= \frac{6+5+5}{2} \\
&= \frac{16}{2} \\
&= 8
\end{aligned}
$$

Note: Problems 12.26–12.28 refer to isosceles triangle XYZ, as illustrated in Problem 12.26.

12.27 Apply Heron's formula to calculate the area of triangle *XYZ*.

Heron's formula is the square root of the product of the semiperimeter and the differences of the semiperimeter and each of the side lengths.

$$A = \sqrt{s(s-x)(s-y)(s-z)}$$
$$= \sqrt{s(s-YZ)(s-XZ)(s-XY)}$$
$$= \sqrt{8(8-6)(8-5)(8-5)}$$
$$= \sqrt{8(2)(3)(3)}$$
$$= \sqrt{144}$$
$$= 12$$

Note: Problems 12.26–12.28 refer to isosceles triangle XYZ, as illustrated in Problem 12.26.

12.28 Verify your solution to Problem 12.27 by dividing triangle *XYZ* into two congruent right triangles and calculating their areas.

The segment connecting *X* and the base $\overline{YZ}$ of the isosceles triangle bisects that base, dividing it in half, and is perpendicular to the base. In the diagram below, $\overline{XW} \perp \overline{YZ}$ and $WY = WZ$.

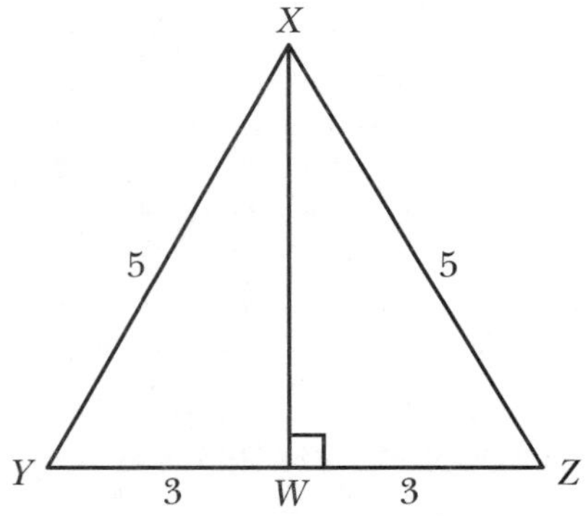

Apply the Pythagorean theorem to calculate *WX*.

$$(WX)^2 + (WY)^2 = (XY)^2$$
$$(WX)^2 + 3^2 = 5^2$$
$$(WX)^2 = 25 - 9$$
$$(WX)^2 = 16$$
$$\sqrt{(WX)^2} = \sqrt{16}$$
$$WX = 4$$

If a triangle is drawn in the coordinate plane, you may have to worry about signed length, meaning the side of a triangle can have a positive or negative value. This triangle is not drawn in the coordinate plane, so its side lengths MUST be positive. In other words, the answer WX = ±4 is wrong.

Both right triangles have a base of 3 and a height of 4. Use the standard triangle area formula to calculate their areas.

$$\text{area}\left(\triangle WXY\right) = \text{area}\left(\triangle WXZ\right) = \frac{1}{2}bh$$
$$= \frac{1}{2}(3)(4)$$
$$= \frac{12}{2}$$
$$= 6$$

Each of the right triangles has an area of 6. Together they have a total area of 12, which verifies the solution to Problem 12.27.

Note: Problems 12.29–12.30 refer to a triangle with side lengths 7, 9, and 10.

12.29 Calculate the semiperimeter of the triangle.

Add the lengths of the sides and divide by 2.

$$s = \frac{7+9+10}{2}$$
$$= \frac{26}{2}$$
$$= 13$$

Note: Problems 12.29–12.30 refer to a triangle with side lengths 7, 9, and 10.

12.30 Calculate the exact area of the triangle.

Apply Heron's formula, setting $x = 7$, $y = 9$, $z = 10$, and $s = 13$ (as calculated in Problem 12.29).

$$A = \sqrt{s(s-x)(s-y)(s-z)}$$
$$= \sqrt{13(13-7)(13-9)(13-10)}$$
$$= \sqrt{13(6)(4)(3)}$$
$$= \sqrt{936}$$
$$= \sqrt{36 \cdot 26}$$
$$= 6\sqrt{26}$$

Note: Problems 12.31–12.32 refer to triangle LMN, illustrated below.

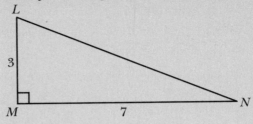

12.31 Calculate the semiperimeter of the triangle.

You are given the lengths of two sides of the right triangle; apply the Pythagorean theorem to calculate the third.

$$(LM)^2 + (MN)^2 = (LN)^2$$
$$3^2 + 7^2 = (LN)^2$$
$$58 = (LN)^2$$
$$\sqrt{58} = LN$$

Calculate the semiperimeter of the triangle.

$$s = \frac{LM + MN + LN}{2}$$
$$= \frac{3 + 7 + \sqrt{58}}{2}$$
$$= \frac{10 + \sqrt{58}}{2}$$
$$= \frac{10}{2} + \frac{\sqrt{58}}{2}$$
$$= 5 + \frac{\sqrt{58}}{2}$$

Note: Problems 12.31–12.32 refer to triangle LMN, illustrated in Problem 12.31.

12.32 Calculate the exact area of the triangle.

According to Problem 12.31, $LN = \sqrt{58}$ and $s = 5 + \sqrt{58}/2$. Apply Heron's formula.

$$A = \sqrt{s(s - LM)(s - MN)(s - LN)}$$
$$= \sqrt{\left(5 + \frac{\sqrt{58}}{2}\right)\left(5 + \frac{\sqrt{58}}{2} - 3\right)\left(5 + \frac{\sqrt{58}}{2} - 7\right)\left(5 + \frac{\sqrt{58}}{2} - \sqrt{58}\right)}$$
$$= \sqrt{\left(5 + \frac{\sqrt{58}}{2}\right)\left(2 + \frac{\sqrt{58}}{2}\right)\left(-2 + \frac{\sqrt{58}}{2}\right)\left(5 + \frac{\sqrt{58}}{2} - \frac{2\sqrt{58}}{2}\right)}$$
$$= \sqrt{\left(5 + \frac{\sqrt{58}}{2}\right)\left(2 + \frac{\sqrt{58}}{2}\right)\left(-2 + \frac{\sqrt{58}}{2}\right)\left(5 - \frac{\sqrt{58}}{2}\right)}$$

Uh oh. That means no calculator and a bunch of arithmetic. Bring it on!

Group the factors to simplify the product.

$$A = \sqrt{\left[\left(5+\frac{\sqrt{58}}{2}\right)\left(5-\frac{\sqrt{58}}{2}\right)\right]\left[\left(2+\frac{\sqrt{58}}{2}\right)\left(-2+\frac{\sqrt{58}}{2}\right)\right]}$$

Multiply the factors within each bracketed group using the FOIL method.

$$A = \sqrt{\left[25-\frac{5\sqrt{58}}{2}+\frac{5\sqrt{58}}{2}-\frac{\sqrt{58}^2}{4}\right]\left[-4+\frac{2\sqrt{58}}{2}-\frac{2\sqrt{58}}{2}+\frac{\sqrt{58}^2}{4}\right]}$$

$$= \sqrt{\left[25-\frac{58}{4}\right]\left[-4+\frac{58}{4}\right]}$$

$$= \sqrt{\left[\frac{100}{4}-\frac{58}{4}\right]\left[-\frac{16}{4}+\frac{58}{4}\right]}$$

$$= \sqrt{\left(\frac{42}{4}\right)\left(\frac{42}{4}\right)}$$

$$= \sqrt{\left(\frac{21}{2}\right)\left(\frac{21}{2}\right)}$$

$$= \frac{21}{2}$$

Note: Problems 12.33–12.34 refer to the diagram below.

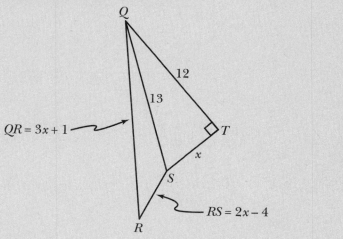

12.33 Calculate the lengths of the sides of triangle QRS.

Note that QST is a right triangle, so apply the Pythagorean theorem to calculate ST (which is also equal to x).

$$\left(QT\right)^2 + \left(ST\right)^2 = \left(QS\right)^2$$

$$12^2 + x^2 = 13^2$$

$$144 + x^2 = 169$$

$$x^2 = 25$$

$$\sqrt{x^2} = \sqrt{25}$$

$$x = 5$$

Now that you know $ST = x = 5$, you can substitute $x = 5$ into the expressions that represent the lengths of QR and RS.

$$QR = 3x + 1 \qquad RS = 2x - 4$$
$$= 3(5) + 1 \qquad\quad = 2(5) - 4$$
$$= 16 \qquad\qquad\quad = 6$$

The lengths of the sides of triangle QRS are $RS = 6$, $QS = 13$, and $QR = 16$.

Note: Problems 12.33–12.34 refer to the diagram in Problem 12.33.

12.34 Apply Heron's formula to calculate the area of triangle QRS. Report the answer accurate to the thousandths place.

According to Problem 12.33, the lengths of the sides of triangle QRS are $RS = 6$, $QS = 13$, and $QR = 16$. Calculate the semiperimeter.

$$s = \frac{RS + QS + QR}{2}$$
$$= \frac{6 + 13 + 16}{2}$$
$$= \frac{35}{2}$$

Apply Heron's formula to calculate the area of triangle QRS.

$$A = \sqrt{s(s - RS)(s - QS)(s - QR)}$$
$$= \sqrt{\left(\frac{35}{2}\right)\left(\frac{35}{2} - 6\right)\left(\frac{35}{2} - 13\right)\left(\frac{35}{2} - 16\right)}$$
$$= \sqrt{\left(\frac{35}{2}\right)\left(\frac{35}{2} - \frac{12}{2}\right)\left(\frac{35}{2} - \frac{26}{2}\right)\left(\frac{35}{2} - \frac{32}{2}\right)}$$
$$= \sqrt{\left(\frac{35}{2}\right)\left(\frac{23}{2}\right)\left(\frac{9}{2}\right)\left(\frac{3}{2}\right)}$$
$$= \sqrt{\frac{21,735}{16}}$$
$$\approx 36.857$$

> The vertex of a central angle lies on the center of a circle and its sides extend outward from the center, intersecting the circle.

Area of a Sector
Surface area of a pizza slice

12.35 Identify the formula used to calculate the area of a sector with central angle θ.

A sector is a portion of a circle bounded by an arc and two radii of the circle. To compute the area of a sector, you need to know the measure of its central angle θ and the radius r of the circle.

$$\text{area}(\text{sector}) = \frac{\text{measure of central angle}}{\text{measure of one full rotation}} \cdot \pi r^2$$

Essentially, this formula multiplies the area of the full circle (πr^2) by a fraction that identifies how much of the circle is contained within the sector. For instance, a semicircle has a central angle of 180°, and its area is 180°/360° = 1/2 of the circle.

Note: Problems 12.36–12.39 refer to the diagram below, in which circle X has radius r = 2.

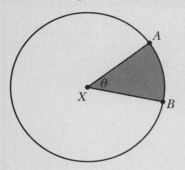

12.36 Calculate the area of circle X.

The area of a circle is the product of π and the squared radius (r^2) of the circle. The problem states that the radius is $r = 2$.

$$A = \pi r^2$$
$$= \pi (2)^2$$
$$= 4\pi$$

Note: Problems 12.36–12.39 refer to the diagram in Problem 12.36. Note that circle X has radius r = 2.

12.37 Calculate the perimeter of the shaded sector, assuming θ is measured in radians.

The sector is bounded by two radii $\left(\overline{AX} \text{ and } \overline{BX}\right)$ and an arc $\left(\overparen{AB}\right)$. Each of the radii has length 2. Apply the formula below to calculate the length of the arc.

$$\text{length}\left(\overparen{AB}\right) = r \cdot \theta$$
$$= 2\theta$$

The perimeter p of the sector is the sum of the lengths of the bounding radii and the arc.

$$p = AX + BX + \text{length}\left(\overparen{AB}\right)$$
$$= 2 + 2 + 2\theta$$
$$= 4 + 2\theta$$

If the central angle of an arc is measured in degrees rather than radians, the following formula calculates the length of that arc; the formula multiplies angle θ (measured in degrees) by $\pi/180$, thereby converting the angle measurement to radians.

$$\text{length}\left(\overparen{AB}\right) = \theta \cdot r \cdot \frac{\pi}{180}$$

Note: Problems 12.36–12.39 refer to the diagram in Problem 12.36. Note that circle X has radius r = 2.

12.38 Calculate the area of the shaded sector, assuming $\theta = \pi/6$.

Apply the sector area formula presented in Problem 12.35. Note that the measure of the central angle is expressed in radians, so you should also use radians to express the measure of one full rotation: 2π.

$$A = \frac{\text{measure of central angle}}{\text{measure of one full rotation}} \cdot \pi r^2$$

$$= \frac{\cancel{\pi}/6}{2\cancel{\pi}} \cdot \pi(2)^2$$

$$= \frac{1/6}{2} \cdot 4\pi$$

$$= \frac{1}{12} \cdot 4\pi$$

$$= \frac{\pi}{3}$$

Note: Problems 12.36–12.39 refer to the diagram in Problem 12.36. Note that circle X has radius r = 2.

12.39 Calculate the *unshaded* region of the circle, assuming $\theta = 40°$.

The unshaded region of the circle is also a sector. Its central angle measures $360° - 40° = 320°$. Because the central angle is expressed in degrees, you should also express the measure of one full rotation in degrees when applying the area formula.

$$A = \frac{\text{measure of central angle}}{\text{measure of one full rotation}} \cdot \pi r^2$$

$$= \frac{320}{360} \cdot \pi(2)^2$$

$$= \frac{8}{9} \cdot 4\pi$$

$$= \frac{32\pi}{9}$$

Note: Problems 12.40–12.42 refer to a circle with radius 6 and a sector of the circle that has a central angle measuring 100°.

12.40 Calculate the area and circumference of the circle.

The area A of a circle is equal to the product of π and the square of the radius.

$$A = \pi r^2$$

$$= \pi(6)^2$$

$$= 36\pi$$

The circumference C of a circle is the product of 2π and the radius.

$$C = 2\pi r$$
$$= 2\pi(6)$$
$$= 12\pi$$

Note: Problems 12.40–12.42 refer to a circle with radius 6 and a sector of the circle that has a central angle measuring 100°.

12.41 Calculate the area of the sector.

Apply the area formula for the sector of a circle.

$$A = \frac{\text{measure of central angle}}{\text{measure of one full rotation}} \cdot \pi r^2$$
$$= \frac{100°}{360°} \cdot \pi(6)^2$$
$$= \frac{5}{18} \cdot 36\pi$$
$$= \frac{180\pi}{18}$$
$$= 10\pi$$

Note: Problems 12.40–12.42 refer to a circle with radius 6 and a sector of the circle that has a central angle measuring 100°.

12.42 Calculate the perimeter of the sector.

The perimeter of a sector consists of two radii and an arc of the circle. The radius of this circle is $r = 6$. To calculate the arc length, apply the formula introduced in Problem 12.37; use the version of the formula intended for central angles measured in degrees.

$$\text{arc length} = \theta \cdot r \cdot \frac{\pi}{180}$$
$$= 100(6)\left(\frac{\pi}{180}\right)$$
$$= \frac{600\pi}{180}$$
$$= \frac{10\pi}{3}$$

Add the lengths of the two radii to the arc length you calculated.

$$\text{perimeter} = 2r + \text{arc length}$$

$$= 2(6) + \frac{10\pi}{3}$$

$$= 12 + \frac{10\pi}{3}$$

$$= \frac{36}{3} + \frac{10\pi}{3}$$

$$= \frac{36 + 10\pi}{3}$$

Note: Problems 12.43–12.45 refer to the diagram below, in which XY = 13 and XZ = 24.

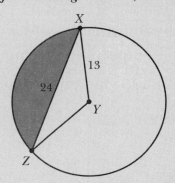

12.43 Calculate the area of the sector, bounded by $\overline{XY}$, $\overline{YZ}$, and $\overarc{XZ}$. Report the answer accurate to the thousandths place.

See Problem 12.28.

Two sides of triangle XYZ are also radii of circle Y, so they have the same length ($XY = YZ = 13$). Therefore, XYZ is an isosceles triangle. Draw $\overline{WY}$, the segment connecting Y to the opposite side, and note that it is perpendicular to—and bisects—$\overline{XZ}$.

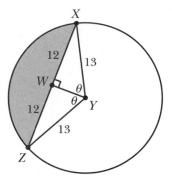

By drawing the perpendicular bisector, you have divided XYZ into two congruent right triangles, WZY and WXY. Furthermore, you know that the measure of angle XYZ is equal to $\theta + \theta$. Apply a trigonometric ratio to calculate θ.

$$\sin\theta = \frac{\text{length of side opposite } \theta}{\text{length of the hypotenuse}}$$

$$\sin\theta = \frac{WX}{XY}$$

$$\sin\theta = \frac{12}{13}$$

$$\theta = \arcsin\frac{12}{13}$$

Therefore, the measure of central angle XYZ is equal to $2\theta = 2\arcsin(12/13)$. Apply the sector area formula. In the solution below, the calculator is set to degrees mode.

$$A = \frac{\text{measure of central angle}}{\text{measure of one full rotation}} \cdot \pi r^2$$

$$= \frac{2 \cdot \arcsin(12/13)}{360} \cdot \pi(13)^2$$

$$= \frac{\arcsin(12/13)}{180} \cdot 169\pi$$

$$= \frac{169\pi \cdot \arcsin(12/13)}{180}$$

$$\approx 198.74488$$

The area of the sector, accurate to the thousandths place, is 198.745.

Note: Problems 12.43–12.45 refer to the diagram in Problem 12.43, in which XY = 13 and XZ = 24.

12.44 Calculate the area of triangle XYZ.

As Problem 12.43 explains, $XY = YZ = 13$. You are given the length of the third side: $XZ = 24$. Calculate the semiperimeter of triangle XYZ.

$$s = \frac{XY + YZ + XZ}{2}$$

$$= \frac{13 + 13 + 24}{2}$$

$$= \frac{50}{2}$$

$$= 25$$

Now apply Heron's formula to calculate the area of triangle XYZ.

$$A = \sqrt{s(s - XY)(s - YZ)(s - XZ)}$$

$$= \sqrt{25(25 - 13)(25 - 13)(25 - 24)}$$

$$= \sqrt{25(12)(12)(1)}$$

$$= \sqrt{3,600}$$

$$= \sqrt{60 \cdot 60}$$

$$= 60$$

Note: Problems 12.43–12.45 refer to the diagram in Problem 12.43, in which XY = 13 and XZ = 24.

12.45 Calculate the area of the shaded region, accurate to the thousandths place.

The shaded region is the portion of the sector investigated in Problem 12.43 that does not include the area of triangle *XYZ*. Subtract the area you calculated in Problem 12.44 from the area you calculated in Problem 12.43.

$$\text{shaded area} = \text{sector area} - \text{area of triangle } XYZ$$
$$\approx 198.74488 - 60$$
$$\approx 138.745$$

Chapter 13
OBLIQUE TRIANGLE LAWS

Oblique = Not a right triangle

In Chapter 12, you apply different formulas to calculate the area of a triangle. Often, the formula you choose is dictated by the information provided in the problem. For example, Heron's formula only applies if you know the lengths of all three sides—which is described as "SSS" or "side-side-side."

In this chapter, you once again select between formulas—the law of sines and the law of cosines—based on the information you are given. Specifically, given any two angles and a side (AAS or ASA), you apply the law of sines to calculate lengths of the remaining sides. Given the lengths of all three sides, or two sides and the included angle measurement, you apply the law of cosines.

There are two important formulas to memorize in this chapter—the law of sines and the law of cosines. Both refer to triangles, and in both formulas, matching upper- and lowercase letters represent opposite elements. In other words, angle A is opposite the side with length a.

The law of sines is a bit easier than the law of cosines, and you'll see the law of sines more often. In fact, you will probably use it in the later stages of problems that you start with the law of cosines.

Two rules of thumb:
(1) Use the law of sines whenever you can
(2) Be careful when you are given two sides and a non-included angle (SSA). You'll see why in Problems 13.9–13.12.

Law of Sines

Given AAS, ASA, and occasionally SSA

13.1 Given the diagram below, in which *A* is opposite *a*, *B* is opposite *b*, and *C* is opposite *c*, state the law of sines.

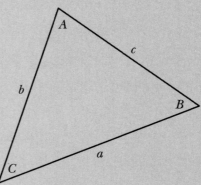

The law of sines states that the ratio of the sine of an angle to the length of the opposite side is constant in a triangle.

$$\frac{\sin A}{a} = \frac{\sin B}{b} = \frac{\sin C}{c}$$

You may also express the law of sines using the reciprocals of the ratios.

$$\frac{a}{\sin A} = \frac{b}{\sin B} = \frac{c}{\sin C}$$

Note: Problems 13.2–13.4 refer to triangle ABC below.

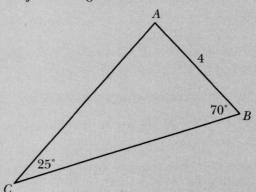

13.2 Calculate *b* and report your answer accurate to the thousandths place.

> The side is not sandwiched between the two angles. In this diagram, $\overline{BC}$ is the side included between angles B and C.

> AC and b are two different ways to write the same length. One names the side according to its endpoints and the other names the side according to the opposite angle.

You are given the measures of two angles and the length of a non-included side: $m\angle B = 70°$, $m\angle C = 25°$, and $c = 4$; in other words, you are given angle-angle-side (AAS). You are asked to calculate *b*, the side opposite angle *B*, which is also written *AC*. Because you know the measure of angle *C* and the length of the opposite side *c*, the ratio (sin *C*)/*c* should appear in the law of sines proportion. You are asked to calculate *AC*, which is equal to *b*, because it is the length of the side opposite angle *B*.

$$\frac{\sin B}{b} = \frac{\sin C}{c}$$

$$\frac{\sin 70°}{b} = \frac{\sin 25°}{4}$$

Cross-multiply and solve for b.

$$(\sin 70°)(4) = (b)(\sin 25°)$$

$$\frac{4 \cdot \sin 70°}{\sin 25°} = b$$

$$8.894008668 \approx b$$

You conclude that $b \approx 8.894$.

Note: Problems 13.2–13.4 refer to the diagram in Problem 13.2.

13.3 Calculate the measure of angle A.

The measures of the interior angles of a triangle have a sum of 180°. You are given two of those measures, so subtract them from 180° to calculate the measure of angle A.

$$m\angle A = 180° - m\angle B - m\angle C$$

$$m\angle A = 180° - 70° - 25°$$

$$m\angle A = 85°$$

Note: Problems 13.2–13.4 refer to the diagram in Problem 13.2.

13.4 Calculate the perimeter of ABC and report your answer accurate to three decimal places.

According to the diagram, the side opposite angle C has length $c = 4$; in Problem 13.2, you calculate b, the length of another side of the triangle. In order to compute the perimeter, you need to calculate the length of the remaining side, a. Apply the law of sines once again.

$$\frac{\sin A}{a} = \frac{\sin C}{c}$$

According to Problem 13.3, the measure of angle A is 85°. Note that you can replace $(\sin C)/c$ with $(\sin B)/b$ in the proportion above, now that you know the value of b. However, because b is a lengthy decimal and c is an integer, $(\sin C)/c$ is preferred.

$$\frac{\sin 85°}{a} = \frac{\sin 25°}{4}$$

$$(4)(\sin 85°) = (a)(\sin 25°)$$

$$\frac{4 \cdot \sin 85°}{\sin 25°} = a$$

Add the lengths of the sides to calculate the perimeter of ABC.

$$\text{perimeter of } ABC = a + b + c$$

$$= \frac{4 \cdot \sin 85°}{\sin 25°} + \frac{4 \cdot \sin 70°}{\sin 25°} + 4$$

$$\approx 22.3227985$$

$$\approx 22.323$$

Substitute the exact values of a and b, the ugly fractions, into the formula. If you use decimals rounded to the thousandths place, it might lead to a rounding error, making your solution slightly inaccurate. If you hate typing fractions into your calculator, you can use the long decimal equivalents, but don't round anything until the final step of the problem.

Problems 13.5–13.7 refer to triangle RST in the diagram below.

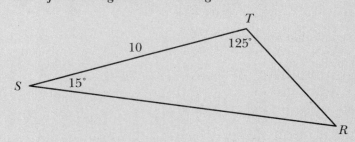

13.5 Calculate the measure of angle R.

The measures of two interior angles of triangle RST are given. Subtract them from 180° to calculate the measure of the remaining angle.

$$m\angle R = 180° - m\angle S - m\angle T$$

$$m\angle R = 180° - 15° - 125°$$

$$m\angle R = 40°$$

Problems 13.5–13.7 refer to triangle RST, illustrated in Problem 13.5.

13.6 Calculate RT, accurate to the thousandths place.

You are given the measure of two angles and the included side, or angle-side-angle (ASA). Apply the law of sines, noting that angle S is opposite side $\overline{RT}$ and angle R is opposite side $\overline{ST}$.

$$\frac{\sin R}{r} = \frac{\sin S}{s}$$

$$\frac{\sin R}{ST} = \frac{\sin S}{RT}$$

One fraction should always contain a known side length opposite a known angle measure (in this case, r and R). The other fraction should contain the side or angle you're calculating and the side or angle opposite it (in this case, you are calculating s = RT and the opposite angle is S).

According to Problem 13.5, the measure of angle R is 40°.

$$\frac{\sin 40°}{10} = \frac{\sin 15°}{RT}$$

$$(RT)(\sin 40°) = (10)(\sin 15°)$$

$$RT = \frac{10 \cdot \sin 15°}{\sin 40°}$$

$$RT \approx 4.027$$

Problems 13.5–13.7 refer to triangle RST, illustrated in Problem 13.5.

13.7 Calculate *RS*, accurate to the thousandths place.

Apply the law of sines, noting that $t = RS$ because side $\overline{RS}$ is opposite angle *T*.

$$\frac{\sin R}{r} = \frac{\sin T}{t}$$

$$\frac{\sin 40°}{10} = \frac{\sin 125°}{RS}$$

$$(RS)(\sin 40°) = (10)(\sin 125°)$$

$$RS = \frac{10 \cdot \sin 125°}{\sin 40°}$$

$$RS \approx 12.744$$

13.8 Calculate *YZ* in the triangle illustrated below. Report your answer accurate to the thousandths place.

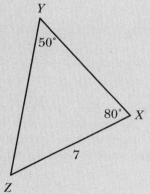

You are given two angles and a non-included side (AAS), so apply the law of sines. Note that you are given the measure of angle *Y* (50°) and the length of the opposite side ($y = 7$). You are asked to calculate $x = YZ$, the length of the side opposite angle *X*.

$$\frac{\sin Y}{y} = \frac{\sin X}{x}$$

$$\frac{\sin 50°}{7} = \frac{\sin 80°}{YZ}$$

$$(\sin 50°)(YZ) = (7)(\sin 80°)$$

$$YZ = \frac{7 \cdot \sin 80°}{\sin 50°}$$

$$YZ \approx 8.999$$

Problems 13.9–13.12 refer to a triangle ABC, such that m∠A = 20°, a = 3, and c = 5.

13.9 Explain why the information provided by the problem classifies this as the ambiguous case for the law of sines.

You are given the measure of one angle and the lengths of two sides of triangle *ABC*. Note that one of the side lengths given (*a* = 3) is opposite the given angle measurement. Therefore, the angle is not formed by the sides, classifying this information as side-side-angle (SSA), rather than side-angle-side (SAS). Unlike the preceding problems in this chapter, you are given only one angle measurement, rather than two.

A SSA triangle is described as the ambiguous case, because the law of sines cannot always reliably predict whether or not an angle is obtuse when only one angle measurement is known. Because of this limitation, you should apply the law of cosines (described in Problems 13.13–13.24) to calculate any angle that may be obtuse, whenever possible. However, in this case, you must apply the law of sines and carefully interpret your answer.

> The angle included between two sides is the letter not named by the sides. In this case, angle B is included between (and formed by) sides a and c. If you had a triangle named MNP, and you were given side lengths n and p, those sides would form angle M, the remaining letter in the name of the triangle.

Problems 13.9–13.12 refer to a triangle ABC, such that m∠A = 20°, a = 3, and c = 5.

13.10 Demonstrate, via an illustration drawn to scale, that angle *C* of triangle *ABC* may be acute or obtuse.

Use a ruler and a protractor to draw two different triangles named *ABC* with side lengths *a* = 3 and *c* = 5. Note that the angle opposite *a* must measure 20°. The diagram below illustrates two different triangles that meet these requirements; in one case, angle *C* is acute, and in the other, angle *C* is obtuse.

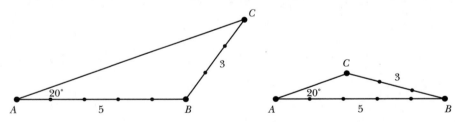

Because two different triangles may (and in this problem actually do) satisfy the SSA conditions described, there may be two different—and equally valid—ways to calculate the measures of the remaining sides and angles of the triangle. Hence, a SSA triangle is considered the ambiguous case for the law of sines. In Problems 13.11–13.12, you calculate the two possible measures of angle *C*.

Problems 13.9–13.12 refer to a triangle ABC, such that m∠A = 20°, a = 3, and c = 5.

13.11 Calculate the measure of angle C, assuming C is acute. Report your answer accurate to the thousandths place.

In the ambiguous case, you apply the law of sines and use arcsine to calculate the measure of an angle. Recall that the restricted range of arcsine is $-\pi/2 \le \theta \le \pi/2$. Because triangles not on the coordinate plane must have positive angle measurements, arcsine will report an angle between 0 and $\pi/2$. In other words, unless the unknown angle is a right angle, it will be acute.

$$\frac{\sin A}{a} = \frac{\sin C}{c}$$
$$\frac{\sin 20°}{3} = \frac{\sin C}{5}$$
$$5(\sin 20°) = 3(\sin C)$$
$$\frac{5 \cdot \sin 20°}{3} = \sin C$$
$$\arcsin\left(\frac{5 \cdot \sin 20°}{3}\right) = C$$
$$\arcsin(0.5700335722) \approx C$$
$$34.75256688 \approx C$$

Therefore, $m\angle C \approx 34.753°$.

Problems 13.9–13.12 refer to a triangle ABC, such that m∠A = 20°, a = 3, and c = 5.

13.12 Calculate the measure of angle C, assuming C is obtuse. Report your answer accurate to the thousandths place.

According to Problem 13.11, angle C has a sine value of approximately 0.5700335722. This is true whether C is acute or obtuse. An angle in standard position with a measure of approximately 34.753° has that sine value, and so does a reference angle with the same measure in the second quadrant. The terminal sides of both angles intersect the unit circle at the same height, so they have the same sine value, as illustrated below.

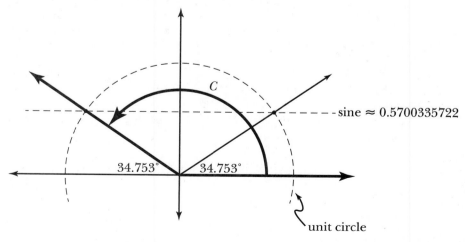

Angle C is the highlighted angle in the diagram, an obtuse angle in standard position that terminates in the second quadrant. Calculate its measure by subtracting the measure of its reference angle from 180° (as explained in Problem 4.7).

$$m\angle C = 180° - \text{measure of reference angle}$$
$$\approx 180° - 34.753°$$
$$\approx 145.247°$$

If C is an obtuse angle in triangle ABC, then it measures 145.247°.

Law of Cosines

Given SAS or SSS

13.13 State the law of cosines, given a triangle ABC such that a is the length of the side opposite angle A, b is the length of the side opposite angle B, and c is the length of the side opposite angle C.

The law of cosines takes one of three forms, based on the information you are given in the problem. In each case, the square of the length of one side is equal to the sum of the squares of the other side lengths minus double the product of those lengths multiplied by the cosine of the angle opposite the original side.

$$a^2 = b^2 + c^2 - 2bc \cdot \cos A$$
$$b^2 = a^2 + c^2 - 2ac \cdot \cos B$$
$$c^2 = a^2 + b^2 - 2ab \cdot \cos C$$

Added bonus: There is no ambiguous case for the law of cosines. If the formula produces an answer, you can be sure that answer is correct.

Problems 13.14–13.17 refer to triangle ABC, illustrated below.

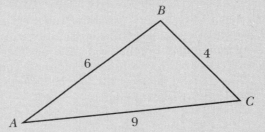

13.14 Explain why you must begin this problem with the law of cosines, not the law of sines.

The law of cosines is usually applied to triangles in which you are given the lengths of two sides and an included angle (side-angle-side or SAS) or the lengths of all three sides (side-side-side or SSS). In this problem, you are given the length of all three sides.

Problems 13.14–13.17 refer to triangle ABC, illustrated in Problem 13.14.

13.15 Identify the angle measure you should calculate first and explain your answer.

When possible, you should use the law of cosines to calculate the largest angle in a triangle, in case it is obtuse. As Problems 13.9–13.12 explain, the law of sines is unable to differentiate between acute and obtuse angles; the law of cosines does not suffer this limitation. The largest angle in a triangle is always opposite the longest side. Because the longest side of this triangle is $b = 9$, you should apply the law of cosines to calculate the measure of angle B.

> Because the restricted range of arccosine is $0 \leq \theta \leq 180°$, it can calculate acute angles (whose measures are greater than 0° but less than 90°) and obtuse angles (whose measures are greater than 90° but less than 180°).

Problems 13.14–13.17 refer to triangle ABC, illustrated in Problem 13.14.

13.16 Calculate the angle you identified in Problem 13.15, and report the answer accurate to the thousandths place.

As Problem 13.15 explains, you should calculate the measure of angle B first. Use the version of the law of cosines that includes B.

$$b^2 = a^2 + c^2 - 2ac \cdot \cos B$$
$$9^2 = 4^2 + 6^2 - 2(4)(6) \cdot \cos B$$
$$81 = 16 + 36 - 48 \cdot \cos B$$
$$81 = 52 - 48 \cdot \cos B$$
$$81 - 52 = -48 \cdot \cos B$$
$$29 = -48 \cdot \cos B$$
$$-\frac{29}{48} = \cos B$$
$$\arccos(-29/48) \approx B$$
$$127.1688997 \approx B$$

The measure of angle B is approximately $127.169°$.

> If the cosine is negative, the angle is obtuse. If the cosine is positive, the angle is acute.

Problems 13.14–13.17 refer to triangle ABC, illustrated in Problem 13.14.

13.17 Calculate the measures of the remaining angles in triangle ABC, accurate to the thousandths place.

According to Problem 13.16, angle B measures approximately $127.1688997°$; the opposite side has length $b = 9$. Although you can apply the law of cosines to calculate another angle measure in the triangle, you should apply the law of sines instead, because it requires less arithmetic to complete. Furthermore, because you have already calculated the measure of the largest angle in triangle ABC, the ambiguous case is no longer a concern.

You can choose to calculate the measure of angle A or C; below, the law of sines is used to calculate the measure of angle A.

$$\frac{\sin A}{a} = \frac{\sin B}{b}$$
$$\frac{\sin A}{4} \approx \frac{\sin 127.1688997°}{9}$$
$$9(\sin A) \approx 4(\sin 127.1688997°)$$
$$\sin A \approx \frac{3.187431915}{9}$$
$$\sin A \approx 0.3541591017$$
$$A \approx \arcsin(0.3541591017)$$
$$A \approx 20.74191647°$$

Angle A measures approximately $20.742°$. Now that you know two of the angle measures, you can subtract them from $180°$ to identify the remaining angle measure in the triangle. Use the unrounded angle measurements to avoid possible rounding errors.

$$m\angle C = 180° - m\angle A - m\angle B$$
$$m\angle C \approx 180° - 20.74191648° - 127.1688997°$$
$$m\angle C \approx 32.089°$$

Problems 13.18–13.21 refer to triangle XYZ, illustrated below.

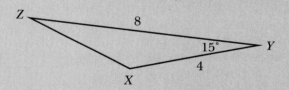

13.18 Calculate XZ using the law of cosines, and report your answer accurate to the thousandths place.

The diagram provides the lengths of two sides and the measure of the angle they form, so you are given side-angle-side or SAS; therefore, the law of cosines should be applied to calculate the missing measurement. Note that side $\overline{XZ}$ is opposite angle Y, so $y = XZ$.

$$y^2 = x^2 + z^2 - 2xz \cdot \cos Y$$
$$y^2 = 8^2 + 4^2 - 2(8)(4) \cdot \cos 15°$$
$$y^2 = 64 + 16 - 64 \cdot \cos 15°$$
$$y^2 \approx 80 - 64(\cos 15°)$$
$$y \approx \sqrt{18.18074712}$$
$$y \approx 4.263888732$$

Therefore, $XZ \approx 4.264$.

Problems 13.18–13.21 refer to triangle XYZ, illustrated in Problem 13.18.

13.19 Identify an angle that must be acute and calculate its measure using the law of sines. Report your answer accurate to the thousandths place.

Two side lengths are given in the diagram: $XY = 4$ and $YZ = 8$. According to Problem 13.18, $XZ \approx 4.264$. The largest angle must be opposite the largest side, so angle X is larger than angles Y and Z. A triangle can have, at most, one obtuse angle. In this case, only angle X could be obtuse; thus, angle Z must be acute.

$$\frac{\sin Y}{y} = \frac{\sin Z}{z}$$

$$\frac{\sin 15°}{4.263888732} \approx \frac{\sin Z}{4}$$

$$4(\sin 15°) \approx 4.263888732(\sin Z)$$

$$\frac{4 \cdot \sin 15°}{4.263888732} \approx \sin Z$$

$$\arcsin(0.2428009372) \approx Z$$

$$14.05191311° \approx Z$$

Angle Z measures approximately $14.052°$.

Problems 13.18–13.21 refer to triangle XYZ, illustrated in Problem 13.18.

13.20 Calculate the measure of the remaining angle, accurate to the thousandths place.

The diagram states that angle Y measures $15°$; according to Problem 13.19, angle Z measures approximately $14.05191311°$. Subtract these values from $180°$ to identify the measure of angle X.

$$m\angle X = 180° - m\angle Y - m\angle Z$$

$$m\angle X \approx 180° - 15° - 14.05191311°$$

$$m\angle X \approx 150.9480869°$$

Angle X measures approximately $150.948°$.

Problems 13.18–13.21 refer to triangle XYZ, illustrated in Problem 13.18.

13.21 Verify your answer to Problem 13.20 using the law of cosines.

Apply the law of cosines to calculate the measure of angle X, given the lengths of the sides of the triangle: $x = 8$, $y \approx 4.263888732$, and $z = 4$.

$$x^2 = y^2 + z^2 - 2yz \cdot \cos X$$

$$8^2 \approx (4.263888732)^2 + 4^2 - 2(4.263888732)(4) \cdot \cos X$$

$$64 \approx 34.18074712 - 34.11110986 \cdot \cos X$$

$$29.81925288 \approx -34.11110986 \cdot \cos X$$

$$\arccos\left(-\frac{29.81925288}{34.11110986}\right) \approx X$$

$$150.9480869° \approx X$$

Problems 13.22–13.24 refer to triangle LMN, illustrated below.

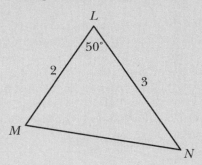

13.22 Calculate the area of triangle *LMN* using only the information given in the diagram, and report your answer accurate to the thousandths place.

You are given the lengths of two sides and the measure of the included angle (SAS); apply the area formula introduced in Problem 12.14.

$$A = \frac{1}{2}mn \sin L$$

$$= \frac{\cancel{1}}{\cancel{2}}(3)\cancel{(2)}\sin(50°)$$

$$= 3 \cdot \sin 50°$$

$$\approx 2.298$$

Problems 13.22–13.24 refer to triangle LMN, illustrated in Problem 13.22.

13.23 Calculate *MN*, and report your answer accurate to the thousandths place.

Given the lengths of two sides of a triangle and the measure of the included angle, you should apply the law of cosines to calculate the length of the side opposite the included angle. Note that *l* = *MN*.

$$l^2 = m^2 + n^2 - 2mn \cdot \cos L$$

$$l^2 = 3^2 + 2^2 - 2(3)(2) \cdot \cos 50°$$

$$l^2 = 9 + 4 - 12 \cdot \cos 50°$$

$$l^2 = 13 - 12 \cdot \cos 50°$$

$$l^2 \approx 5.286548684$$

$$l \approx \sqrt{5.286548684}$$

$$l \approx 2.299249591$$

Thus, *MN* ≈ 2.299.

Problems 13.22–13.24 refer to triangle LMN, illustrated in Problem 13.22.

13.24 Verify your answer to Problem 13.22 using Heron's formula.

Heron's formula, first introduced in Problem 12.25, allows you to calculate the area of a triangle based on the lengths of its sides. You are given $LM = 2$ and $LN = 3$; according to Problem 13.23, $MN \approx 2.299249591$. Begin by calculating the semiperimeter, s.

$$s = \frac{LM + LN + MN}{2}$$
$$\approx \frac{2 + 3 + 2.299249591}{2}$$
$$\approx \frac{7.299249591}{2}$$
$$\approx 3.649624796$$

Now apply Heron's formula to calculate the area A of triangle LMN.

$$A = \sqrt{s(s - LM)(s - LN)(s - MN)}$$
$$\approx \sqrt{3.649624796(3.649624796 - 2)(3.649624796 - 3)(3.649624796 - 2.299249591)}$$
$$\approx \sqrt{3.649624796(1.649624796)(0.649624796)(1.350375205)}$$
$$\approx \sqrt{5.281416806} \longleftarrow$$
$$\approx 2.298$$

This is the same area value you calculated in Problem 13.22, thus verifying your answer.

> Some calculators use more decimals than they actually show on screen, so the number under your square root may not end in "806" like the book's does. That's okay, as long as you still end up with the same answer rounded to the thousandths place: 2.298.

Chapter 14

VECTORS

Arrows on the coordinate plane

Vectors are directed line segments in the coordinate plane that possess both magnitude and direction. In this chapter, you plot vectors in the coordinate plane, identify equivalent and opposite vectors, express vectors in component form, calculate vector magnitude, and investigate unit vectors. Although Chapters 14–16 deal exclusively with vectors, many of the concepts you explore in this chapter are echoed in later concepts, including complex numbers and the parametric representation of graphs.

If you have had your fill of triangles, you're in luck. The final five chapters of this book are a startling change of pace. For example, in this chapter you study vectors, which are basically rays lying in the coordinate plane. Like line segments, they have a length you can calculate, but unlike line segments, they have direction. One end of a vector is represented by an arrowhead, while the other is just a point.

Plotting Vectors
Using initial and terminal points

Note: Problems 14.1–14.5 refer to vector **v**, *which has initial point P = (–2,3) and terminal point Q = (5,1).*

14.1   Plot **v** in the coordinate plane.

Vector **v** begins at point *P* = (–2,3) and extends to point *Q* = (5,1), where it terminates, as illustrated in the following graph. Note that the initial and terminal points are not explicitly depicted on the graph of the vector.

> In other words, just draw an arrow whose pointy end is the terminal point. Don't draw dots on the graph to represent the initial and terminal points of the vector.

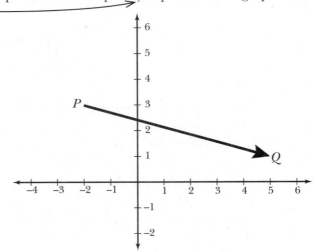

Note: Problems 14.1–14.5 refer to vector **v**, *which has initial point P = (–2,3) and terminal point Q = (5,1).*

14.2   There are two ways to name a vector. In this problem, the vector is defined using the single letter **v**. Identify the other, equally valid, name for the vector.

In this problem, vector **v** is assigned a name, but you can also identify a vector according to its initial and terminal points. Because **v** has initial point *P* and terminal point *Q*, you can refer to **v** as $\overrightarrow{PQ}$. (Some textbooks use a slightly different arrow to denote a vector: $\overline{PQ}$.) The first letter in the name must be the initial point of the vector, so $\overrightarrow{QP}$ is not the same vector as **v**; they point in opposite directions.

Note: Problems 14.1–14.5 refer to vector **v**, *which has initial point P = (–2,3) and terminal point Q = (5,1).*

14.3   Verify that the vector with initial point *R* = (0,5) and terminal point *S* = (7,3) is equivalent to **v** and explain your answer.

Equivalent vectors have the same length and direction. However, you do not need to calculate the lengths (or magnitudes) of two vectors to determine whether or not they are equivalent. You simply need to compare the horizontal and vertical changes exhibited by both vectors.

If you travel along **v** from P to Q, you move 7 units to the right and 2 units down. To calculate the change in x (horizontal change), subtract the x-coordinate of P from the x-coordinate of Q. Similarly, calculate the change in y (vertical change) by subtracting the corresponding y-coordinates.

$$\text{change in } x \text{ for } \overrightarrow{PQ}: 5-(-2)=5+2=7$$

$$\text{change in } y \text{ for } \overrightarrow{PQ}: 1-3=-2$$

Now calculate the changes in x and y that occur as you travel from the initial point to the terminal point of $\overrightarrow{RS}$.

$$\text{change in } x \text{ for } \overrightarrow{RS}: 7-0=7$$

$$\text{change in } y \text{ for } \overrightarrow{RS}: 3-5=-2$$

Although **v** and $\overrightarrow{RS}$ appear in different locations on the coordinate plane, they both describe a vector that travels 7 units to the right and 2 units down. Thus, the vectors are equivalent.

---

*Note: Problems 14.1–14.5 refer to vector **v**, which has initial point P = (–2,3) and terminal point Q = (5,1).*

14.4 Let **w** represent a vector that is equivalent to **v** and has initial point (–1,0). Identify the terminal point of **w**.

Let (x, y) represent the terminal point of **w**. As Problem 14.3 explains, equivalent vectors exhibit the same changes in x and y as you travel from their initial to their terminal points. Furthermore, Problem 14.3 states that **v** exhibits a +7 change in x and a –2 change in y.

Let (x, y) represent the terminal point of **w**. Calculate the changes in x and y for vector **w** by subtracting the coordinates of its initial point $(-1, 0)$ from the corresponding coordinates of its terminal point (x, y). Recall that the change in x is equal to 7; the change in y is equal to –2.

$$\text{change in } x \text{ for } \mathbf{w}: x-(-1)=7 \qquad \text{change in } y \text{ for } \mathbf{w}: y-0=-2$$
$$x+1=7 \qquad\qquad\qquad y=-2$$
$$x=6$$

Thus, the terminal point of **w** is (6,–2). As you travel along the vector to this coordinate from initial point (–1,0), you move 7 units right and 2 units down.

---

*Note: Problems 14.1–14.5 refer to vector **v**, which has initial point P = (–2,3) and terminal point Q = (5,1).*

14.5 Assume **z** is the opposite vector of **v**. Identify the initial and terminal points of **z**.

Vector **v** has initial point P and terminal point Q. Reverse those points to identify the opposite vector **z**. In other words, the initial point of **z** is Q and the terminal point is P. A vector and its opposite have the same length but opposite directions.

Note: In Problems 14.6–14.7, M = (0,6) and N = (–3,–1).

14.6 Plot vector $\overrightarrow{MN}$.

The vector begins at initial point $M = (0,6)$ and extends toward its terminal point $N = (–3,–1)$, as illustrated in the following graph.

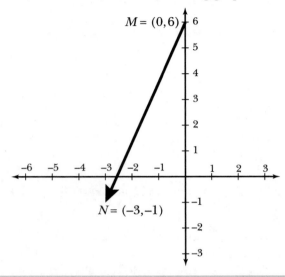

There is more than one correct answer to this problem.

Note: In Problems 14.6–14.7, M = (0,6) and N = (–3,–1).

14.7 Construct a vector **v** that is equivalent to $\overrightarrow{MN}$ and plot it on the coordinate plane.

In the diagram below, $\overrightarrow{MN}$ is illustrated as a dotted vector. It is equivalent to **v**, which has initial point $(3, 2)$ and terminal point $(0, –5)$. Both vectors travel left 3 units and down 7 units as you trace the path from their initial to terminal points.

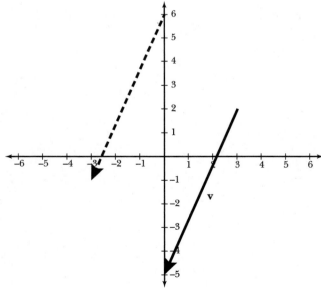

There is no single correct solution to this problem; any vector with a terminal point 3 units left of, and 7 units below, its initial point is a valid solution.

14.8 Given $P = (1.5, -3.5)$, $Q = (-4, 9.5)$, $R = (13, 6)$, and $S = (7.5, 19)$, determine whether or not $\overrightarrow{PQ}$ and $\overrightarrow{RS}$ are equivalent vectors and explain your answer.

Subtract the x- and y- coordinates of each vector's initial point from the coordinates of its terminal point to determine whether the vectors demonstrate the same changes in x and y.

$$\text{change in } x \text{ for } \overrightarrow{PQ}: \ -4 - 1.5 = -5.5$$
$$\text{change in } y \text{ for } \overrightarrow{PQ}: \ 9.5 - (-3.5) = 13$$

$$\text{change in } x \text{ for } \overrightarrow{RS}: \ 7.5 - 13 = -5.5$$
$$\text{change in } y \text{ for } \overrightarrow{RS}: \ 19 - 6 = 13$$

Both vectors travel left 5.5 units and up 13 units from their initial to terminal points, so they are equivalent.

14.9 Given $\overrightarrow{AB}$, the vector illustrated below, identify the terminal point of the equivalent vector **v**, which has initial point $(9, -8)$.

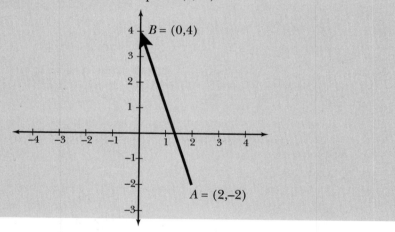

Vector $\overrightarrow{AB}$ has initial point A = $(2, -2)$ and terminal point $(0, 4)$. Thus, it exhibits a horizontal change of $0 - 2 = -2$ and a vertical change of $4 - (-2) = 6$. You are given an initial point of $(9, -8)$ for equivalent vector **v**. Let (x, y) represent its terminal point. Recall that equivalent vectors exhibit the same horizontal and vertical changes.

change in x for **v**: $x - 9 = -2$ change in y for **v**: $y - (-8) = 6$
$$x = 9 - 2$$
$$x = 7$$
$$y + 8 = 6$$
$$y = -2$$

The terminal point of **v** is $(7, -2)$.

Note: In Problems 14.10–14.11, assume that $\overrightarrow{AB}$ and $\overrightarrow{CD}$ are equivalent vectors, such that $A = (-4, 11)$, $B = (8, -3)$, $C = (x - 1, 3x)$, and $D = (4x, y + 2)$.

14.10 Calculate x.

> Make sure to distribute the negative sign to both terms in the parentheses: x and -1.

Calculate the horizontal change exhibited by each vector.

$$\text{change in } x \text{ for } \overrightarrow{AB}: \ 8-(-4)=8+4=12$$

$$\text{change in } x \text{ for } \overrightarrow{CD}: \ 4x-(x-1)=4x-x+1=3x+1$$

Equivalent vectors have equivalent horizontal changes, so $12 = 3x + 1$. Solve this equation for x.

$$12 = 3x + 1$$
$$11 = 3x$$
$$\frac{11}{3} = x$$

Note: In Problems 14.10–14.11, assume that $\overrightarrow{AB}$ and $\overrightarrow{CD}$ are equivalent vectors, such that $A = (-4, 11)$, $B = (8, -3)$, $C = (x - 1, 3x)$, and $D = (4x, y + 2)$.

14.11 Calculate y.

Calculate the vertical change exhibited by each vector.

$$\text{change in } y \text{ for } \overrightarrow{AB}: \ -3-11=-14$$

$$\text{change in } y \text{ for } \overrightarrow{CD}: \ y+2-3x$$

Equivalent vectors have equivalent vertical changes, so $-14 = y + 2 - 3x$. According to Problem 14.10, $x = 11/3$. Substitute that value into the equation and solve for y.

$$-14 = y+2-\cancel{3}\left(\frac{11}{\cancel{3}}\right)$$
$$-14 = y+2-11$$
$$-14 = y-9$$
$$-5 = y$$

Component Form
Move the initial point to the origin

14.12 Given a vector $\mathbf{v}$ in standard position, identify its initial point and explain how to express it in component form.

> If vector v goes right 5 units and up 9 units as you travel from its initial point to its terminal point, then $v = \langle 5, 9 \rangle$.

A vector $\mathbf{v}$ in standard position has an initial point of $(0,0)$—the origin of the coordinate plane. The coordinates of its terminal point are the components of the vector. For example, if $\mathbf{v}$ has terminal point (a, b), then the component form of the vector is $\mathbf{v} = \langle a, b \rangle$.

Note: In Problems 14.13–14.16, v = < 2, –1 >.

14.13 Identify the initial and terminal points of **v**, assuming **v** is in standard position.

> These pointy grouping symbols (that look like less than and greater than symbols) indicate that the vector is in component form.

A vector in component form describes the horizontal and vertical changes of the vector without tying it to a specific position on the coordinate plane. Thus, **v** may be located anywhere in the coordinate plane, but you know that its terminal point is 2 units right of, and 1 unit below, its initial point.

This problem states that **v** is in standard position, so its initial point is (0,0). Its terminal point has the same coordinates as the component form of **v**. Therefore, the terminal point of **v** is (2,–1).

> Some textbooks refer to vectors in standard position as "standard vectors."

Note: In Problems 14.13–14.16, v = < 2, –1 >.

14.14 Plot **v** in standard position.

According to Problem 14.13, **v** has initial point (0,0) and terminal point (2,–1).

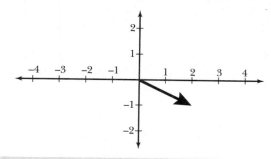

Note: In Problems 14.13–14.16, v = < 2, –1 >.

14.15 Assume that **w** is the opposite of **v**. Express **w** in component form.

If a vector has component form $<a, b>$, its opposite vector has component form $<-a, -b>$. Thus, **w** = <–2, 1>.

Note: In Problems 14.13–14.16, v = < 2, –1 >.

14.16 Plot **w** in standard position on the coordinate plane.

Problem 14.15 states that **w** = <–2, 1>. If **w** is in standard position, then its initial point is (0,0) and its terminal point is (–2,1).

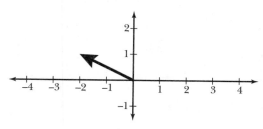

Note: In Problems 14.17–14.18, V = (4,7) and W = (1,6).

14.17 Express $\overrightarrow{VW}$ in component form.

The component form of a vector describes its horizontal and vertical changes as you travel from the initial to the terminal point. Thus, you apply the technique introduced in Problem 14.3, subtracting the x- and y-coordinates of the initial point from the corresponding coordinates of the terminal point.

In other words, if a vector has initial point (a, b) and terminal point (c, d), then the component form of that vector is $\langle c - a, d - b \rangle$. In this problem, $a = 4$, $b = 7$, $c = 1$, and $d = 6$.

$$\overrightarrow{VW} = \langle 1 - 4, 6 - 7 \rangle$$
$$= \langle -3, -1 \rangle$$

Note: In Problems 14.17–14.18, V = (4,7) and W = (1,6).

14.18 Plot $\overrightarrow{VW}$ in its original position and in standard position to visually verify that the vectors are equivalent.

According to Problem 14.17, the component form of the vector is $\langle -3, -1 \rangle$, so in standard position, the vector travels from an initial point of $(0,0)$ to a terminal point of $(-3,-1)$.

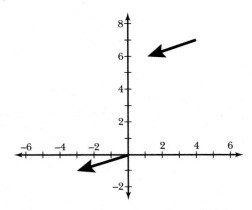

14.19 Given $S = (-10,6)$ and $T = (3,-3)$, express $\overrightarrow{ST}$ in component form.

Apply the technique described in Problem 14.17, subtracting the coordinates of S from the corresponding coordinates of T.

$$\overrightarrow{ST} = \langle 3 - (-10), -3 - 6 \rangle$$
$$= \langle 13, -9 \rangle$$

14.20 Given $Q = \left(\dfrac{5}{2}, -2\right)$ and $R = \left(-8, \dfrac{5}{3}\right)$ express $\overrightarrow{QR}$ in component form.

Subtract the components of Q from the corresponding components of R. Use least common denominators to reduce the fractions to lowest terms.

$$\overrightarrow{QR} = \left\langle -8 - \frac{5}{2}, \frac{5}{3} - (-2) \right\rangle$$

$$= \left\langle -\frac{16}{2} - \frac{5}{2}, \frac{5}{3} + \frac{6}{3} \right\rangle$$

$$= \left\langle -\frac{21}{2}, \frac{11}{3} \right\rangle$$

*Note: Problems 14.21–14.22 refer to **w**, the vector plotted below.*

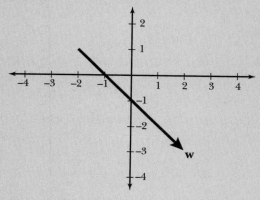

14.21 Express **w** in component form.

Vector **w** has initial point (–2,1) and terminal point (2,–3).

$$\mathbf{w} = \left\langle 2 - (-2), -3 - 1 \right\rangle$$

$$= \left\langle 2 + 2, -3 - 1 \right\rangle$$

$$= \left\langle 4, -4 \right\rangle$$

*Note: Problems 14.21–14.22 refer to **w**, the vector plotted in Problem 14.21.*

14.22 Plot **w** in standard position.

According to Problem 14.21, **w** = <4,–4>. Therefore, in standard position, **w** has initial point (0,0) and terminal point (4,–4).

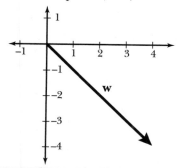

14.23 Determine whether vectors **r** and **s**, plotted below, are equivalent.

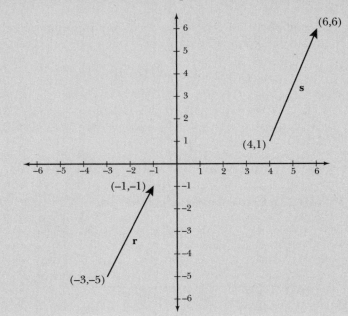

Vector **r** has initial point (–3,–5) and terminal point (–1,–1); **s** has initial point (4,1) and terminal point (6,6). Express both vectors in component form.

$$\mathbf{r} = \langle -1-(-3), -1-(-5) \rangle \qquad \mathbf{s} = \langle 6-4, 6-1 \rangle$$
$$= \langle -1+3, -1+5 \rangle \qquad\qquad = \langle 2,5 \rangle$$
$$= \langle 2,4 \rangle$$

Both vectors exhibit the same horizontal change—they travel 2 units to the right between their initial and terminal points. However, **r** and **s** exhibit different vertical changes, so they are not equivalent vectors. Two vectors are equivalent if and only if they have the same component form, and $\langle 2,4 \rangle \neq \langle 2,5 \rangle$.

In other words, if two vectors have the same component form, then they are equivalent. Furthermore, if two vectors are equivalent, then they must have the same component form.

14.24 Given $P = \left(2c^2 - 3c, 4d - d^2\right)$ and $Q = \left(7c - 1, d^2 + 6d - 10\right)$, such that c and d are real numbers, express $\overrightarrow{PQ}$ in component form.

Subtract the components of initial point P from the corresponding components of terminal point Q and combine like terms.

$$\overrightarrow{PQ} = \left\langle 7c - 1 - \left(2c^2 - 3c\right), d^2 + 6d - 10 - \left(4d - d^2\right) \right\rangle$$
$$= \left\langle 7c - 1 - 2c^2 + 3c, d^2 + 6d - 10 - 4d + d^2 \right\rangle$$
$$= \left\langle -2c^2 + 10c - 1, 2d^2 + 2d - 10 \right\rangle$$

Magnitude

How long is the vector?

Note: Problems 14.25–14.26 refer to a vector in component form: v = <3,4>.

14.25 Calculate $\|\mathbf{v}\|$.

The notation $\|\mathbf{v}\|$ represents the magnitude, or length, of vector **v**. Given a vector **v** expressed in component form $\mathbf{v} = \langle a, b \rangle$, the magnitude of **v** is defined according to the following formula.

$$\|\mathbf{v}\| = \sqrt{a^2 + b^2}$$

In this problem, the components of **v** are $a = 3$ and $b = 4$. Substitute these values into the magnitude formula.

$$\begin{aligned} \|\mathbf{v}\| &= \sqrt{3^2 + 4^2} \\ &= \sqrt{9 + 16} \\ &= \sqrt{25} \\ &= 5 \end{aligned}$$

The magnitude of **v** is 5, so the vector has a length of 5.

Note: Problems 14.25–14.26 refer to a vector in component form: v = <3,4>.

14.26 Verify your answer to Problem 14.25 by plotting **v** in the coordinate plane and applying the Pythagorean theorem.

Plot **v** in standard position in the coordinate plane, with initial point (0,0) and terminal point (3,4). If you draw a vertical segment connecting the terminal point to the *x*-axis, you construct a right triangle whose legs have lengths 3 and 4.

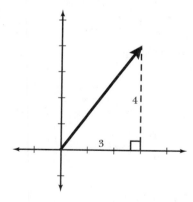

Apply the Pythagorean theorem to calculate the length of the vector, the hypotenuse of the right triangle.

$$3^2 + 4^2 = \left(\|\mathbf{v}\|\right)^2$$
$$9 + 16 = \left(\|\mathbf{v}\|\right)^2$$
$$\sqrt{25} = \sqrt{\left(\|\mathbf{v}\|\right)^2}$$
$$5 = \|\mathbf{v}\|$$

It is no coincidence that the solutions to Problems 14.25–14.26 are equivalent, as the vector magnitude formula is derived from the Pythagorean theorem.

14.27 Calculate the magnitude of $\mathbf{w} = <-12, 5>$.

Apply the vector magnitude formula introduced in Problem 12.25.

$$\|\mathbf{w}\| = \sqrt{(-12)^2 + 5^2}$$
$$= \sqrt{144 + 25}$$
$$= \sqrt{169}$$
$$= 13$$

14.28 Calculate the magnitude of $\mathbf{r} = <-6, -3>$.

Remember to square both components when you apply the magnitude formula. Magnitude is always a positive real number value.

$$\|\mathbf{r}\| = \sqrt{(-6)^2 + (-3)^2}$$
$$= \sqrt{36 + 9}$$
$$= \sqrt{45}$$
$$= 3\sqrt{5}$$

14.29 Given $\mathbf{s} = <4, -11>$, calculate $\|\mathbf{s}\|$.

The magnitude of a vector is equal to the square root of the sum of its components' squares.

$$\|\mathbf{s}\| = \sqrt{4^2 + (-11)^2}$$
$$= \sqrt{16 + 121}$$
$$= \sqrt{137}$$

14.30 Given $\mathbf{v} = \left\langle \dfrac{4}{3}, -\dfrac{7}{2} \right\rangle$, calculate $\|\mathbf{v}\|$.

Substitute the components of $\mathbf{v}$ into the magnitude formula and use a least common denominator to reduce the expression to lowest terms.

$$
\begin{aligned}
\|\mathbf{v}\| &= \sqrt{\left(\frac{4}{3}\right)^2 + \left(-\frac{7}{2}\right)^2} \\
&= \sqrt{\frac{16}{9} + \frac{49}{4}} \\
&= \sqrt{\frac{16}{9} \cdot \frac{4}{4} + \frac{49}{4} \cdot \frac{9}{9}} \\
&= \sqrt{\frac{64}{36} + \frac{441}{36}} \\
&= \sqrt{\frac{505}{36}} \\
&= \frac{\sqrt{505}}{6}
\end{aligned}
$$

14.31 Given $A = (2, -9)$ and $B = (-4, -1)$, calculate $\left\| \overrightarrow{AB} \right\|$.

Begin by rewriting the vector in component form: $\overrightarrow{AB} = \left\langle -6, 8 \right\rangle$. Substitute the components into the magnitude formula.

$$
\begin{aligned}
\left\| \overrightarrow{AB} \right\| &= \sqrt{(-6)^2 + (8)^2} \\
&= \sqrt{36 + 64} \\
&= \sqrt{100} \\
&= 10
\end{aligned}
$$

<-4 - 2, -1 - (-9)>
= <-6, -1 + 9>
= <-6, 8>

14.32 Given $P = (8, -5)$ and $Q = (4, -14)$, calculate $\left\| \overrightarrow{PQ} \right\|$.

Express the vector in component form: $\overrightarrow{PQ} = \left\langle 4 - 8, -14 - (-5) \right\rangle = \left\langle -4, -9 \right\rangle$. Next, apply the magnitude formula.

$$
\begin{aligned}
\left\| \overrightarrow{PQ} \right\| &= \sqrt{(-4)^2 + (-9)^2} \\
&= \sqrt{16 + 81} \\
&= \sqrt{97}
\end{aligned}
$$

Note: In Problems 14.33–14.34, $\mathbf{v} = \langle 4,4 \rangle$ and $\mathbf{w} = \langle -5, -\sqrt{7} \rangle$.

14.33 Verify that $\|\mathbf{v}\| = \|\mathbf{w}\|$.

Compute the magnitude of each vector individually.

$$\|\mathbf{v}\| = \sqrt{4^2 + 4^2} \qquad \|\mathbf{w}\| = \sqrt{(-5)^2 + \left(-\sqrt{7}\right)^2}$$
$$= \sqrt{16 + 16} \qquad\qquad = \sqrt{25 + 7}$$
$$= \sqrt{32} \qquad\qquad\quad = \sqrt{32}$$
$$= 4\sqrt{2} \qquad\qquad\quad = 4\sqrt{2}$$

> Vectors are only equivalent if they have the same component form. In other words, $v = \langle a,b \rangle$ and $w = \langle c,d \rangle$ are equivalent only if $a = c$ and $b = d$.

Note: In Problems 14.33–14.34, $\mathbf{v} = \langle 4,4 \rangle$ and $\mathbf{w} = \langle -5, -\sqrt{7} \rangle$.

14.34 Determine whether $\mathbf{v}$ and $\mathbf{w}$ are equivalent vectors and explain your answer.

Equivalent vectors always have the same magnitude, but vectors with the same magnitude are not always equivalent. In order for two vectors to be equivalent, they must have the same length and direction. While $\mathbf{v}$ and $\mathbf{w}$ have the same length, they do not have the same direction. If you plot $\mathbf{v}$ in standard position, its terminal point is (4,4), which is located in the first quadrant. Note that $\mathbf{w}$, plotted in standard position, terminates in the third quadrant.

14.35 Given $\mathbf{r} = \langle 3x, -6 \rangle$ and $\|\mathbf{r}\| = 10$, calculate x.

Apply the magnitude formula, noting that the magnitude of the vector is equal to 10.

$$\|\mathbf{r}\| = \sqrt{(3x)^2 + (-6)^2}$$
$$10 = \sqrt{9x^2 + 36}$$
$$(10)^2 = \left(\sqrt{9x^2 + 36}\right)^2$$
$$100 = 9x^2 + 36$$
$$100 = 9\left(x^2 + 4\right)$$
$$\frac{100}{9} = x^2 + 4$$

Solve for x.

$$\frac{100}{9} - 4 = x^2$$
$$\frac{100}{9} - \frac{36}{9} = x^2$$
$$\frac{64}{9} = x^2$$
$$\pm\sqrt{\frac{64}{9}} = x$$
$$\pm\frac{8}{3} = x$$

> If you substitute either $x = +8/3$ or $x = -8/3$ into the x-component of r, the magnitude of r is 10.

14.36 Given $\mathbf{v} = \langle a+1, a-1 \rangle$ and $\|\mathbf{v}\| = \sqrt{34}$, calculate a.

Apply the magnitude formula to $\mathbf{v}$, noting that its magnitude is $\sqrt{34}$.

$$\|\mathbf{v}\| = \sqrt{(a+1)^2 + (a-1)^2}$$
$$\sqrt{34} = \sqrt{(a^2 + 2a + 1) + (a^2 - 2a + 1)}$$
$$\sqrt{34} = \sqrt{2a^2 + 2}$$

Square both sides of the equation and solve for a.

$$\left(\sqrt{34}\right)^2 = \left(\sqrt{2a^2 + 2}\right)^2$$
$$34 = 2a^2 + 2$$
$$\frac{34}{2} = \frac{2a^2}{2} + \frac{2}{2}$$
$$17 = a^2 + 1$$
$$16 = a^2$$
$$\pm\sqrt{16} = a$$
$$\pm 4 = a$$

If $a = -4$ or $a = +4$, then $\left\|\langle a+1, a-1 \rangle\right\| = \sqrt{34}$.

Unit Vectors

Vectors with a magnitude of 1

14.37 Describe the relationship between a vector $\mathbf{v}$ and its unit vector $\hat{\mathbf{v}}$.

Given a vector $\mathbf{v}$, the corresponding unit vector $\hat{\mathbf{v}}$ has the same direction as $\mathbf{v}$ with a magnitude of 1.

> Just like a unit circle has a radius of 1, a unit vector has a length of 1.

Note: In Problems 14.38–14.39, $\mathbf{v}$ = <4, 3>.

14.38 Given $\mathbf{v}$ = <4,3>, express its unit vector $\hat{\mathbf{v}}$ in component form.

Begin by calculating the magnitude of $\mathbf{v}$.

$$\|\mathbf{v}\| = \sqrt{4^2 + 3^2}$$
$$= \sqrt{16 + 9}$$
$$= \sqrt{25}$$
$$= 5$$

To compute the unit vector, apply this formula: $\hat{\mathbf{v}} = \dfrac{\mathbf{v}}{\|\mathbf{v}\|}$. In other words, divide each component of **v** by the magnitude of **v**.

$$\hat{\mathbf{v}} = \left\langle \frac{4}{\|\mathbf{v}\|}, \frac{3}{\|\mathbf{v}\|} \right\rangle$$

$$= \left\langle \frac{4}{5}, \frac{3}{5} \right\rangle$$

*Note: In Problems 14.38–14.39, **v** = <4, 3>.*

14.39 Verify that the unit vector $\hat{\mathbf{v}}$ identified in Problem 14.38 has a magnitude of 1.

According to problem 14.38, $\hat{\mathbf{v}} = \left\langle \dfrac{4}{5}, \dfrac{3}{5} \right\rangle$. Apply the magnitude formula.

$$\|\hat{\mathbf{v}}\| = \sqrt{\left(\frac{4}{5}\right)^2 + \left(\frac{3}{5}\right)^2}$$

$$= \sqrt{\frac{16}{25} + \frac{9}{25}}$$

$$= \sqrt{\frac{25}{25}}$$

$$= \sqrt{1}$$

$$= 1$$

14.40 Given **w** = <–2,–5>, express the corresponding unit vector $\hat{\mathbf{w}}$ in component form.

Begin by calculating the magnitude of **w**.

$$\|\mathbf{w}\| = \sqrt{(-2)^2 + (-5)^2}$$

$$= \sqrt{4 + 25}$$

$$= \sqrt{29}$$

Divide each of the components of **w** by the magnitude of the vector and rationalize the denominators.

$$\hat{\mathbf{w}} = \left\langle \frac{-2}{\sqrt{29}}, \frac{-5}{\sqrt{29}} \right\rangle$$

$$\hat{\mathbf{w}} = \left\langle -\frac{2\sqrt{29}}{29}, -\frac{5\sqrt{29}}{29} \right\rangle$$

Unless your instructor doesn't mind square roots in the denominator

14.41 Express the standard unit vectors **i** and **j** in component form.

The standard unit vectors **i** and **j** (which may also be written $\hat{\mathbf{i}}$ and $\hat{\mathbf{j}}$ or $\hat{i}$ and $\hat{j}$) represent unit vectors in the horizontal and vertical directions, respectively. In other words, **i** is a horizontal vector with length 1 and **j** is a vertical vector with length 1. In component form, **i** = <1, 0> and **j** = <0, 1>.

14.42 Write vector **v** = <8, 3> in terms of standard unit vectors.

In standard positition, **v** stretches from the origin to the point (8,3). It moves a total of 8 units to the right and 3 units up. Each of those units in the horizontal and vertical directions can be represented by individual unit vectors: 8 horizontal vectors **i** and 3 vertical vectors **j**.

$$\mathbf{v} = 8\mathbf{i} + 3\mathbf{j}$$

If a vector is in component form <*a, b*>, you can write it in terms of standard unit vectors by multiplying the *x*-component by **i**, multiplying the *y*-component by **j**, and then adding those products together.

$$<a, b> = a\mathbf{i} + b\mathbf{j}$$

14.43 Write vector $\mathbf{w} = \left\langle 5, -\sqrt{2} \right\rangle$ in terms of standard unit vectors.

Apply the technique described in Problem 14.42: Multiply the horizontal component by the standard unit vector **i** = <1, 0>, multiply the vertical component by the standard unit vector **j** = <0, 1>, and add the products.

$$\mathbf{w} = 5\mathbf{i} + \left(-\sqrt{2}\right)(\mathbf{j})$$
$$\mathbf{w} = 5\mathbf{i} - \sqrt{2}\mathbf{j}$$

*Note: In Problems 14.44–14.45, **r** = <1, 1> and **s** = <4, 4>.*

14.44 Identify the direction in which **r** travels by graphing it in standard position and calculating its direction angle, the measure of the acute angle it forms with the positive *x*-axis.

Consider the diagram below, in which **r** is the hypotenuse of a right triangle. The horizontal and vertical legs of the triangle are, respectively, the *x*- and *y*-components of **r**. In other words, both legs have a length of 1.

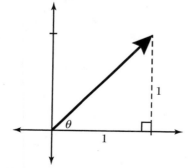

Recall that the tangent value of an angle in a right triangle is equal to the length of the side opposite that angle divided by the length of the side adjacent to that angle. Use this trigonometric ratio to calculate θ.

$$\tan\theta = \frac{\text{length of side opposite } \theta}{\text{length of side adjacent to } \theta}$$

$$\tan\theta = \frac{1}{1}$$

$$\tan\theta = 1$$

$$\theta = \arctan(1)$$

$$\theta = \frac{\pi}{4} \text{ or } 45°$$

You conclude that **r** has a direction of 45°, measured counterclockwise from the positive x-axis when **r** is in standard position.

> *More simply, the vector has a direction angle of 45°.*

Note: In Problems 14.44–14.45, r = <1, 1> and s = <4, 4>.

14.45 Verify that **s** has the same direction angle as **r**.

In the diagram below, **s** is plotted in standard position, forming the hypotenuse of a right triangle whose legs have a length of 4.

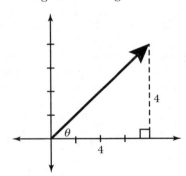

Apply the technique described in Problem 14.44; use the tangent ratio to calculate the measure of θ.

$$\tan\theta = \frac{\text{length of side opposite } \theta}{\text{length of side adjacent to } \theta}$$

$$\tan\theta = \frac{4}{4}$$

$$\tan\theta = 1$$

$$\theta = \arctan(1)$$

$$\theta = \frac{\pi}{4} \text{ or } 45°$$

You conclude that **s**, like **r**, has a direction angle of 45°.

14.46 Based on your answers to Problems 14.44–14.45, identify a vector that has the same direction as **v** = <*a*, *b*>.

Vectors **r** and **s** in Problems 14.44–14.45 have the same direction angle: 45°. In fact, all vectors in the first quadrant with equal *x*- and *y*-components have direction angle 45°.

If you multiply both components of a vector by the same positive real number, the result is a vector in the same direction. In this case, multiplying each of the components of **r** = <1, 1> by 4 produces the vector **s** = <4, 4> in the same direction as **r**. Therefore, if you multiply the components of **v** = <*a*, *b*> by a positive real number *c*, then vector <*ac*, *bc*> has the same direction as **v**. ←

> If *c* is negative, then <*ac*, *bc*> and **v** travel in opposite directions.

14.47 Explain why **v** = <−2, 5> and **w** = <−4, 10> have the same direction and visually verify your answer by plotting both vectors on the same coordinate plane.

According to Problem 14.46, multiplying both components of a vector by the same positive real number produces a vector in the same direction. In this case, multiplying the components of **v** by 2 generates **w**.

$$<(-2)(2), (5)(2)> = <-4, 10>$$

Plot **v** and **w** to visually verify that the vectors have the same direction. In the diagram below, the shorter vector **v** overlaps longer vector **w**, so both vectors have the same direction.

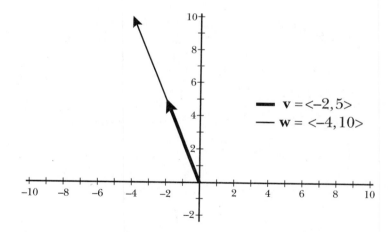

*Note: In Problems 14.48–14.49, **v** = <12, –5>.*

14.48 Create a vector of length 2 that has the same direction as **v** = <12, –5>.

Begin by calculating **v̂**, a unit vector in the same direction as **v**. Apply the technique described in Problems 14.38 and 14.40, which requires that you first compute the magnitude of the vector.

$$\|\mathbf{v}\| = \sqrt{12^2 + (-5)^2}$$
$$= \sqrt{144 + 25}$$
$$= \sqrt{169}$$
$$= 13$$

Divide each component of **v** by the magnitude to identify the unit vector in the same direction as **v**.

$$\hat{\mathbf{v}} = \left\langle \frac{12}{13}, -\frac{5}{13} \right\rangle$$

Now that you have identified a unit vector with the same direction as **v**, you need to identify a vector in that direction with a length of 2. Simply multiply each component of **v̂** by the desired magnitude, in this case 2.

$$2\hat{\mathbf{v}} = \left\langle 2\left(\frac{12}{13}\right), 2\left(-\frac{5}{13}\right) \right\rangle$$
$$= \left\langle \frac{24}{13}, -\frac{10}{13} \right\rangle$$

Remember, multiplying both components of a vector by a positive real number produces a vector in the same direction as the original vector, so <24/13, –10/13> has the same direction as **v** and **v̂**.

*Note: In Problems 14.48–14.49, **v** = <12, –5>.*

14.49 Verify that the vector you identified in Problem 14.48 has a magnitude of 2.

According to Problem 14.48, <24/13, –10/13> has a magnitude of 2. Apply the magnitude formula to verify this claim.

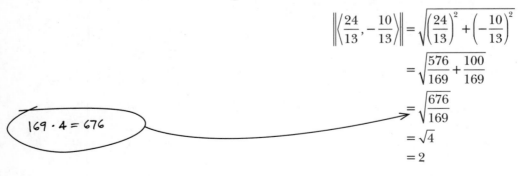

$$\left\| \left\langle \frac{24}{13}, -\frac{10}{13} \right\rangle \right\| = \sqrt{\left(\frac{24}{13}\right)^2 + \left(-\frac{10}{13}\right)^2}$$
$$= \sqrt{\frac{576}{169} + \frac{100}{169}}$$
$$= \sqrt{\frac{676}{169}}$$
$$= \sqrt{4}$$
$$= 2$$

$169 \cdot 4 = 676$

14.50 Create a vector of length 6 that has the same direction as $\mathbf{w} = \langle -1, 2\sqrt{2} \rangle$.

Calculate the magnitude of $\mathbf{w}$.

$$\begin{aligned} \|\mathbf{w}\| &= \sqrt{(-1)^2 + \left(2\sqrt{2}\right)^2} \\ &= \sqrt{1 + 4 \cdot 2} \\ &= \sqrt{1 + 8} \\ &= \sqrt{9} \\ &= 3 \end{aligned}$$

Divide each component of $\mathbf{w}$ by its magnitude to identify the unit vector $\hat{\mathbf{w}}$ in the same direction as $\mathbf{w}$.

$$\hat{\mathbf{w}} = \left\langle -\frac{1}{3}, \frac{2\sqrt{2}}{3} \right\rangle$$

You are directed to identify a vector in the same direction with a magnitude of 6. Thus, you should multiply both components of the unit vector by 6 to complete the problem.

$$\begin{aligned} 6\hat{\mathbf{w}} &= \left\langle 6\left(-\frac{1}{3}\right), 6\left(\frac{2\sqrt{2}}{3}\right) \right\rangle \\ &= \left\langle -\frac{6}{3}, \frac{12\sqrt{2}}{3} \right\rangle \\ &= \left\langle -2, 4\sqrt{2} \right\rangle \end{aligned}$$

14.51 Create a vector of length d that has the same direction as $\mathbf{v} = \langle a, b \rangle$.

Begin by calculating the magnitude of $\mathbf{v}$.

$$\|\mathbf{v}\| = \sqrt{a^2 + b^2}$$

Divide each component of $\mathbf{v}$ by its magnitude to identify a unit vector $\hat{\mathbf{v}}$ in the same direction as $\mathbf{v}$.

$$\hat{\mathbf{v}} = \left\langle \frac{a}{\sqrt{a^2 + b^2}}, \frac{b}{\sqrt{a^2 + b^2}} \right\rangle$$

You are instructed to create a vector of length d, so multiply each of the components by that constant.

$$\left\langle \frac{ad}{\sqrt{a^2 + b^2}}, \frac{bd}{\sqrt{a^2 + b^2}} \right\rangle$$

This is just like Problems 14.48 and 14.50, but you're using variables instead of numbers.

Chapter 15
BASIC VECTOR OPERATIONS
Add/subtract vectors and multiply by scalars

In Chapter 14, you are introduced to vectors, including how they are plotted in the coordinate plane, how they can be expressed in component form, and how they can be transformed into vectors of any length in the same direction. All of those concepts and operations involved a single vector.

In this chapter, you will explore vector operations, which involve multiple vectors and real number values. You begin by combining vectors via addition and subtraction, eventually extending both of those operations through scalar multiplication. At the end of this chapter, you return to the familiar topic of component form, but with a complicating twist: How can you determine the component form of a vector if you are not given the coordinates of its initial or terminal points?

You start this chapter by adding two vectors together using a very curious technique: You glue the terminal point of the first vector to the initial point of the second vector. Once you feel comfortable with vector addition (and subtraction, which is almost the same thing) in the coordinate plane, you graduate to the shortcut technique. Trust me, you will love the shortcut. Less graphing + easy formula = happy you!

Next, it's on to scalar multiplication, which is very straightforward. In fact, you unknowingly applied it in Chapter 14. Finally, you'll dust off sine and cosine to calculate the components of a vector that's floating somewhere in the coordinate plane, not in standard position. Basically, given the magnitude and direction of a vector, you'll be able to write it in component form.

Adding and Subtracting Graphically

Head-to-tail technique

15.1 Describe the technique used to add vectors graphically in the coordinate plane.

The sum of two vectors is also a vector, and it is a simple matter to visualize that sum in the coordinate plane through a technique known informally as the head-to-tail method. To add vector **w** to vector **v**, place the initial point of **w** at the terminal point of **v**. The sum **v** + **w** is called the resultant vector; it shares an initial point with **v** and a terminal point with **w**.

> In other words, move w so it starts exactly where v ends.

15.2 Given **v** = <2,0>, calculate **v** + **v** by plotting the resultant vector in the coordinate plane.

Vector **v** = <2,0> is a horizontal vector with a magnitude of 2 and a direction angle of 0°; it points directly to the right, due east from its initial point. If you graph **v** in standard position, its initial point is (0,0) and its terminal point is (2,0).

In order to calculate **v** + **v**, graph **v** in standard position and then plot a second copy of **v**, such that the second vector begins where the first ends. As the following diagram illustrates, the second copy of **v** has an initial point of (2,0) and a terminal point of (4,0).

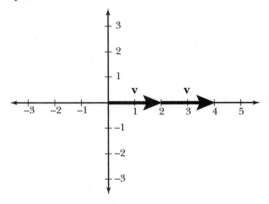

The resultant vector, illustrated in the following diagram, represents **v** + **v**. It begins at the initial point of the first vector (0,0) and ends at the terminal point of the second vector (4,0).

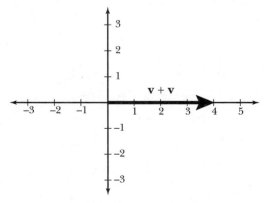

Note that the resultant vector is in standard position, because its initial point is the origin. Therefore, the coordinates of its terminal point are also its components: **v** + **v** = <4, 0>.

15.3 Given **r** = <3, 0> and **s** = <0, 5>, calculate **r** + **s** by plotting the resultant vector in the coordinate plane.

Plot **r** in standard position; it has initial point (0, 0) and terminal point (3, 0). Next, plot **s** so that its initial point coincides with the terminal point of **r**. In other words, from its initial point (3, 0), **s** travels a horizontal distance of 0 and a vertical distance of 5 units up.

Just pretend that (3,0)—the terminal point of r—is the origin, and plot s as you normally would.

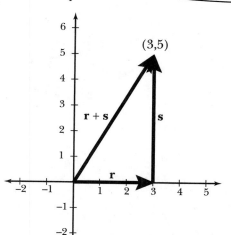

The resultant vector **r** + **s** has initial point (0, 0) and terminal point (3, 5). Therefore, **r** + **s** = <3, 5>.

*Note: In Problems 15.4–15.5, **v** = <–4, –1> and **w** = <9, –2>.*

15.4 Given **v** = <–4, –1> and **w** = <9, –2>, calculate **v** + **w** by plotting the resultant vector in the coordinate plane.

Plot **v** in standard position and then place the initial point of **w** at the terminal point of **v**.

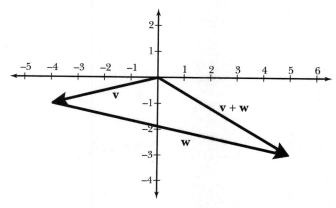

The resultant vector has an initial point of (0, 0) and a terminal point of (5, –3). Thus, **v** + **w** = <5, –3>.

Commutative means the order in which you add doesn't matter:

$v + w = w + v.$

Note: In Problems 15.4–15.5, v = <−4,−1> and w = <9,−2>.

15.5 Demonstrate that vector addition is commutative by plotting **w** + **v** in the coordinate plane and verifying that the resultant vector is equal to the sum calculated in Problem 15.4.

Plot **w** in standard position and then place the initial point of **v** at the terminal point of **w**.

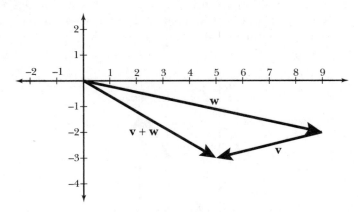

The resultant vector has an initial point of $(0,0)$ and a terminal point of $(5,-3)$. Thus, **w** + **v** = <5,−3>.

15.6 Given $A = (0,3)$, $B = (4,2)$, $C = (−1,−5)$, and $D = (−7,1)$ calculate $\overrightarrow{AB} + \overrightarrow{CD}$ by plotting the resultant vector in the coordinate plane.

Because $\overrightarrow{CD} = <−6,6>$, you should travel 6 units left and 6 units up from (4, −1).

Express both of the vectors in component form by subtracting the components of their initial points from the components of their terminal points.

$$\overrightarrow{AB} = \langle 4 - 0, 2 - 3 \rangle \qquad \overrightarrow{CD} = \langle -7 - (-1), 1 - (-5) \rangle$$
$$= \langle 4, -1 \rangle \qquad\qquad = \langle -7 + 1, 1 + 5 \rangle$$
$$\qquad\qquad\qquad = \langle -6, 6 \rangle$$

Plot $\overrightarrow{AB}$ in standard position. Its terminal point, $(4,-1)$, is the initial point of $\overrightarrow{CD}$.

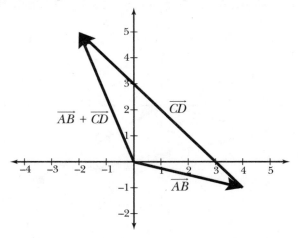

You conclude that $\overrightarrow{AB} + \overrightarrow{CD} = \langle -2, 5 \rangle$.

15.7   Given **r** = <0,5> and **s** = <6,4>, calculate **r** − **s** by plotting the resultant vector in the coordinate plane.

In order to plot the resultant vector for vector subtraction, rewrite the subtraction expression using addition. In other words, rather than subtract **s** from **r**, you add −**s** to **r**. Multiply both components of **s** by −1 to calculate −**s**.

$$-\mathbf{s} = \langle -1 \cdot 6, -1 \cdot 4 \rangle$$
$$= \langle -6, -4 \rangle$$

Plot **r** in standard position and then place the initial point of −**s** at the terminal point of **r**.

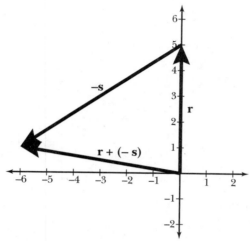

You conclude that **r** − **s** = <−6, 1>.

Note: In Problems 15.8–15.9, **v** = −2**i** − **j** *and* **w** = 4**i** − 3**j**.

15.8   Calculate **v** + **w** by plotting the resultant vector in the coordinate plane.

Note that **v** and **w** are expressed in terms of standard unit vectors **i** = <1,0> and **j** = <0,1>. To rewrite each in component form, use the coefficients of **i** and **j** as the *x*- and *y*-components, respectively: **v** = <−2,−1> and **w** = <4,−3>. Plot **v** in standard position and then place the initial point of **w** on the terminal point of **v**, as illustrated in the following diagram.

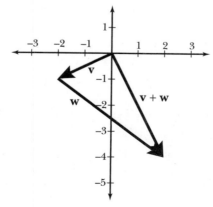

The resultant vector, **v** + **w**, has component form <2,−4>.

Note: In Problems 15.8–15.9, v = –2i – j and w = 4i – 3j.

15.9 Calculate **v** – **w** by plotting the resultant vector in the coordinate plane.

According to Problem 15.8, **v** = <–2,–1> and **w** = <4,–3>. Multiply the components of **w** by –1 in order to generate the vector –**w**.

$$-\mathbf{w} = \langle -1 \cdot 4, -1(-3) \rangle$$
$$= \langle -4, 3 \rangle$$

> Just like
> $10 + (-5) = 10 - 5 = 5$

The vector sum **v** + (–**w**) is equivalent to the vector difference **v** – **w**. Plot **v** in standard position and then place the initial point of –**w** on the terminal point of **v**.

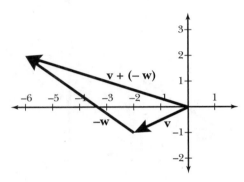

You conclude that **v** – **w** = <–6, 2>.

15.10 Given **r** = <3, 2> and **s** = <1, –4>, calculate **r** – **s** by plotting the resultant vector in the coordinate plane.

Multiply the components of **s** by –1 to generate the vector –**s**.

$$-\mathbf{s} = \langle -1 \cdot 1, -1(-4) \rangle$$
$$= \langle -1, 4 \rangle$$

The vector **r** – **s** is equivalent to the sum **r** + (–**s**).

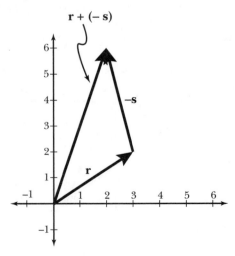

You conclude that **r** – **s** = <2, 6>.

15.11 Given **u** = <−3,−1>, **v** = <−4,−5>, **w** = <−2,−8>, calculate **u** + **v** − **w**.

Begin by multiplying the components of **w** by −1 to generate vector −**w**.

$$-\mathbf{w} = \langle -1(-2), (-1)(-8) \rangle$$
$$= \langle 2, 8 \rangle$$

Plot **u** in standard position and place the initial point of **v** at the terminal point of **u**. Then, place the initial point of −**w** at the terminal point of **v**. The resultant vector stretches from the origin to (−5,2), as illustrated in the following diagram.

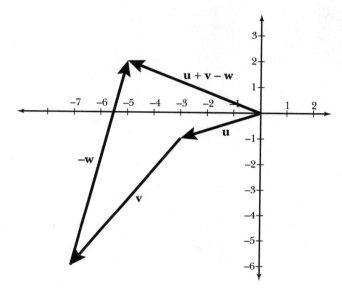

You conclude that **u** + **v** − **w** = <−5,2>.

Adding and Subtracting Algebraically
Calculate <a, b> + <c, d>

Note: In Problems 15.12–15.13, **v** = <a, b> *and* **w** = <c, d>.

15.12 Calculate **v** + **w**.

To compute the sum of two vectors in component form algebraically, add the corresponding components:

$$\mathbf{v} + \mathbf{w} = <a + c, b + d>$$

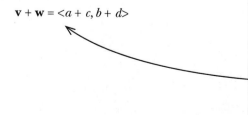

In other words, without having to graph the vectors

Add the x-components together to get the x-component of the sum, and do the same thing for the y-components.

Note: In Problems 15.12–15.13, v = <a, b> and w = <c, d>.

15.13 Calculate **v** – **w**.

In Problem 15.12, you add corresponding *x*- and *y*-components. In this problem, you subtract them.

$$\mathbf{v} - \mathbf{w} = <a - c, b - d>$$

Note: In Problems 15.14–15.16, r = <3, 1> and s = <6, 8>.

15.14 Calculate **r** + **s** by plotting the resultant vector in the coordinate plane.

Plot **r** in standard position and then place the initial point of **s** at the terminal point of **r**. The following diagram plots the resultant vector, **r** + **s**.

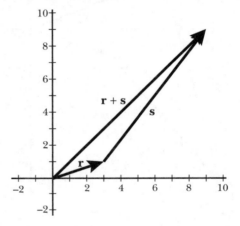

The terminal point of the resultant vector is (9,9), so you conclude that **r** + **s** = <9, 9>.

Note: In Problems 15.14–15.16, r = <3, 1> and s = <6, 8>.

15.15 Verify your answer to Problem 15.14 by calculating **r** + **s** algebraically.

Computing the sums of vectors algebraically is a simple task: Add the *x*-components of **r** and **s** to compute the *x*-component of the sum **r** + **s**. Similarly, add the *y*-components of the individual vectors to compute the *y*-component of **r** + **s**.

$$\mathbf{r} + \mathbf{s} = \langle 3 + 6, 1 + 8 \rangle$$
$$= \langle 9, 9 \rangle$$

The solution **r** + **s** = <9, 9> matches the solution to Problem 15.14, thus verifying the answer.

*Note: In Problems 15.14–15.16, **r** = <3, 1> and **s** = <6, 8>.*

15.16 Calculate **r** − **s**.

Subtract the components of **s** from the components of **r**, applying the technique described in Problem 15.13.

$$\mathbf{r} - \mathbf{s} = \langle 3 - 6, 1 - 8 \rangle$$
$$= \langle -3, -7 \rangle$$

15.17 Calculate the sum: <11, −9> + <−3, −4>.

Combine the corresponding components of each vector.

$$\langle 11, -9 \rangle + \langle -3, -4 \rangle = \langle 11 + (-3), -9 + (-4) \rangle$$
$$= \langle 11 - 3, -9 - 4 \rangle$$
$$= \langle 8, -13 \rangle$$

15.18 Calculate the difference: <x − 3, y + 5> − <9 − 7x, 8y>.

Subtract the components of <9 − 7x, 8y> from the corresponding components of <x − 3, y + 5>.

$$\langle x - 3, y + 5 \rangle - \langle 9 - 7x, 8y \rangle = \langle x - 3 - (9 - 7x), y + 5 - 8y \rangle$$
$$= \langle x - 3 - 9 + 7x, y + 5 - 8y \rangle$$
$$= \langle x + 7x - 3 - 9, y - 8y + 5 \rangle$$
$$= \langle 8x - 12, -7y + 5 \rangle$$

You conclude that <x − 3, y + 5> − <9 − 7x, 8y> = <8x − 12, −7y + 5>.

15.19 Given **v** = <−10, 2>, A = (7, 6), and B = (−9, 5), calculate $\overrightarrow{AB}$ − **v**.

Begin by expressing $\overrightarrow{AB}$ in component form.

$$\overrightarrow{AB} = \langle -9 - 7, 5 - 6 \rangle$$
$$= \langle -16, -1 \rangle$$

Review Problem 15.6 if you forget how to do this.

Now that both vectors are in component form, calculate the difference.

$$\overrightarrow{AB} - \mathbf{v} = \langle -16, -1 \rangle - \langle -10, 2 \rangle$$
$$= \langle -16 - (-10), -1 - 2 \rangle$$
$$= \langle -16 + 10, -3 \rangle$$
$$= \langle -6, -3 \rangle$$

15.20 Given $L = (5, -1)$, $M = (8, 4)$, $P = (-3, 3)$, and $Q = (2, 5)$, calculate $\left\| \overrightarrow{LM} + \overrightarrow{PQ} \right\|$.

Express the vectors in component form.

$$\overrightarrow{LM} = \langle 8 - 5, 4 - (-1) \rangle \qquad \overrightarrow{PQ} = \langle 2 - (-3), 5 - 3 \rangle$$
$$= \langle 8 - 5, 4 + 1 \rangle \qquad\qquad = \langle 2 + 3, 5 - 3 \rangle$$
$$= \langle 3, 5 \rangle \qquad\qquad\qquad = \langle 5, 2 \rangle$$

Calculate the sum of the vectors.

$$\overrightarrow{LM} + \overrightarrow{PQ} = \langle 3, 5 \rangle + \langle 5, 2 \rangle$$
$$= \langle 3 + 5, 5 + 2 \rangle$$
$$= \langle 8, 7 \rangle$$

Note that you are asked to calculate the magnitude of the vector sum.

$$\left\| \overrightarrow{LM} + \overrightarrow{PQ} \right\| = \left\| \langle 8, 7 \rangle \right\|$$
$$= \sqrt{8^2 + 7^2}$$
$$= \sqrt{64 + 49}$$
$$= \sqrt{113}$$

> You're calculating magnitude because there are double vertical bars around the vector in the original problem. Look at Problem 14.25 if you need to review the magnitude formula.

Scalar Multiplication
Calculate c <a, b>

15.21 Express $c\langle a, b \rangle$ in component form, assuming a, b, and c are real numbers.

The vector $\langle a, b \rangle$ is multiplied by a real number c; that real number is called a scalar. To multiply a vector in component form by a scalar value, multiply each of the components by the scalar. In this example, multiply both a and b by c.

$$c\langle a, b \rangle = \langle ca, cb \rangle$$

> A vector has a value (magnitude) and a direction. A scalar only has a value. However, that value changes the SCALE of the vector. For example, multiplying a vector by the scalar 2 doubles its length.

15.22 Express the vector in component form: $3\langle 4, 7 \rangle$.

Multiply each of the components of vector $\langle 4, 7 \rangle$ by 3.

$$3\langle 4, 7 \rangle = \langle 3(4), 3(7) \rangle$$
$$= \langle 12, 21 \rangle$$

15.23 Express the vector in component form: $-5 \langle -1, 8 \rangle$.

Distribute -5 to each of the components of the vector.

$$-5 \langle -1, 8 \rangle = \langle -5(-1), -5(8) \rangle$$
$$= \langle 5, -40 \rangle$$

Note: In Problems 15.24–15.27, $\mathbf{v} = <2, 1>$.

15.24 Calculate $\|\mathbf{v}\|$.

Apply the vector magnitude formula presented in Problem 14.25.

$$\|\mathbf{v}\| = \sqrt{2^2 + 1^2}$$
$$= \sqrt{4 + 1}$$
$$= \sqrt{5}$$

Note: In Problems 15.24–15.27, $\mathbf{v} = <2, 1>$.

15.25 Express $4\mathbf{v}$ in component form.

Multiply the x- and y-components of $\mathbf{v}$ by the scalar 4.

$$4\mathbf{v} = 4 \langle 2, 1 \rangle$$
$$= \langle 4(2), 4(1) \rangle$$
$$= \langle 8, 4 \rangle$$

Note: In Problems 15.24–15.27, $\mathbf{v} = <2, 1>$.

15.26 Based on your answers to Problems 15.24–15.25, explain how scalar multiplication affects the magnitude of a vector.

In Problem 15.24, you determine that the magnitude of $\mathbf{v}$ is $\sqrt{5}$; in Problem 15.25, you determine that $4\mathbf{v} = <8, 4>$. Calculate the magnitude of $4\mathbf{v}$ and compare it to the magnitude of $\mathbf{v}$.

$$\|4\mathbf{v}\| = \|\langle 8, 4 \rangle\|$$
$$= \sqrt{8^2 + 4^2}$$
$$= \sqrt{64 + 16}$$
$$= \sqrt{80}$$
$$= 4\sqrt{5}$$

Vector $4\mathbf{v}$ has magnitude $4\sqrt{5}$, which is 4 times the magnitude of $\mathbf{v}$: $\|4\mathbf{v}\| = 4\|\mathbf{v}\|$. This demonstrates one property of scalar multiplication: The product of a scalar and a vector's magnitude is equal to the magnitude of the product of the scalar and the vector.

In this problem, four times the length of v is equal to the length of vector 4v.

Note: In Problems 15.24–15.27, **v** *= <2, 1>.*

15.27 Visually verify that 4**v** is four times as long as **v**.

Note that 4**v** = **v** + **v** + **v** + **v**. Use the head-to-tail technique demonstrated in Problems 15.1–15.11 to add those four vectors together. As the following diagram illustrates, the sum is a vector whose length must be four times the length of **v**, because it is literally comprised of four copies of that vector.

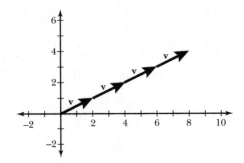

15.28 Prove that multiplying **v** = <*a, b*> by the scalar *c* creates a vector with *c* times the magnitude of **v**.

This problem asks you to verify that $\|c\,\mathbf{v}\| = c\|\mathbf{v}\|$. In Problems 15.24–15.27, you verify this is true for a specific vector and a specific scalar (<2, 1> and 4, respectively); in this problem you prove that the same relationship between magnitude and scalar multiplication holds true for all vectors and scalars. Apply the magnitude formula to both sides of the equation $\|c\,\mathbf{v}\| = c\|\mathbf{v}\|$.

$$\|c\,\mathbf{v}\| = c\|\mathbf{v}\|$$
$$\|\langle ca, cb\rangle\| = c\|\langle a, b\rangle\|$$
$$\sqrt{(ca)^2 + (cb)^2} = c\left(\sqrt{a^2 + b^2}\right)$$
$$\sqrt{c^2 a^2 + c^2 b^2} = c\sqrt{a^2 + b^2}$$

Factor c^2 out of both terms on the left side of the equation and simplify the radical expression.

$$\sqrt{c^2\left(a^2 + b^2\right)} = c\sqrt{a^2 + b^2}$$
$$c\sqrt{a^2 + b^2} = c\sqrt{a^2 + b^2}$$

Multiplying **v** = <*a, b*> by the scalar *c* creates a vector with length $c\sqrt{a^2 + b^2}$, which is *c* times the length of **v**.

Note: In Problems 15.29–15.31, $\mathbf{u}$ = *<2, –4> and* $\mathbf{v}$ = *<7, –5>.*

15.29 Calculate $2\mathbf{u} + 3\mathbf{v}$.

Multiply the components of $\mathbf{u}$ by 2, multiply the components of $\mathbf{v}$ by 3, and add the vectors.

$$
\begin{aligned}
2\mathbf{u} + 3\mathbf{v} &= 2\langle 2, -4 \rangle + 3\langle 7, -5 \rangle \\
&= \langle 4, -8 \rangle + \langle 21, -15 \rangle \\
&= \langle 4 + 21, -8 - 15 \rangle \\
&= \langle 25, -23 \rangle
\end{aligned}
$$

Note: In Problems 15.29–15.31, $\mathbf{u}$ = *<2, –4> and* $\mathbf{v}$ = *<7, –5>.*

15.30 Calculate $4\mathbf{v} - \dfrac{1}{2}\mathbf{u}$.

Multiply the components of $\mathbf{v}$ by 4, multiply the components of $\mathbf{u}$ by 1/2, and subtract the vectors.

$$
\begin{aligned}
4\mathbf{v} - \frac{1}{2}\mathbf{u} &= 4\langle 7, -5 \rangle - \left(\frac{1}{2}\right)\langle 2, -4 \rangle \\
&= \langle 28, -20 \rangle - \langle 1, -2 \rangle \\
&= \langle 28 - 1, -20 - (-2) \rangle \\
&= \langle 28 - 1, -20 + 2 \rangle \\
&= \langle 27, -18 \rangle
\end{aligned}
$$

> Or you could multiply the components of u by –1/2 and add them to 4v.

Note: In Problems 15.29–15.31, $\mathbf{u}$ = *<2, –4> and* $\mathbf{v}$ = *<7, –5>.*

15.31 Identify the component form of the unit vector in the direction of $\mathbf{w} = 4\mathbf{u} - \mathbf{v}$.

Begin by calculating $\mathbf{w}$.

$$
\begin{aligned}
\mathbf{w} &= 4\langle 2, -4 \rangle - \langle 7, -5 \rangle \\
&= \langle 8, -16 \rangle - \langle 7, -5 \rangle \\
&= \langle 8 - 7, -16 - (-5) \rangle \\
&= \langle 8 - 7, -16 + 5 \rangle \\
&= \langle 1, -11 \rangle
\end{aligned}
$$

Now calculate the magnitude of $\mathbf{w}$.

$$
\begin{aligned}
\|\mathbf{w}\| &= \sqrt{1^2 + (-11)^2} \\
&= \sqrt{1 + 121} \\
&= \sqrt{122}
\end{aligned}
$$

According to Problem 14.38, given a vector $\mathbf{w}$, the corresponding unit vector is $\hat{\mathbf{w}} = \mathbf{w}/\|\mathbf{w}\|$. It is equal to the component form of the vector divided by its magnitude.

$$\hat{\mathbf{w}} = \frac{\mathbf{w}}{\|\mathbf{w}\|}$$

$$= \frac{\langle 1, -11 \rangle}{\sqrt{122}}$$

$$= \left\langle \frac{1}{\sqrt{122}}, -\frac{11}{\sqrt{122}} \right\rangle$$

Rationalize the denominators of the components.

$$\hat{\mathbf{w}} = \left\langle \frac{\sqrt{122}}{122}, -\frac{11\sqrt{122}}{122} \right\rangle$$

Note: In Problems 15.32–15.34, r = <3x, 8> and s = <x – 1, 5>.

15.32 Calculate 4**s** – 5**r**.

Multiply each of the factors by the appropriate scalar values. Note that the x-component of **s** is $x – 1$, and both of those terms must be multiplied by 4 when you compute 4**s**.

Alternative: Instead of multiplying r by –5 and adding 4s + (–5r), you could multiply r by 5 and subtract 5r from 4s.

$$4\mathbf{s} - 5\mathbf{r} = 4\langle x - 1, 5 \rangle + (-5)\langle 3x, 8 \rangle$$

$$= \langle 4(x - 1), 20 \rangle + \langle -15x, -40 \rangle$$

$$= \langle 4x - 4, 20 \rangle + \langle -15x, -40 \rangle$$

$$= \langle 4x - 4 - 15x, 20 - 40 \rangle$$

$$= \langle -11x - 4, -20 \rangle$$

Note: In Problems 15.32–15.34, r = <3x, 8> and s = <x – 1, 5>.

15.33 Calculate **r** – 2**s**.

When you compute –2**s**, make sure to distribute –2 to both terms in the x-component: $-2(x - 1) = -2x + 2$.

$$\mathbf{r} - 2\mathbf{s} = \langle 3x, 8 \rangle + (-2)\langle x - 1, 5 \rangle$$

$$= \langle 3x, 8 \rangle + \langle -2x + 2, -10 \rangle$$

$$= \langle 3x - 2x + 2, 8 - 10 \rangle$$

$$= \langle x + 2, -2 \rangle$$

Note: In Problems 15.32–15.34, **r** *= <3x, 8> and* **s** *= <x – 1, 5>.*

15.34 Given $\|\mathbf{r} - 2\mathbf{s}\| = 2\sqrt{2}$, calculate x.

According to Problem 15.33, $\mathbf{r} - 2\mathbf{s} = \langle x + 2, -2\rangle$. Apply the magnitude formula, noting that the magnitude of the vector is equal to $2\sqrt{2}$.

$$\|\mathbf{r} - 2\mathbf{s}\| = \sqrt{(x+2)^2 + (-2)^2}$$
$$2\sqrt{2} = \sqrt{(x^2 + 4x + 4) + 4}$$
$$2\sqrt{2} = \sqrt{x^2 + 4x + 8}$$

Square both sides of the equation to eliminate the square root symbols.

$$\left(2\sqrt{2}\right)^2 = \left(\sqrt{x^2 + 4x + 8}\right)^2$$
$$2^2\left(\sqrt{2}\right)^2 = x^2 + 4x + 8$$
$$4(2) = x^2 + 4x + 8$$
$$8 = x^2 + 4x + 8$$
$$0 = x^2 + 4x + 8 - 8$$
$$0 = x^2 + 4x$$

Squaring both sides could produce extra solutions that aren't true, so you check the solutions at the end of the problem.

Factor the equation: $0 = x(x + 4)$. Apply the zero-product property to solve for x.

$$x = 0 \qquad x + 4 = 0$$
$$x = -4$$

See Problems 10.12–10.13 to review the zero-product property.

There are two potential solutions: $x = 0$ and $x = -4$. Substitute each into $\mathbf{r} - 2\mathbf{s} = \langle x + 2, -2\rangle$ and calculate the magnitude of the resulting vectors to verify that each equals $2\sqrt{2}$.

Check $x = 0$	Check $x = -4$
$\mathbf{r} - 2\mathbf{s} = \langle x + 2, -2\rangle$	$\mathbf{r} - 2\mathbf{s} = \langle x + 2, -2\rangle$
$\mathbf{r} - 2\mathbf{s} = \langle 0 + 2, -2\rangle$	$\mathbf{r} - 2\mathbf{s} = \langle -4 + 2, -2\rangle$
$\mathbf{r} - 2\mathbf{s} = \langle 2, -2\rangle$	$\mathbf{r} - 2\mathbf{s} = \langle -2, -2\rangle$
$\|\mathbf{r} - 2\mathbf{s}\| = \sqrt{2^2 + (-2)^2}$	$\|\mathbf{r} - 2\mathbf{s}\| = \sqrt{(-2)^2 + (-2)^2}$
$\|\mathbf{r} - 2\mathbf{s}\| = \sqrt{4 + 4}$	$\|\mathbf{r} - 2\mathbf{s}\| = \sqrt{4 + 4}$
$\|\mathbf{r} - 2\mathbf{s}\| = \sqrt{8}$	$\|\mathbf{r} - 2\mathbf{s}\| = \sqrt{8}$
$\|\mathbf{r} - 2\mathbf{s}\| = 2\sqrt{2}$	$\|\mathbf{r} - 2\mathbf{s}\| = 2\sqrt{2}$

You conclude that $\|\mathbf{r} - 2\mathbf{s}\| = 2\sqrt{2}$ when $x = 0$ or $x = -4$.

Note: In Problems 15.35–15.36, v = <–3, –2y> and w = <–5, 4 – 3y>.

15.35 Calculate 6w – 7v.

Multiply the components of **w** by 6, multiply the components of **v** by –7, and combine the vectors.

$$6\mathbf{w} - 7\mathbf{v} = 6\langle -5, 4-3y \rangle + (-7)\langle -3, -2y \rangle$$
$$= \langle -30, 6(4-3y) \rangle + \langle 21, 14y \rangle$$
$$= \langle -30, 24-18y \rangle + \langle 21, 14y \rangle$$
$$= \langle -30+21, 24-18y+14y \rangle$$
$$= \langle -9, 24-4y \rangle$$

Note: In Problems 15.35–15.36, v = <–3, –2y> and w = <–5, 4 – 3y>.

15.36 Given $A = (-5, 11)$ and $B = (-14, -3)$, and assuming 6w – 7v is equivalent to $\overrightarrow{AB}$, calculate y.

Equivalent vectors have the same component form, so begin by expressing $\overrightarrow{AB}$ in terms of its components.

$$\overrightarrow{AB} = \langle -14-(-5), -3-11 \rangle$$
$$= \langle -14+5, -14 \rangle$$
$$= \langle -9, -14 \rangle$$

According to Problem 15.35, 6w – 7v = <–9, 24 – 4y>. Note that its x-component is already equal to the x-component of $\overrightarrow{AB}$. In order for the vectors to be equivalent, the y-components must be equal as well.

$$y\text{-component of } 6\mathbf{w} - 7\mathbf{v} = y\text{-component of } \overrightarrow{AB}$$
$$24 - 4y = -14$$
$$-4y = -14 - 24$$
$$-4y = -38$$
$$y = \frac{38}{4}$$
$$y = \frac{19}{2}$$

Identifying Components Given Magnitude and Direction

Instead of coordinates

Note: Problems 15.37–15.38 refer to **v** = *<x, y> with direction angle θ, as illustrated in the diagram below.*

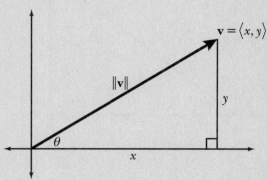

15.37 Evaluate cos θ, sin θ, and tan θ.

The length of the side adjacent to θ is x, the length of the side opposite θ is y, and the length of the hypotenuse is $\|\mathbf{v}\|$. Use these values to calculate cos θ, sin θ, and tan θ.

$$\cos\theta = \frac{\text{length of side adjacent to } \theta}{\text{length of hypotenuse}} = \frac{x}{\|\mathbf{v}\|}$$

$$\sin\theta = \frac{\text{length of side opposite } \theta}{\text{length of hypotenuse}} = \frac{y}{\|\mathbf{v}\|}$$

$$\tan\theta = \frac{\text{length of side opposite } \theta}{\text{length of side adjacent to } \theta} = \frac{y}{x}$$

Note: Problems 15.37–15.38 refer to **v** = *<x, y> with direction angle θ, as illustrated in Problem 15.37.*

15.38 Identify the component form of **v** in terms of $\|\mathbf{v}\|$.

Note that the component form of the vector is **v** = <x, y>. According to Problem 15.37, cos θ = x/$\|\mathbf{v}\|$ and sin θ = y/$\|\mathbf{v}\|$. Cross-multiply to solve these trigonometric equations for x and y.

$$\cos\theta = \frac{x}{\|\mathbf{v}\|} \qquad\qquad \sin\theta = \frac{y}{\|\mathbf{v}\|}$$

$$\frac{\cos\theta}{1} = \frac{x}{\|\mathbf{v}\|} \qquad\qquad \frac{\sin\theta}{1} = \frac{y}{\|\mathbf{v}\|}$$

$$1 \cdot x = \|\mathbf{v}\|\cos\theta \qquad\qquad 1 \cdot y = \|\mathbf{v}\|\sin\theta$$

$$x = \|\mathbf{v}\|\cos\theta \qquad\qquad y = \|\mathbf{v}\|\sin\theta$$

Substitute these values into **v** = <x, y>.

$$\mathbf{v} = \left\langle x, y \right\rangle$$

$$\mathbf{v} = \left\langle \|\mathbf{v}\|\cos\theta, \|\mathbf{v}\|\sin\theta \right\rangle$$

> This is the formula that calculates the component form of a vector v with direction angle θ and magnitude $\|v\|$.

*Note: Problems 15.39–15.40 refer to **v** = <x, y> with direction angle θ, as illustrated in the diagram below.*

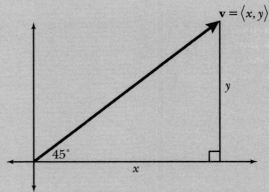

15.39 Given $\|\mathbf{v}\| = 6\sqrt{2}$ and $\theta = 45°$, apply the vector component formula you generated in Problem 15.38 to express **v** in component form.

Substitute the magnitude and the direction angle into the formula $\mathbf{v} = \langle \|\mathbf{v}\| \cos\theta, \|\mathbf{v}\| \sin\theta \rangle$. Note that $45° = \pi/4$ radians, and $\cos 45° = \sin 45° = \sqrt{2}/2$.

$$\mathbf{v} = \langle \|\mathbf{v}\| \cos\theta, \|\mathbf{v}\| \sin\theta \rangle$$
$$= \langle 6\sqrt{2}\,(\cos 45°), 6\sqrt{2}\,(\sin 45°) \rangle$$
$$= \left\langle 6\sqrt{2} \cdot \frac{\sqrt{2}}{2}, 6\sqrt{2} \cdot \frac{\sqrt{2}}{2} \right\rangle$$
$$= \left\langle \frac{6\sqrt{4}}{2}, \frac{6\sqrt{4}}{2} \right\rangle$$
$$= \left\langle \frac{6 \cdot \cancel{2}}{\cancel{2}}, \frac{6 \cdot \cancel{2}}{\cancel{2}} \right\rangle$$
$$= \langle 6, 6 \rangle$$

You conclude that **v** = <6, 6>.

*Note: Problems 15.39–15.40 refer to **v** = <x, y> with direction angle θ, as illustrated in Problem 15.39.*

15.40 Verify your answer to Problem 15.39 by applying the properties of a 45°–45°–90° triangle to calculate x and y.

The right triangle in the figure has angles that measure 45°, 45°, and 90°. Therefore, the length of the hypotenuse is equal to $\sqrt{2}$ times the length of the legs. In this triangle, the hypotenuse has length $6\sqrt{2}$ and the legs have lengths x and y.

You know two angle measurements of the triangle: a right angle and a 45° angle at the origin. All three angles have to add up to 180°, so the third angle measures 45°.

$$\text{length of leg} \cdot \sqrt{2} = \text{length of hypotenuse}$$
$$\text{length of leg} \cdot \sqrt{2} = 6\sqrt{2}$$
$$\text{length of leg} = \frac{6\sqrt{2}}{\sqrt{2}}$$
$$\text{length of leg} = 6$$

Recall that the lengths of the legs of a 45°–45°–90° triangle are equal, so $x = y = 6$. These are also the components of the vector: $\mathbf{v} = <x, y> = <6, 6>$. This verifies your solution to Problem 15.39.

15.41 Given a vector $\mathbf{w}$ with direction angle $\theta = 180°$ and magnitude $\|\mathbf{w}\| = 2$, express $\mathbf{w}$ in component form.

Substitute the angle and magnitude into the vector component formula generated in Problem 15.38. Note that $180° = \pi$ radians; thus, $\cos 180° = -1$ and $\sin 180° = 0$.

$$\mathbf{w} = \left\langle \|\mathbf{w}\|\cos\theta, \|\mathbf{w}\|\sin\theta \right\rangle$$
$$= \left\langle 2 \cdot \cos 180°, 2 \cdot \sin 180° \right\rangle$$
$$= \left\langle 2(-1), 2(0) \right\rangle$$
$$= \left\langle -2, 0 \right\rangle$$

15.42 Given a vector $\mathbf{r}$ with direction angle $\theta = 135°$ and magnitude $\|\mathbf{r}\| = 7\sqrt{3}$, express $\mathbf{r}$ in component form.

Note that $135° = 3\pi/4$ radians, so $\cos 135° = -\sqrt{2}/2$ and $\sin 135° = \sqrt{2}/2$. Substitute these values, as well as the magnitude of $\mathbf{r}$, into the vector component formula.

$$\mathbf{r} = \left\langle \|\mathbf{r}\|\cos\theta, \|\mathbf{r}\|\sin\theta \right\rangle$$
$$= \left\langle 7\sqrt{3}\left(-\frac{\sqrt{2}}{2}\right), 7\sqrt{3}\left(\frac{\sqrt{2}}{2}\right) \right\rangle$$
$$= \left\langle -\frac{7\sqrt{6}}{2}, \frac{7\sqrt{6}}{2} \right\rangle$$

15.43 Given a vector **s** with direction angle $\theta = 60°$ and magnitude $\|\mathbf{s}\| = 8$, express **s** in component form.

Note that $60° = \pi/3$ radians, so $\cos 60° = 1/2$ and $\sin 60° = \sqrt{3}/2$. Substitute these values into the vector component formula.

$$\mathbf{s} = \left\langle \|\mathbf{s}\|\cos 60°, \|\mathbf{s}\|\sin 60° \right\rangle$$

$$= \left\langle 8\left(\frac{1}{2}\right), 8\left(\frac{\sqrt{3}}{2}\right) \right\rangle$$

$$= \left\langle \frac{8}{2}, \frac{8\sqrt{3}}{2} \right\rangle$$

$$= \left\langle 4, 4\sqrt{3} \right\rangle$$

*Note: In Problems 15.44–15.45, **v** has direction angle 30°, **w** has direction angle 270°, $\|\mathbf{v}\| = 3$, and $\|\mathbf{w}\| = 5$.*

15.44 Express **v** and **w** in component form.

Apply the technique demonstrated in Problems 15.41–15.43 to express both vectors in component form. Note that $30° = \pi/6$ radians and $270° = 3\pi/2$ radians; both angles appear on the standard unit circle.

$$\mathbf{v} = \left\langle \|\mathbf{v}\|\cos\theta, \|\mathbf{v}\|\sin\theta \right\rangle \qquad \mathbf{w} = \left\langle \|\mathbf{w}\|\cos\theta, \|\mathbf{w}\|\sin\theta \right\rangle$$

$$= \left\langle 3(\cos 30°), 3(\sin 30°) \right\rangle \qquad = \left\langle 5(\cos 270°), 5(\sin 270°) \right\rangle$$

$$= \left\langle 3\left(\frac{\sqrt{3}}{2}\right), 3\left(\frac{1}{2}\right) \right\rangle \qquad = \left\langle 5(0), 5(-1) \right\rangle$$

$$= \left\langle \frac{3\sqrt{3}}{2}, \frac{3}{2} \right\rangle \qquad = \left\langle 0, -5 \right\rangle$$

*Note: In Problems 15.44–15.45, **v** has direction angle 30°, **w** has direction angle 270°, ‖**v**‖ = 3, and ‖**w**‖ = 5.*

15.45 Determine whether the following statement is true and explain your answer.

$$\|\mathbf{v}\| + \|\mathbf{w}\| = \|\mathbf{v} + \mathbf{w}\|$$

This question asks whether the length of a vector sum is equal to the sum of the lengths of the individual vectors. In Problem 15.44, you express both vectors in component form. Begin this problem by calculating the sum **v** + **w**.

$$\mathbf{v} + \mathbf{w} = \left\langle \frac{3\sqrt{3}}{2}, \frac{3}{2} \right\rangle + \langle 0, -5 \rangle$$

$$= \left\langle \frac{3\sqrt{3}}{2} + 0, \frac{3}{2} - 5 \right\rangle$$

$$= \left\langle \frac{3\sqrt{3}}{2}, \frac{3}{2} - \frac{10}{2} \right\rangle$$

$$= \left\langle \frac{3\sqrt{3}}{2}, -\frac{7}{2} \right\rangle$$

Calculate the magnitude of **v** + **w**.

$$\|\mathbf{v} + \mathbf{w}\| = \sqrt{\left(\frac{3\sqrt{3}}{2} \right)^2 + \left(-\frac{7}{2} \right)^2}$$

$$= \sqrt{\frac{9 \cdot 3}{4} + \frac{49}{4}}$$

$$= \sqrt{\frac{27}{4} + \frac{49}{4}}$$

$$= \sqrt{\frac{76}{4}}$$

$$= \sqrt{19}$$

Notice that the magnitude of the sum is not equal to the sum of the individual magnitudes.

$$\|\mathbf{v}\| + \|\mathbf{w}\| \neq \|\mathbf{v} + \mathbf{w}\|$$
$$3 + 5 \neq \sqrt{19}$$
$$8 \neq \sqrt{19}$$

The magnitude of a vector sum is equal to the sum of the magnitudes of the individual vectors only when those vectors have the same direction angle. Vectors **v** and **w** do not have the same direction.

Chapter 16
ADVANCED VECTOR OPERATIONS

All about the dot product

In Chapter 15, you explore vector addition, which combines two vectors to produce another vector. Mathematically speaking, this means that the vectors are considered "closed" under addition. Simply stated, a set (like vectors) is closed under an operation when performing an operation (like addition) on that set produces another member of that set.

This is not an entirely new concept. The set of integers is closed under addition, subtraction, and multiplication—if you add, subtract, or multiply two integers, the result is another integer. Note that integers are not closed over division, because dividing two integers can produce a fraction: The operations $1 \div 2$ and $3 \div 5$ produce fractions, not integers.

In this chapter, you explore the dot product, an operation under which the set of vectors is *not* closed. The dot product of two vectors is *not* a vector; it is a real number.

You spend the first half of the chapter getting familiar with the dot product, answering questions like "How do I calculate it?", "Is it commutative?", "Is it distributive?", "Does it prefer dogs or cats as pets?" ... that sort of thing.

By the end of the chapter, you'll be using the dot product to calculate the angle between vectors, to identify orthogonal (perpendicular) vectors, to create vector projections (and actually understand what that means), and to apply all of this in a real life situation. That's a non-threatening way of saying "word problems," but don't panic.

Dot Product
Looks like multiplication, but it's not

16.1 Given $\mathbf{v} = <a, b>$ and $\mathbf{w} = <c, d>$, explain how to calculate the dot product $\mathbf{v} \cdot \mathbf{w}$.

To calculate the dot product of two vectors, add the product of their x-components to the product of their y-components.

$$\mathbf{v} \cdot \mathbf{w} = ac + bd$$

Even though the dot here looks like multiplication, it stands for the "dot product."

Note: In Problems 16.2–16.4, $\mathbf{v} = <1, 4>$ and $\mathbf{w} = <5, 2>$.

16.2 Calculate $\mathbf{v} \cdot \mathbf{w}$.

Apply the formula presented in Problem 16.1, which adds the products of the x-components and the y-components.

$$\langle 1, 4 \rangle \cdot \langle 5, 2 \rangle = 1(5) + 4(2)$$
$$= 5 + 8$$
$$= 13$$

Note: In Problems 16.2–16.4, $\mathbf{v} = <1, 4>$ and $\mathbf{w} = <5, 2>$.

16.3 Calculate $3\mathbf{v} \cdot \mathbf{w}$.

Before you apply the dot product, multiply $\mathbf{v}$ by the scalar 3.

$$3\mathbf{v} \cdot \mathbf{w} = 3\langle 1, 4 \rangle \cdot \langle 5, 2 \rangle$$
$$= \langle 3(1), 3(4) \rangle \cdot \langle 5, 2 \rangle$$
$$= \langle 3, 12 \rangle \cdot \langle 5, 2 \rangle$$

Now apply the dot product formula.

$$= 3(5) + 12(2)$$
$$= 15 + 24$$
$$= 39$$

You conclude that $3\mathbf{v} \cdot \mathbf{w} = 39$.

Note: In Problems 16.2–16.4, v = <1, 4> and w = <5, 2>.

16.4 Calculate $2\mathbf{v} \cdot (-3\mathbf{w})$.

Multiply both components of **v** by 2, multiply both components of **w** by –3, and then compute the dot product.

$$2\mathbf{v} \cdot (-3\mathbf{w}) = 2\langle 1, 4 \rangle \cdot (-3)\langle 5, 2 \rangle$$
$$= \langle 2(1), 2(4) \rangle \cdot \langle -3(5), -3(2) \rangle$$
$$= \langle 2, 8 \rangle \cdot \langle -15, -6 \rangle$$
$$= 2(-15) + 8(-6)$$
$$= -30 - 48$$
$$= -78$$

Note: In Problems 16.5–16.6, you verify that the dot product is commutative using the vectors r = <–3, 7> and s = <–2, 10>.

16.5 Calculate $\mathbf{r} \cdot \mathbf{s}$.

In other words, r · s = s · r.

Add the product of the *x*-components to the product of the *y*-components.

$$\mathbf{r} \cdot \mathbf{s} = \langle -3, 7 \rangle \cdot \langle -2, 10 \rangle$$
$$= -3(-2) + 7(10)$$
$$= 6 + 70$$
$$= 76$$

Note: In Problems 16.5–16.6, you verify that the dot product is commutative using the vectors r = <–3, 7> and s = <–2, 10>.

16.6 Calculate $\mathbf{s} \cdot \mathbf{r}$.

According to Problem 16.5, $\mathbf{r} \cdot \mathbf{s} = 76$. If the dot product is commutative, then $\mathbf{s} \cdot \mathbf{r}$ will also equal 76.

$$\mathbf{s} \cdot \mathbf{r} = \langle -2, 10 \rangle \cdot \langle -3, 7 \rangle$$
$$= -2(-3) + 10(7)$$
$$= 6 + 70$$
$$= 76$$

Only the order of the components changes: -2(-3) instead of -3(-2), 10(7) instead of 7(10). Because multiplication is commutative, the order doesn't affect the products.

The order in which you list the vectors does not affect the value of the dot product—you still add the products of the *x*-components to the products of the *y*-components, and the result is the same. Therefore, the dot product is commutative.

Note: In Problems 16.7–16.8, you verify that the dot product of a vector with itself is equal to the square of the vector's magnitude.

16.7 Given $\mathbf{v} = <-4, 6>$, verify that $\mathbf{v} \cdot \mathbf{v} = \|\mathbf{v}\|^2$.

In this problem, you demonstrate that the relationship $\mathbf{v} \cdot \mathbf{v} = \|\mathbf{v}\|^2$ is true for the specific vector $\mathbf{v} = <-4, 6>$; in Problem 16.8, you prove that the relationship is true for any vector.

Calculate the dot product $\mathbf{v} \cdot \mathbf{v}$ on the left side of the equation, and apply the vector magnitude formula to calculate $\|\mathbf{v}\|$ on the right side of the equation.

$$\mathbf{v} \cdot \mathbf{v} = \|\mathbf{v}\|^2$$
$$\langle -4, 6 \rangle \cdot \langle -4, 6 \rangle = \left(\sqrt{(-4)^2 + (6)^2}\right)^2$$
$$-4(-4) + 6(6) = \left(\sqrt{16 + 36}\right)^2$$
$$16 + 36 = \left(\sqrt{52}\right)^2$$
$$52 = 52$$

The square and square root cancel out, leaving 52 on the right side of the equation.

Note: In Problems 16.7–16.8, you verify that the dot product of a vector with itself is equal to the square of the vector's magnitude.

16.8 Prove that $\mathbf{v} \cdot \mathbf{v} = \|\mathbf{v}\|^2$ for any general vector $\mathbf{v} = <a, b>$.

Apply the same technique you used in Problem 16.7. In the solution below, note that the dot product $\mathbf{v} \cdot \mathbf{v}$ is equal to the expression inside the radical symbol.

$$\mathbf{v} \cdot \mathbf{v} = \|\mathbf{v}\|^2$$
$$\langle a, b \rangle \cdot \langle a, b \rangle = \left(\sqrt{a^2 + b^2}\right)^2$$
$$a(a) + b(b) = \left(\sqrt{a^2 + b^2}\right)^2$$
$$a^2 + b^2 = a^2 + b^2$$

Note: In Problems 16.9–16.10, you use vectors $\mathbf{u} = <-3, 5>$, $\mathbf{v} = <-4, 2>$, and $\mathbf{w} = <-9, -1>$ to demonstrate that the dot product is distributive over addition: $\mathbf{u} \cdot (\mathbf{v} + \mathbf{w}) = \mathbf{u} \cdot \mathbf{v} + \mathbf{u} \cdot \mathbf{w}$.

16.9 Calculate $\mathbf{u} \cdot (\mathbf{v} + \mathbf{w})$.

In this problem, you calculate the dot product of $\mathbf{u}$ with the vector sum $\mathbf{v} + \mathbf{w}$. In Problem 16.10, you verify that this value is equal to the sum of the dot products $\mathbf{u} \cdot \mathbf{v}$ and $\mathbf{u} \cdot \mathbf{w}$. In other words, the dot product can be distributed over a sum (or a difference). Begin by calculating $\mathbf{v} + \mathbf{w}$.

$$\mathbf{u} \cdot (\mathbf{v} + \mathbf{w}) = \langle -3, 5 \rangle \cdot (\langle -4, 2 \rangle + \langle -9, -1 \rangle)$$
$$= \langle -3, 5 \rangle \cdot \langle -4 + (-9), 2 + (-1) \rangle$$
$$= \langle -3, 5 \rangle \cdot \langle -13, 1 \rangle$$

Calculate the dot product.

$$= -3(-13) + 5(1)$$
$$= 39 + 5$$
$$= 44$$

*Note: In Problems 16.9–16.10, you use vectors **u** = <–3, 5>, **v** = <–4, 2>, and **w** = <–9, –1> to demonstrate that the dot product is distributive over addition: **u** · (**v** + **w**) = **u** · **v** + **u** · **w**.*

16.10 Demonstrate that **u** · (**v** + **w**) = **u** · **v** + **u** · **w**.

According to Problem 16.9, **u** · (**v** + **w**) = 44. In this problem, you calculate the sum of the dot products **u** · **v** and **u** · **w** to verify that the dot product with a sum is equal to the sum of the dot products.

$$\mathbf{u} \cdot \mathbf{v} + \mathbf{u} \cdot \mathbf{w} = \langle -3, 5 \rangle \cdot \langle -4, 2 \rangle + \langle -3, 5 \rangle \cdot \langle -9, -1 \rangle$$
$$= -3(-4) + 5(2) + (-3)(-9) + 5(-1)$$
$$= 12 + 10 + 27 - 5$$
$$= 22 + 27 - 5$$
$$= 49 - 5$$
$$= 44$$

You conclude that **u** · (**v** + **w**) = **u** · **v** + **u** · **w** = 44.

16.11 Evaluate the expression: $\left\langle 3, \dfrac{1}{2} \right\rangle \cdot \left(\left\langle 0, \dfrac{3}{4} \right\rangle + \left\langle 2, \dfrac{1}{8} \right\rangle \right).$

As Problems 16.9–16.10 demonstrate, the dot product is distributive over a vector sum. Thus, you can calculate the dot product of <3, 1/2> with each of the terms in parentheses and then calculate the sum.

$$\left\langle 3, \frac{1}{2} \right\rangle \cdot \left(\left\langle 0, \frac{3}{4} \right\rangle + \left\langle 2, \frac{1}{8} \right\rangle \right) = \left\langle 3, \frac{1}{2} \right\rangle \cdot \left\langle 0, \frac{3}{4} \right\rangle + \left\langle 3, \frac{1}{2} \right\rangle \cdot \left\langle 2, \frac{1}{8} \right\rangle$$
$$= 3(0) + \frac{1}{2}\left(\frac{3}{4}\right) + 3(2) + \frac{1}{2}\left(\frac{1}{8}\right)$$
$$= 0 + \frac{3}{8} + 6 + \frac{1}{16}$$

Compute the sum using the least common denominator, 16.

$$= \frac{3}{8} \cdot \frac{2}{2} + \frac{6}{1} \cdot \frac{16}{16} + \frac{1}{16}$$
$$= \frac{6}{16} + \frac{96}{16} + \frac{1}{16}$$
$$= \frac{103}{16}$$

Note: In Problems 16.12–16.14, you use the vectors v = <7, 2> *and* w = <–4, 3> *to investigate how scalar multiplication affects a dot product.*

16.12 Demonstrate that $5(\mathbf{v} \cdot \mathbf{w}) \neq 5\mathbf{v} \cdot 5\mathbf{w}$.

As Problems 16.9–16.10 demonstrate, the dot product is distributive over a vector sum. However, scalar multiplication is not distributive over a dot product. In this problem, you verify that multiplying the dot product $\mathbf{v} \cdot \mathbf{w}$ by 5 is not equivalent to the dot product of $5\mathbf{v}$ and $5\mathbf{w}$. Simplify both sides of the equation individually.

$$5(\langle 7, 2 \rangle \cdot \langle -4, 3 \rangle) \neq 5 \langle 7, 2 \rangle \cdot 5 \langle -4, 3 \rangle$$
$$5(7(-4) + 2(3)) \neq \langle 5(7), 5(2) \rangle \cdot \langle 5(-4), 5(3) \rangle$$
$$5(-28 + 6) \neq \langle 35, 10 \rangle \cdot \langle -20, 15 \rangle$$
$$5(-22) \neq 35(-20) + 10(15)$$
$$-110 \neq -700 + 150$$
$$-110 \neq -550$$

Note: In Problems 16.12–16.14, you use the vectors v = <7, 2> *and* w = <–4, 3> *to investigate how scalar multiplication affects a dot product.*

16.13 Demonstrate that $5(\mathbf{v} \cdot \mathbf{w}) = 5\mathbf{v} \cdot \mathbf{w}$.

In Problem 16.12, you verify that scalar multiplication is not distributive over a dot product. However, multiplying a dot product of two vectors by a scalar is equivalent to the dot product of those vectors when exactly one of them—not both—is multiplied by the scalar. According to Problem 16.12, $5(\mathbf{v} \cdot \mathbf{w}) = -110$; in this problem you verify that $5\mathbf{v} \cdot \mathbf{w} = -110$ as well.

$$\begin{aligned} 5\mathbf{v} \cdot \mathbf{w} &= 5 \langle 7, 2 \rangle \cdot \langle -4, 3 \rangle \\ &= \langle 5(7), 5(2) \rangle \cdot \langle -4, 3 \rangle \\ &= \langle 35, 10 \rangle \cdot \langle -4, 3 \rangle \\ &= 35(-4) + 10(3) \\ &= -140 + 30 \\ &= -110 \end{aligned}$$

You conclude that $5(\mathbf{v} \cdot \mathbf{w}) = 5\mathbf{v} \cdot \mathbf{w} = -110$.

In other words, if you multiply ONE vector by the scalar and then calculate the dot product, you get the same answer as if you'd taken the dot product of the vectors first and then multiplied by the scalar.

Note: In Problems 16.12–16.14, you use the vectors v = <7, 2> *and* w = <–4, 3> *to investigate how scalar multiplication affects a dot product.*

16.14 Demonstrate that $5(\mathbf{v} \cdot \mathbf{w}) = \mathbf{v} \cdot 5\mathbf{w}$.

In Problem 16.13, you verify that multiplying $\mathbf{v}$ by the scalar 5 and then computing the dot product with $\mathbf{w}$ is equivalent to multiplying the dot product $\mathbf{v} \cdot \mathbf{w}$ by the scalar 5. In this problem, you demonstrate that multiplying the other vector ($\mathbf{w}$) by the scalar 5 has the same effect on the dot product. Recall that $5(\mathbf{v} \cdot \mathbf{w}) = -110$.

$$\mathbf{v} \cdot 5\mathbf{w} = \langle 7, 2 \rangle \cdot 5 \langle -4, 3 \rangle$$
$$= \langle 7, 2 \rangle \cdot \langle 5(-4), 5(3) \rangle$$
$$= \langle 7, 2 \rangle \cdot \langle -20, 15 \rangle$$
$$= 7(-20) + 2(15)$$
$$= -140 + 30$$
$$= -110$$

Thus, $5(\mathbf{v} \cdot \mathbf{w}) = \mathbf{v} \cdot 5\mathbf{w} = -110.$ ←

> Moral of the story: If a dot product is multiplied by a scalar, you can multiply either one of the vectors by the scalar and then calculate the dot product to get the same result.

16.15 Evaluate the expression: $2\big(\langle 10, -4 \rangle \cdot \langle -2, 6 \rangle\big)$.

You can either choose to calculate the dot product and then multiply by the scalar 2 or, as Problems 16.12–16.14 explain, multiply exactly one of the vectors by the scalar 2 and then compute the dot product. In the following solution, the vector <–2, 6> is multiplied by 2 and then the dot product is computed.

$$2\big(\langle 10, -4 \rangle \cdot \langle -2, 6 \rangle\big) = \langle 10, -4 \rangle \cdot 2 \langle -2, 6 \rangle$$
$$= \langle 10, -4 \rangle \cdot \langle -4, 12 \rangle$$
$$= 10(-4) + (-4)(12)$$
$$= -40 - 48$$
$$= -88$$

16.16 Evaluate the expression: $\dfrac{1}{3}\left(\left\langle -\dfrac{1}{4}, \dfrac{5}{6} \right\rangle \cdot \left\langle \dfrac{2}{3}, -\dfrac{3}{4} \right\rangle\right)$.

In the following solution, the dot product is computed and the resulting value is multiplied by 1/3.

$$\frac{1}{3}\left[\left\langle -\frac{1}{4}, \frac{5}{6} \right\rangle \cdot \left\langle \frac{2}{3}, -\frac{3}{4} \right\rangle\right] = \frac{1}{3}\left[-\frac{1}{4}\left(\frac{2}{3}\right) + \frac{5}{6}\left(-\frac{3}{4}\right) \right]$$
$$= \frac{1}{3}\left[-\frac{2}{12} - \frac{15}{24} \right]$$
$$= \frac{1}{3}\left[-\frac{4}{24} - \frac{15}{24} \right]$$
$$= \frac{1}{3}\left[-\frac{19}{24} \right]$$
$$= -\frac{19}{72}$$

Measuring Angles Between Vectors

The dot product in disguise

16.17 Identify the alternate dot product formula, which expresses a dot product in terms of the magnitude of the vectors.

According to Problem 16.1, given vectors $\mathbf{v} = \langle a, b\rangle$ and $\mathbf{w} = \langle c, d\rangle$, the dot product is defined according to the following formula.

$$\mathbf{v} \cdot \mathbf{w} = ac + bd$$

You can also calculate the dot product according to an alternate formula, which is based on the magnitudes of the vectors and the cosine of the angle θ between those vectors.

$$\mathbf{v} \cdot \mathbf{w} = \|\mathbf{v}\| \|\mathbf{w}\| \cos\theta$$

> The alternate version is handy when you don't know the components of the vectors.

Note: In Problems 16.18–16.21, v = <0, 5> and w = <2, 2>.

16.18 Calculate $\mathbf{v} \cdot \mathbf{w}$ using the standard dot product formula.

To apply the standard dot product formula, you must know the components of the vectors.

$$\begin{aligned}
\mathbf{v} \cdot \mathbf{w} &= \langle 0, 5\rangle \cdot \langle 2, 2\rangle \\
&= 0(2) + 5(2) \\
&= 0 + 10 \\
&= 10
\end{aligned}$$

Note: In Problems 16.18–16.21, v = <0, 5> and w = <2, 2>.

16.19 Plot $\mathbf{v}$ and $\mathbf{w}$ on the coordinate plane and measure their direction angles in degrees.

Both vectors have initial point $(0, 0)$; $\mathbf{v}$ has terminal point $(0, 5)$ and $\mathbf{w}$ has terminal point $(2, 2)$.

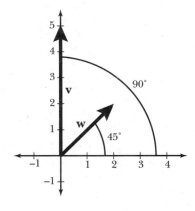

Vector **v** lies on the positive *y*-axis, which forms a right angle with the positive *x*-axis. Therefore, the direction angle of **v** is 90°. The direction angle of w is 45°. To calculate this value, apply the technique demonstrated in Problems 14.44–14.45.

$$\tan\theta = \frac{2}{2}$$

$$\tan\theta = 1$$

$$\theta = \arctan 1$$

$$\theta = 45°$$

> The tangent of the direction angle is equal to the y-component divided by the x-component.

Note: In Problems 16.18–16.21, v = <0, 5> and w = <2, 2>.

16.20 Use the diagram you created in Problem 16.19 to calculate the angle θ between **v** and **w**.

As the following diagram illustrates, the measure of the angle between the vectors is equal to the difference of the larger and smaller direction angles.

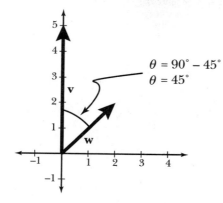

$$\theta = 90° - 45°$$
$$\theta = 45°$$

Note: In Problems 16.18–16.21, v = <0, 5> and w = <2, 2>.

16.21 Verify your answer to Problem 16.20 by applying the alternate dot product formula.

> The alternate dot product formula comes from Problem 16.17.

According to Problem 16.18, **v** · **w** = 10. If you substitute this value and the magnitudes of **v** and **w** into the alternate dot product formula, you can solve for θ to verify that the measure of the angle between **v** and **w** is 45°. Begin by calculating the magnitudes of the vectors.

$$\|\mathbf{v}\| = \sqrt{0^2 + 5^2} \qquad \|\mathbf{w}\| = \sqrt{2^2 + 2^2}$$
$$= \sqrt{0 + 25} \qquad\quad = \sqrt{4 + 4}$$
$$= \sqrt{25} \qquad\qquad = \sqrt{8}$$
$$= 5 \qquad\qquad\quad = 2\sqrt{2}$$

Substitute **v** · **w** = 10, $\|\mathbf{v}\| = 5$, and $\|\mathbf{w}\| = 2\sqrt{2}$ into the alternate dot product formula. Then, isolate $\cos\theta$ on the right side of the equation.

$$\mathbf{v} \cdot \mathbf{w} = \|\mathbf{v}\| \|\mathbf{w}\| \cos\theta$$

$$10 = (5)\left(2\sqrt{2}\right)(\cos\theta)$$

$$10 = 10\sqrt{2} \cos\theta$$

$$\frac{\cancel{10}}{\cancel{10}\sqrt{2}} = \cos\theta$$

$$\frac{1}{\sqrt{2}} = \cos\theta$$

Rationalize the denominator and apply the arccosine function to solve for θ.

$$\frac{1 \cdot \sqrt{2}}{\sqrt{2} \cdot \sqrt{2}} = \cos\theta$$

$$\frac{\sqrt{2}}{2} = \cos\theta$$

> *According to the unit circle, $\cos 45° = \cos \pi/4 = \sqrt{2}/2$.*

$$\arccos\left(\frac{\sqrt{2}}{2}\right) = \theta$$

$$45° = \theta$$

The alternate dot product formula verifies the solution to Problem 16.20—the angle between $\mathbf{v}$ and $\mathbf{w}$ measures 45°.

16.22 Calculate the dot product of vectors $\mathbf{r}$ and $\mathbf{s}$, which form an angle measuring 30° and have magnitudes of 2 and 9.

The alternate dot product states that $\mathbf{r} \cdot \mathbf{s}$ is equal to the product of the magnitudes of the vectors and the cosine of the angle between the vectors. In this problem, you do not know which vector has a magnitude of 2 and which has a magnitude of 9, but that is irrelevant. The following solution assumes that $\|\mathbf{r}\| = 2$ and $\|\mathbf{s}\| = 9$, but the dot product is unchanged if $\|\mathbf{s}\| = 2$ and $\|\mathbf{r}\| = 9$.

$$\mathbf{r} \cdot \mathbf{s} = \|\mathbf{r}\| \|\mathbf{s}\| \cos\theta$$

$$= (2)(9)(\cos 30°)$$

$$= 18\left(\frac{\sqrt{3}}{2}\right)$$

$$= \frac{18}{2}\sqrt{3}$$

$$= 9\sqrt{3}$$

16.23 Calculate the dot product of vectors **u** and **v**, which form an angle measuring 60° and have magnitudes of $\frac{1}{3}$ and $\frac{1}{4}$.

Substitute the magnitudes of the vectors and $\theta = 60°$ into the alternate dot product formula to calculate **u** · **v**.

$$\mathbf{u} \cdot \mathbf{v} = \|\mathbf{u}\| \, \|\mathbf{v}\| \cos\theta$$
$$= \left(\frac{1}{3}\right)\left(\frac{1}{4}\right)(\cos 60°)$$
$$= \frac{1}{12}\cos 60°$$
$$= \frac{1}{12}\left(\frac{1}{2}\right)$$
$$= \frac{1}{24}$$

16.24 Calculate the dot product of vectors **w** and **z**, which form an angle measuring 120° and have magnitudes of $4\sqrt{3}$ and 7.

Substitute $\theta = 120°$, $\|\mathbf{w}\| = 4\sqrt{3}$, and $\|\mathbf{z}\| = 7$ into the alternate dot product formula.

$$\mathbf{w} \cdot \mathbf{z} = \|\mathbf{w}\| \, \|\mathbf{z}\| \cos\theta$$
$$= \left(4\sqrt{3}\right)(7)(\cos 120°)$$
$$= 28\sqrt{3}\,\cos 120°$$
$$= 28\sqrt{3}\left(-\frac{1}{2}\right)$$
$$= -\frac{28}{2}\sqrt{3}$$
$$= -14\sqrt{3}$$

16.25 Calculate the angle formed by $\mathbf{r} = \left\langle 1, \sqrt{3} \right\rangle$ and $\mathbf{s} = \left\langle \sqrt{3}, 1 \right\rangle$.

Calculate the magnitudes of the vectors.

$$\|\mathbf{r}\| = \sqrt{1^2 + \left(\sqrt{3}\right)^2} \qquad \|\mathbf{s}\| = \sqrt{\left(\sqrt{3}\right)^2 + 1^2}$$
$$= \sqrt{1+3} \qquad\qquad = \sqrt{3+1}$$
$$= \sqrt{4} \qquad\qquad = \sqrt{4}$$
$$= 2 \qquad\qquad\quad = 2$$

Now calculate the dot product of the vectors.

$$\mathbf{r} \cdot \mathbf{s} = 1\left(\sqrt{3}\right) + \sqrt{3}\,(1)$$
$$= \sqrt{3} + \sqrt{3}$$
$$= 2\sqrt{3}$$

Apply the alternate dot product formula and solve for θ.

$$\mathbf{r} \cdot \mathbf{s} = \|\mathbf{r}\| \, \|\mathbf{s}\| \cos \theta$$
$$2\sqrt{3} = 2(2)(\cos \theta)$$
$$2\sqrt{3} = 4 \cos \theta$$
$$\frac{1}{4}\left(2\sqrt{3}\right) = \frac{1}{4}\left(4 \cos \theta\right)$$
$$\frac{2\sqrt{3}}{4} = \cos \theta$$
$$\frac{\sqrt{3}}{2} = \cos \theta$$
$$\arccos\left(\frac{\sqrt{3}}{2}\right) = \theta$$
$$30° = \theta$$

According to the unit circle, $\cos 30° = \cos \pi/6 = \sqrt{3}/2$.

Thus, vectors $\mathbf{r}$ and $\mathbf{s}$ form a $30°$ angle.

16.26 Calculate the angle formed by $\mathbf{w} = \langle -4, -3 \rangle$ and $\mathbf{z} = \langle 6, -1 \rangle$, and report your answer as a degree measurement accurate to the thousandths place.

Calculate $\mathbf{w} \cdot \mathbf{z}$, $\|\mathbf{w}\|$, and $\|\mathbf{z}\|$.

$$\mathbf{w} \cdot \mathbf{z} = -4(6) + (-3)(-1) \qquad \|\mathbf{w}\| = \sqrt{(-4)^2 + (-3)^2} \qquad \|\mathbf{z}\| = \sqrt{6^2 + (-1)^2}$$
$$= -24 + 3 \qquad\qquad = \sqrt{16 + 9} \qquad\qquad = \sqrt{36 + 1}$$
$$= -21 \qquad\qquad\quad = \sqrt{25} \qquad\qquad\quad = \sqrt{37}$$
$$= 5$$

Substitute these values into the alternate dot product formula and solve for θ. Use a scientific or graphing calculator set to degrees mode and calculate the measure of the angle.

$$\mathbf{w} \cdot \mathbf{z} = \|\mathbf{w}\| \, \|\mathbf{z}\| \cos \theta$$
$$-21 = (5)\left(\sqrt{37}\right)(\cos \theta)$$
$$-\frac{21}{5\sqrt{37}} = \cos \theta$$
$$\arccos\left(-\frac{21}{5\sqrt{37}}\right) = \theta$$
$$\arccos(-0.6904757467) \approx \theta$$
$$133.668° \approx \theta$$

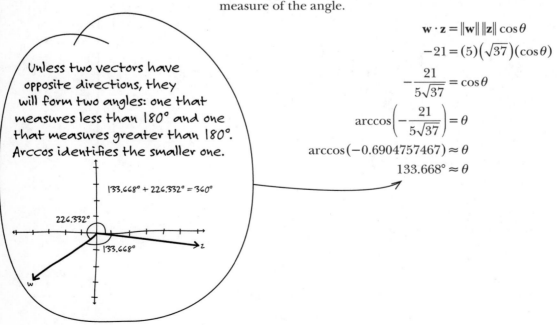

Unless two vectors have opposite directions, they will form two angles: one that measures less than 180° and one that measures greater than 180°. Arccos identifies the smaller one.

133.668° + 226.332° = 360°

226.332°

133.668°

z

w

Note: In Problems 16.27–16.30, **u** *= <–5,–12> and* **v** *= <–8, 10>.*

16.27 Calculate the angle θ between vectors **u** and **v**, and report your answer accurate to the thousandths place.

Calculate $\mathbf{u} \cdot \mathbf{v}$, $\|\mathbf{u}\|$, and $\|\mathbf{v}\|$.

$$\begin{aligned}
\mathbf{u} \cdot \mathbf{v} &= -5(-8)+(-12)(10) & \|\mathbf{u}\| &= \sqrt{(-5)^2 + (-12)^2} & \|\mathbf{v}\| &= \sqrt{(-8)^2 + 10^2} \\
&= 40-120 & &= \sqrt{25+144} & &= \sqrt{64+100} \\
&= -80 & &= \sqrt{169} & &= \sqrt{164} \\
& & &= 13 & &= 2\sqrt{41}
\end{aligned}$$

Substitute these values into the alternate dot product formula.

$$\mathbf{u} \cdot \mathbf{v} = \|\mathbf{u}\|\,\|\mathbf{v}\|\cos\theta$$
$$-80 = (13)\left(2\sqrt{41}\right)(\cos\theta)$$
$$\frac{-80}{26\sqrt{41}} = \cos\theta$$
$$\arccos\left(-0.480534652\right) \approx \theta$$
$$118.720° \approx \theta$$

Note: In Problems 16.27–16.30, **u** *= <–5,–12> and* **v** *= <–8, 10>.*

16.28 Identify $\hat{\mathbf{u}}$ and $\hat{\mathbf{v}}$, unit vectors that have the same direction angles as **u** and **v**.

According to Problem 16.27, $\|\mathbf{u}\| = 13$ and $\|\mathbf{v}\| = 2\sqrt{41}$. Divide each component of **u** and **v** by the magnitude of its vector to identify the corresponding unit vectors.

$$\begin{aligned}
\hat{\mathbf{u}} &= \left\langle \frac{-5}{\|\mathbf{u}\|}, \frac{-12}{\|\mathbf{u}\|} \right\rangle & \hat{\mathbf{v}} &= \left\langle \frac{-8}{\|\mathbf{v}\|}, \frac{10}{\|\mathbf{v}\|} \right\rangle \\
&= \left\langle -\frac{5}{13}, -\frac{12}{13} \right\rangle & &= \left\langle -\frac{8}{2\sqrt{41}}, \frac{10}{2\sqrt{41}} \right\rangle \\
& & &= \left\langle -\frac{4}{\sqrt{41}}, \frac{5}{\sqrt{41}} \right\rangle \\
& & &= \left\langle -\frac{4\sqrt{41}}{41}, \frac{5\sqrt{41}}{41} \right\rangle
\end{aligned}$$

Note: In Problems 16.27–16.30, u = <–5, –12> and v = <–8, 10>.

16.29 Demonstrate that the dot product of the unit vectors you identified in Problem 16.28 is equal to the cosine of the angle between them.

If $\hat{\mathbf{u}}$ and $\hat{\mathbf{v}}$ are unit vectors, then they each have a magnitude of 1. Substitute these values into the alternate dot product formula.

$$\hat{\mathbf{u}} \cdot \hat{\mathbf{v}} = \|\hat{\mathbf{u}}\| \, \|\hat{\mathbf{v}}\| \cos\theta$$
$$= (1)(1)(\cos\theta)$$
$$= \cos\theta$$

Note: In Problems 16.27–16.30, u = <–5, –12> and v = <–8, 10>.

16.30 Verify that the angle between **u** and **v** is equal to the angle between unit vectors $\hat{\mathbf{u}}$ and $\hat{\mathbf{v}}$.

Calculate the dot product of the unit vectors.

$$\hat{\mathbf{u}} \cdot \hat{\mathbf{v}} = \left\langle -\frac{5}{13}, -\frac{12}{13} \right\rangle \cdot \left\langle -\frac{4\sqrt{41}}{41}, \frac{5\sqrt{41}}{41} \right\rangle$$
$$= -\frac{5}{13}\left(-\frac{4\sqrt{41}}{41}\right) + \left(-\frac{12}{13}\right)\left(\frac{5\sqrt{41}}{41}\right)$$
$$= \frac{20\sqrt{41}}{533} - \frac{60\sqrt{41}}{533}$$
$$= -\frac{40\sqrt{41}}{533}$$

As Problem 16.29 explains, the dot product of the unit vectors is equal to the cosine of the angle θ between them.

$$\hat{\mathbf{u}} \cdot \hat{\mathbf{v}} = \cos\theta$$
$$-\frac{40\sqrt{41}}{533} = \cos\theta$$
$$\arccos\left(-\frac{40\sqrt{41}}{533}\right) = \theta$$
$$\arccos(-0.480534652) \approx \theta$$
$$118.720° \approx \theta$$

The angle between vectors **u** and **v** (calculated in Problem 16.27) is equal to the angle between unit vectors $\hat{\mathbf{u}}$ and $\hat{\mathbf{v}}$. Both measure approximately 118.720°.

Orthogonal Vectors
Perpendicular vectors

16.31 Describe the geometric relationship between orthogonal vectors.

Orthogonal vectors are perpendicular to each other. Therefore, intersecting orthogonal vectors intersect at right angles.

Note: In Problems 16.32–16.36, you explore four different techniques to verify that vectors **u** = <3, 3> *and* **v** = <5, –5> *are orthogonal.*

16.32 Plot **u** and **v**, and use their direction angles to verify that the vectors are orthogonal.

Plot **u** and **v** in the coordinate plane. You will measure the angle between them by calculating the angles each vector forms with the positive *x*-axis.

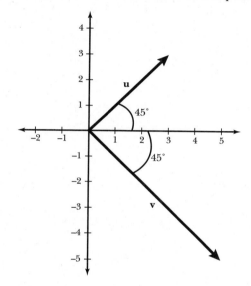

Apply the technique demonstrated in Problem 16.19 to calculate the direction angle θ of vector **u**—the measure of the angle formed by **u** and the positive *x*-axis.

$$\tan\theta = \frac{3}{3}$$
$$\tan\theta = 1$$
$$\theta = \arctan 1$$
$$\theta = 45°$$

Similarly, the angle between **v** and the positive *x*-axis measures 45°. The angle between **u** and **v** is the sum of those angles: 45° + 45° = 90°. Lines— and vectors—that intersect at 90° angles are perpendicular, so **u** and **v** are orthogonal.

Note: In Problems 16.32–16.36, you explore four different techniques to verify that vectors **u** *= <3, 3> and* **v** *= <5, –5> are orthogonal.*

16.33 Calculate the slopes of the lines that coincide with **u** and **v**, and use them to verify that the vectors are orthogonal.

"Coincide with" means "overlap."

The slope of a vector in component form—and the slope of the line that coincides with that vector—is equal to the *y*-component of the vector divided by the *x*-component. Therefore, the slope of the line that coincides with **u** is 3/3 = 1. The slope of the line that coincides with **v** is –5/5 = –1.

Perpendicular lines have slopes that are negative reciprocals of each other. Because 1 and –1 are negative reciprocals, they represent the slopes of perpendicular lines. You conclude that **u** and **v** are orthogonal.

Note: In Problems 16.32–16.36, you explore four different techniques to verify that vectors **u** *= <3, 3> and* **v** *= <5, –5> are orthogonal.*

16.34 Apply the standard dot product formula to verify that **u** and **v** are orthogonal.

According to the standard dot product formula, $<a, b> \cdot <c, d> = ac + bd$.

$$\mathbf{u} \cdot \mathbf{v} = \langle 3, 3 \rangle \cdot \langle 5, -5 \rangle$$
$$= 3(5) + 3(-5)$$
$$= 15 - 15$$
$$= 0$$

You conclude that **u** and **v** are orthogonal because they have a dot product of 0. In fact, any pair of nonzero vectors that have a dot product of 0 are orthogonal.

And vice versa: If vectors are orthogonal, they always have a dot product of 0. This only works for NONZERO vectors. The next problem explains why.

Note: In Problems 16.32–16.36, you explore four different techniques to verify that vectors **u** *= <3, 3> and* **v** *= <5, –5> are orthogonal.*

16.35 Determine whether **u** and **w** = <0, 0> are orthogonal vectors, and explain your answer.

If you calculate the dot product of **u** and **w**, the result is 0.

$$\mathbf{u} \cdot \mathbf{w} = \langle 3, 3 \rangle \cdot \langle 0, 0 \rangle$$
$$= 3(0) + 3(0)$$
$$= 0 + 0$$
$$= 0$$

In fact, the dot product of any vector and the zero vector <0, 0> is equal to 0. However, you cannot conclude that **u** and **w** are orthogonal based upon the dot product value alone. Recall that orthogonal vectors are perpendicular, so the angle formed by the vectors should equal 90°.

Solve the alternate dot product formula for $\cos\theta$ to better understand why **u** and **w** do not form a 90° angle.

$$\mathbf{u}\cdot\mathbf{w} = \|\mathbf{u}\|\,\|\mathbf{w}\|\cos\theta$$

$$\frac{\mathbf{u}\cdot\mathbf{w}}{\|\mathbf{u}\|\,\|\mathbf{w}\|} = \cos\theta \longleftarrow$$

Some textbooks (and teachers) prefer this version of the formula.

Calculate the magnitudes of **u** and **w**.

$$\|\mathbf{u}\| = \sqrt{3^2 + 3^2} \qquad \|\mathbf{w}\| = \sqrt{0^2 + 0^2}$$
$$= \sqrt{9+9} \qquad\qquad = \sqrt{0}$$
$$= \sqrt{18} \qquad\qquad = 0$$
$$= 3\sqrt{2}$$

Substitute the dot product and vector magnitudes into the modified dot product formula.

$$\frac{\mathbf{u}\cdot\mathbf{w}}{\|\mathbf{u}\|\,\|\mathbf{w}\|} = \cos\theta$$

$$\frac{0}{\left(3\sqrt{2}\right)(0)} = \cos\theta$$

$$\frac{0}{0} = \cos\theta$$

Dividing by 0 is prohibited. Because you cannot calculate θ, you cannot conclude that θ is a right angle. Therefore, **u** and **w** are not orthogonal.

Note: In Problems 16.32–16.36, you explore four different techniques to verify that vectors **u** = <3, 3> *and* **v** = <5, –5> *are orthogonal.*

16.36 Calculate the angle between **u** and **v** to verify that the vectors are orthogonal.

Recall that $\mathbf{u}\cdot\mathbf{v} = 0$, and according to Problem 16.35, $\|\mathbf{u}\| = 3\sqrt{2}$. Calculate the magnitude of **v**.

$$\|\mathbf{v}\| = \sqrt{5^2 + (-5)^2}$$
$$= \sqrt{25+25}$$
$$= \sqrt{50}$$
$$= 5\sqrt{2}$$

Apply the modified dot product formula introduced in Problem 16.35 to compute the angle θ between **u** and **v**.

$$\frac{\mathbf{u} \cdot \mathbf{v}}{\|\mathbf{u}\| \|\mathbf{v}\|} = \cos\theta$$

$$\frac{0}{\left(3\sqrt{2}\right)\left(5\sqrt{2}\right)} = \cos\theta$$

$$0 = \cos\theta$$

$$\arccos 0 = \theta$$

$$90° = \theta$$

Vectors **u** and **v** form a 90° angle, so they are orthogonal.

16.37 Determine whether the vectors <2,−3> and <6,4> are orthogonal.

Calculate the dot product of the vectors.

$$\langle 2, -3 \rangle \cdot \langle 6, 4 \rangle = 2(6) + (-3)(4)$$
$$= 12 - 12$$
$$= 0$$

Because their dot product is equal to 0, vectors <2,−3> and <6,4> are orthogonal.

16.38 Given $A = (8,-2)$, $B = (6,1)$, $C = (-3,-7)$, and $D = (4,-4)$, determine whether $\overrightarrow{AB}$ and $\overrightarrow{CD}$ are orthogonal.

Subtract corresponding coordinates to express the vectors in component form.

$$\overrightarrow{AB} = \langle 6 - 8, 1 - (-2) \rangle \qquad \overrightarrow{CD} = \langle 4 - (-3), -4 - (-7) \rangle$$
$$= \langle 6 - 8, 1 + 2 \rangle \qquad\qquad = \langle 4 + 3, -4 + 7 \rangle$$
$$= \langle -2, 3 \rangle \qquad\qquad\quad = \langle 7, 3 \rangle$$

Calculate the dot product of the vectors.

$$\overrightarrow{AB} \cdot \overrightarrow{CD} = \langle -2, 3 \rangle \cdot \langle 7, 3 \rangle$$
$$= -14 + 9$$
$$= -5$$

Vectors $\overrightarrow{AB}$ and $\overrightarrow{CD}$ are not orthogonal because their dot product is not equal to 0.

16.39 Calculate the value of c for which vectors $\mathbf{a} = <4, -2c>$ and $\mathbf{b} = <9, 3>$ are orthogonal.

If $\mathbf{a}$ and $\mathbf{b}$ are orthogonal vectors, then $\mathbf{a} \cdot \mathbf{b} = 0$. Compute the dot product and set it equal to 0.

$$\mathbf{a} \cdot \mathbf{b} = 0$$
$$\langle 4, -2c \rangle \cdot \langle 9, 3 \rangle = 0$$
$$4(9) + (-2c)(3) = 0$$
$$36 - 6c = 0$$

Solve the equation for c.

$$36 = 6c$$
$$\frac{36}{6} = c$$
$$6 = c$$

16.40 Given $A = (3, 4)$, $B = (12, -10)$, and $\mathbf{z} = <5c, -2>$, calculate the value of c for which $\overrightarrow{AB}$ and $\mathbf{z}$ are orthogonal.

Express $\overrightarrow{AB}$ in component form.

$$\overrightarrow{AB} = \langle 12 - 3, -10 - 4 \rangle$$
$$= \langle 9, -14 \rangle$$

Because the vectors are orthogonal, you can set their dot product equal to 0.

$$\overrightarrow{AB} \cdot \mathbf{z} = 0$$
$$\langle 9, -14 \rangle \cdot \langle 5c, -2 \rangle = 0$$
$$9(5c) + (-14)(-2) = 0$$
$$45c + 28 = 0$$

Solve for c.

$$45c = -28$$
$$c = -\frac{28}{45}$$

Vector Projections and Work
Create specific orthogonal vectors

16.41 Given the diagram below, plot vectors **a** and **b** such that **a** + **b** = **v**. Explain why the solution is not unique.

In other words, explain why there isn't one correct answer to this problem.

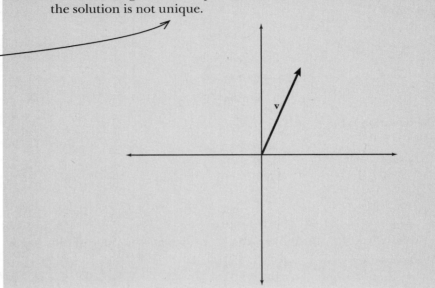

As Problems 15.1–15.3 demonstrate, you add vectors in the coordinate plane using the head-to-tail method. The terminal point of the first vector acts as the initial point of the second vector. In this problem, you are asked to create two vectors, **a** and **b**, that have a sum of **v**. Plot any vector in standard position and label this vector **a**. Then, plot a new vector **b** with an initial point at the terminal point of **a** and the same terminal point as **v**.

The following diagram represents one solution to this problem, but any vectors **a** and **b** that meet the criteria described above are equally valid solutions.

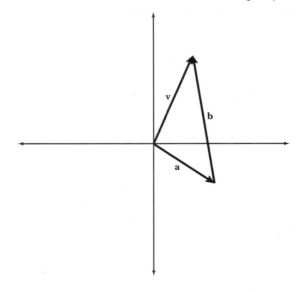

*Note: Problems 16.42–16.45 refer to vectors **v** and **w** in the diagram below.*

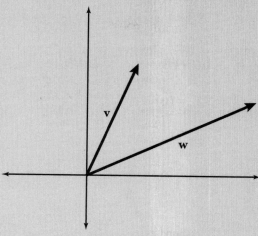

16.42 Plot the unique vector **b** on the coordinate plane, such that: (1) **b** is orthogonal to **w**, (2) the initial point of **b** lies on **w**, and (3) **b** and **v** share the same terminal point.

The figure below illustrates **b**, the vector that is perpendicular to **w**, has an initial point lying on **w**, and terminates at the same point as **v**.

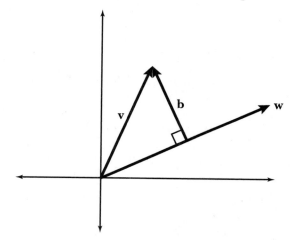

> Problem 16.41 tells you that there are many ways to create two vectors that add up to v. In Problems 16.42–16.44, you create two perpendicular vectors whose sum is v, and one of those vectors will overlap vector w.

Note: Problems 16.42–16.45 refer to the diagram in Problem 16.42.

16.43 Based on your answer to Problem 16.42, identify vector **a** such that **a** = proj$_w$**v**.

The notation "proj$_w$ **v**" is read "the projection of vector **v** onto vector **w**." Because you are projecting **v** onto **w**, the vector **a** you are seeking lies along vector **w** (but **a** and **w** do not have the same magnitude). As the diagram below illustrates, vectors **a** and **w** share the same initial point, but the terminal point of **a** is the initial point of vector **b**, which you identified in Problem 16.42.

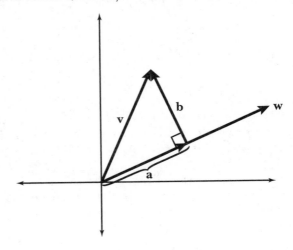

Note: Problems 16.42–16.45 refer to the diagram in Problem 16.42.

16.44 Based on your answers to Problems 16.42–16.43, describe the vector sum **a** + **b**.

According to the geometric technique of vector addition, **a** + **b** = **v**. Although there are infinite pairs of vectors that have a sum of **v**, which you demonstrated in Problem 16.41, these vectors have two unique properties: they are orthogonal and one of the vectors lies along the vector **w** that was specified in the problem.

Note: Problems 16.42–16.45 refer to the diagram in Problem 16.42.

16.45 Identify the formula that expresses **a** = proj$_w$ **v** in component form.

Although it is important to understand the geometric derivation of the projection vector **a** = proj$_w$ **v**, the graph alone is insufficient to calculate the components of **a**. Instead, you should apply the following formula.

$$\text{proj}_w \mathbf{v} = \left(\frac{\mathbf{v} \cdot \mathbf{w}}{\|\mathbf{w}\|^2} \right) \mathbf{w}$$

Look at the fraction in this formula. It's equal to a dot product (which is a number) divided by a magnitude squared (which is also a number). Basically, this formula multiplies vector w by some crazy number in parentheses. In other words, the projection of v onto w is just a scalar multiple of w! It has the same direction as w, but not the same length.

> *Note: In Problems 16.46–16.47,* **v** *= <3, 0> and* **w** *= <2, 2>.*

16.46 Let **a** represent the projection of **v** onto **w**. Express **a** in component form.

Substitute the vectors into the formula presented in Problem 16.45 to calculate **a** = proj$_w$ **v**.

$$\mathbf{a} = \text{proj}_w \mathbf{v}$$

$$= \left(\frac{\mathbf{v} \cdot \mathbf{w}}{\|\mathbf{w}\|^2} \right) \mathbf{w}$$

$$= \left(\frac{\langle 3, 0 \rangle \cdot \langle 2, 2 \rangle}{\|\langle 2, 2 \rangle\|^2} \right) \langle 2, 2 \rangle$$

$$= \left(\frac{3(2) + 0(2)}{\left[\sqrt{2^2 + 2^2} \right]^2} \right) \langle 2, 2 \rangle$$

$$= \left(\frac{6 + 0}{\left[\sqrt{8} \right]^2} \right) \langle 2, 2 \rangle$$

$$= \frac{6}{8} \langle 2, 2 \rangle$$

$$= \frac{3}{4} \langle 2, 2 \rangle$$

Note that **a** = proj$_w$ **v** is equal to the vector **w** multiplied by the scalar 3/4.

$$= \left\langle \frac{3}{4} \cdot 2, \frac{3}{4} \cdot 2 \right\rangle$$

$$= \left\langle \frac{6}{4}, \frac{6}{4} \right\rangle$$

$$= \left\langle \frac{3}{2}, \frac{3}{2} \right\rangle$$

You conclude that the projection of **v** onto **w**, in component form, is **a** = <3/2, 3/2>.

Note: In Problems 16.46–16.47, v = <3, 0> and w = <2, 2>.

16.47 Plot **v**, **w**, **a** = proj$_w$ **v**, and **b** in the coordinate plane, given **a** + **b** = **v**.

Construct vector **b** as directed in Problem 16.42. It should be orthogonal to **a** = <3/2, 3/2>, have an initial point on **w**, and terminate at the same point as **v**.

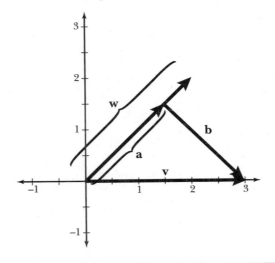

16.48 Given **v** = <–2, 1> and **w** = <–6, –2>, identify vectors **a** and **b**, such that **a** = proj$_w$ **v** and **a** + **b** = **v**.

Apply the vector projection formula to calculate **a**.

$$\mathbf{a} = \left(\frac{\mathbf{v} \cdot \mathbf{w}}{\|\mathbf{w}\|^2} \right) \mathbf{w}$$

$$= \left(\frac{\langle -2, 1 \rangle \cdot \langle -6, -2 \rangle}{\|\langle -6, -2 \rangle\|^2} \right) \langle -6, -2 \rangle$$

$$= \left(\frac{-2(-6) + 1(-2)}{\left[\sqrt{(-6)^2 + (-2)^2} \right]^2} \right) \langle -6, -2 \rangle$$

$$= \left(\frac{12 - 2}{\left[\sqrt{36 + 4} \right]^2} \right) \langle -6, -2 \rangle$$

$$= \left(\frac{10}{40} \right) \langle -6, -2 \rangle$$

$$= \frac{1}{4} \langle -6, -2 \rangle$$

Perform scalar multiplication and reduce the fractions to lowest terms.

$$= \left\langle \frac{1}{4}(-6), \frac{1}{4}(-2) \right\rangle$$

$$= \left\langle -\frac{6}{4}, -\frac{2}{4} \right\rangle$$

$$= \left\langle -\frac{3}{2}, -\frac{1}{2} \right\rangle$$

Recall that $\mathbf{a} + \mathbf{b} = \mathbf{v}$. Substitute $\mathbf{v} = <-2, 1>$ and $\mathbf{a} = <-3/2, -1/2>$ into the equation and solve for $\mathbf{b}$.

$$\mathbf{a} + \mathbf{b} = \mathbf{v}$$

$$\left\langle -\frac{3}{2}, -\frac{1}{2} \right\rangle + \mathbf{b} = \left\langle -2, 1 \right\rangle$$

$$\mathbf{b} = \left\langle -2, 1 \right\rangle - \left\langle -\frac{3}{2}, -\frac{1}{2} \right\rangle$$

$$\mathbf{b} = \left\langle -2 - \left(-\frac{3}{2} \right), 1 - \left(-\frac{1}{2} \right) \right\rangle$$

$$\mathbf{b} = \left\langle -\frac{4}{2} + \frac{3}{2}, \frac{2}{2} + \frac{1}{2} \right\rangle$$

$$\mathbf{b} = \left\langle -\frac{1}{2}, \frac{3}{2} \right\rangle$$

16.49 Calculate the work done moving a particle from point $A = (0,0)$ to point $B = (4,1)$ on the coordinate plane with a force of 6. Assume that the force is applied in the same direction the particle is moving.

The displacement of an object describes the movement of that object from its initial to final location. In this problem, the displacement vector $\mathbf{d}$ has initial point $A = (0,0)$ and final point $B = (4,1)$. In component form, $\mathbf{d} = <4,1>$. The magnitude of the displacement vector is the distance traveled by the particle along that vector.

$$\|\mathbf{d}\| = \|\langle 4, 1 \rangle\|$$

$$= \sqrt{4^2 + 1^2}$$

$$= \sqrt{17}$$

Force is also expressed as a vector: $\mathbf{F}$. The magnitude of the force vector is the total force applied; in this problem, $\|\mathbf{F}\| = 6$. Note that the force is applied in the same direction as the displacement. When force is applied in the same direction as displacement, the formula for work done is $W = \|\mathbf{F}\| \|\mathbf{d}\|$.

$$W = \|\mathbf{F}\| \|\mathbf{d}\|$$

$$= 6\left(\sqrt{17} \right)$$

$$= 6\sqrt{17}$$

> The technique you use does not account for friction, so the answer is more theoretical than realistic. That's why the problem asks you to move an atomic particle along a line instead of pushing a massive couch along a steep ramp. The smaller the mass of the object, the more realistic your answer is.

> The particle moves along a line with slope 1/4, so the force is applied along the same slope. Imagine that the particle is sitting on that line and you (in microscopic form) are pushing the particle along the line as well, up a teeny little ramp.

*Note: In Problems 16.50–16.52, you calculate the total work done to move a particle along displacement vector **d** = <3,4> by applying force vector **F** = <4,2>.*

16.50 Calculate the component of force along the displacement vector.

Unlike Problem 16.49, the force on this particle is not applied in the same direction the particle is moving. Your objective is to calculate **a** = proj$_d$**F**, the vector representing the force applied along the displacement vector. The magnitude of this vector is the component of **F** along **d**.

$$\text{proj}_d \mathbf{F} = \left(\frac{\mathbf{F} \cdot \mathbf{d}}{\|\mathbf{d}\|^2} \right) \mathbf{d}$$

$$= \left(\frac{\langle 4, 2 \rangle \cdot \langle 3, 4 \rangle}{\left[\sqrt{3^2 + 4^2} \right]^2} \right) \langle 3, 4 \rangle$$

$$= \left(\frac{4(3) + 2(4)}{\left[\sqrt{9 + 16} \right]^2} \right) \langle 3, 4 \rangle$$

$$= \left(\frac{12 + 8}{25} \right) \langle 3, 4 \rangle$$

$$= \frac{20}{25} \langle 3, 4 \rangle$$

$$= \frac{4}{5} \langle 3, 4 \rangle$$

$$= \left\langle \frac{12}{5}, \frac{16}{5} \right\rangle$$

Thus, the vector **a** = <12/5, 16/5> represents the force applied along the displacement vector. Calculate the magnitude of **a** to determine the amount of force applied to the particle along **d**.

$$\|\mathbf{a}\| = \sqrt{\left(\frac{12}{5} \right)^2 + \left(\frac{16}{5} \right)^2}$$

$$= \sqrt{\frac{144}{25} + \frac{256}{25}}$$

$$= \sqrt{\frac{400}{25}}$$

$$= \sqrt{16}$$

$$= 4$$

The component of force along the displacement vector is 4.

Note: In Problems 16.50–16.52, you calculate the total work done to move a particle along displacement vector d = <3,4> by applying force vector F = <4,2>.

16.51 Apply the dot product to calculate the total work done.

The total work done to move an object with force vector **F** along displacement vector **d** is equal to the dot product of the vectors.

$$W = \mathbf{F} \cdot \mathbf{d}$$
$$= \langle 4, 2 \rangle \cdot \langle 3, 4 \rangle$$
$$= 4(3) + 2(4)$$
$$= 12 + 8$$
$$= 20$$

Even if F and d don't share the same direction.

Note: In Problems 16.50–16.52, you calculate the total work done to move a particle along displacement vector d = <3,4> by applying force vector F = <4,2>.

16.52 Verify your answer to Problem 16.51 by calculating the product $W = \left\| \text{proj}_\mathbf{d} \mathbf{F} \right\| \left\| \mathbf{d} \right\|$.

According to Problem 16.51, total work done to move a particle along **d** using force **F** is equal to the dot product: **F** · **d** = 20. Problem 16.49 presents an equally valid, if slightly more complicated, method to calculate work done. It states that the total work done is equal to the product of the magnitudes of the displacement vector **d** and the force applied in the same direction. In other words, $W = \left\| \text{proj}_\mathbf{d} \mathbf{F} \right\| \left\| \mathbf{d} \right\|$.

According to Problem 16.50, $\left\| \text{proj}_\mathbf{d} \mathbf{F} \right\| = 4$. Calculate $\left\| \mathbf{d} \right\|$.

$$\left\| \mathbf{d} \right\| = \sqrt{3^2 + 4^2}$$
$$= \sqrt{9 + 16}$$
$$= \sqrt{25}$$
$$= 5$$

proj$_\mathbf{d}$ F represents force in the same direction as d.

Calculate $W = \left\| \text{proj}_\mathbf{d} \mathbf{F} \right\| \left\| \mathbf{d} \right\|$.

$$W = \left\| \text{proj}_\mathbf{d} \mathbf{F} \right\| \left\| \mathbf{d} \right\|$$
$$= 4(5)$$
$$= 20$$

$W = 20$ is the same solution generated in Problem 16.51.

Chapter 17

PARAMETRIC EQUATIONS AND POLAR COORDINATES

Different ways to map the coordinate plane

Throughout this book, you have explored the coordinate plane. Split horizontally by the x-axis and vertically by the y-axis, it pairs a simple coordinate (x,y) with any point in two-dimensional space. It is powerful in its simplicity, but it is not the only way to map points in the coordinate plane.

In this chapter, you will explore parametrically defined curves, in which a third variable called the parameter works in tandem with the familiar variables x and y to define points in the plane. You will also plot points using polar coordinates, which eschew x and y and define the coordinate plane in an entirely new way.

Don't worry—it's not as though you'll never see the normal coordinate plane again. In fact, a lot of the problems in this chapter focus on translating parametric and polar graphs into old-fashioned "rectangular" form, which is a fancy way of describing the coordinate plane you're used to.

Spend some extra time studying polar form. Even though the notation feels uncomfortable at first, polar form allows you to create extremely complex graphs with ease. Also, polar form makes a surprise cameo appearance in Chapter 18.

Parametric Equations

Two equations that describe one curve

Note: Problems 17.1–17.4 refer to the curve defined by parametric equations x = t + 3 and y = 2t – 1.

17.1 Identify the point on the curve that corresponds with $t = -3$.

A parametrically defined graph uses individual equations to represent the coordinates of each point. In this problem, the x- and y-coordinates of each point are defined in terms of a third variable t called the parameter. Substitute $t = -3$ into both of the parametric equations to identify the coordinates of one point on the graph of the parametric curve.

$$x = t + 3 \qquad\qquad y = 2t - 1$$
$$= -3 + 3 \qquad\qquad = 2(-3) - 1$$
$$= 0 \qquad\qquad\qquad = -6 - 1$$
$$\qquad\qquad\qquad\qquad = -7$$

"Rectangular coordinates" are the coordinates you've been using for the entire book, where x represents horizontal distance from the y-axis and y represents vertical distance from the x-axis.

Thus, the parameter $t = -3$ corresponds with the rectangular coordinates $(x, y) = (0, -7)$ on the parametric curve.

Note: Problems 17.1–17.4 refer to the curve defined by parametric equations x = t + 3 and y = 2t – 1.

17.2 Identify points on the parametric curve by substituting the following values into the parameter: –2, –1, 0, 1, and 2.

Apply the technique demonstrated in Problem 17.1 to substitute the values –2, –1, 0, 1, and 2 into t. In the chart below, each value of t is substituted into both parametric equations, producing a coordinate pair (x,y) that represents a point on the graph of the parametric curve.

t	$x = t + 3$	$y = 2t - 1$	(x, y)
-2	$x = -2 + 3$ $= 1$	$y = 2(-2) - 1$ $= -4 - 1$ $= -5$	$(1, -5)$
-1	$x = -1 + 3$ $= 2$	$y = 2(-1) - 1$ $= -2 - 1$ $= -3$	$(2, -3)$
0	$x = 0 + 3$ $= 3$	$y = 2(0) - 1$ $= 0 - 1$ $= -1$	$(3, -1)$
1	$x = 1 + 3$ $= 4$	$y = 2(1) - 1$ $= 2 - 1$ $= 1$	$(4, 1)$
2	$x = 2 + 3$ $= 5$	$y = 2(2) - 1$ $= 4 - 1$ $= 3$	$(5, 3)$

Note: Problems 17.1–17.4 refer to the curve defined by parametric equations x = t + 3 and y = 2t – 1.

17.3 Graph the parametric curve.

In Problem 17.2, you substitute five values of t into the parametric equations to create five points on the parametric curve: (1,–5), (2,–3), (3,–1), (4,1), and (5,3). As the graph below illustrates, those points trace a linear path.

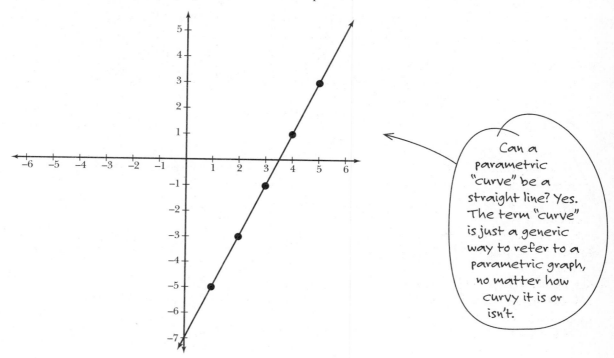

Can a parametric "curve" be a straight line? Yes. The term "curve" is just a generic way to refer to a parametric graph, no matter how curvy it is or isn't.

Note that the graph also passes through the point (0,–7) that you identified in Problem 17.1.

Note: Problems 17.1–17.4 refer to the curve defined by parametric equations x = t + 3 and y = 2t – 1.

17.4 Eliminate the parameter to express the parametric curve in rectangular form.

To rewrite a parametric curve in rectangular form, you need to eliminate the parameter, leaving behind only the variables x and y. In this problem, you should solve one of the parametric equations for t and substitute it into the other parametric equation.

If you subtract 3 from both sides of the parametric equation $x = t + 3$, the result is $x - 3 = t$. Replace t in the other parametric equation with the equivalent value $x - 3$.

$$y = 2t - 1$$
$$y = 2(x - 3) - 1$$
$$y = 2x - 6 - 1$$
$$y = 2x - 7$$

The parametric equations $x = t + 3$ and $y = 2t - 1$ describe the linear equation $y = 2x - 7$.

> *Note: Problems 17.5–17.7 refer to the curve defined by parametric equations $x = t^2 - 3$ and $y = 4 - t$.*

17.5 Identify points on the parametric curve by substituting the following values into the parameter: –2, –1, 0, 1, and 2.

This problem asks you to generate a table of values, substituting $t = -2, -1, 0, 1,$ and 2 into the parametric equations to produce points on the graph.

t	$x = t^2 - 3$	$y = 4 - t$	(x, y)
–2	$x = (-2)^2 - 3$ $= 4 - 3$ $= 1$	$y = 4 - (-2)$ $= 4 + 2$ $= 6$	$(1, 6)$
–1	$x = (-1)^2 - 3$ $= 1 - 3$ $= -2$	$y = 4 - (-1)$ $= 4 + 1$ $= 5$	$(-2, 5)$
0	$x = (0)^2 - 3$ $= 0 - 3$ $= -3$	$y = 4 - 0$ $= 4$	$(-3, 4)$
1	$x = (1)^2 - 3$ $= 1 - 3$ $= -2$	$y = 4 - 1$ $= 3$	$(-2, 3)$
2	$x = (2)^2 - 3$ $= 4 - 3$ $= 1$	$y = 4 - 2$ $= 2$	$(1, 2)$

The curve passes through points $(1, 6)$, $(-2, 5)$, $(-3, 4)$, $(-2, 3)$, and $(1, 2)$.

Note: Problems 17.5–17.7 refer to the curve defined by parametric equations $x = t^2 - 3$ and $y = 4 - t$.

17.6 Graph the curve.

According to Problem 17.5, the curve passes through points $(1, 6)$, $(-2, 5)$, $(-3, 4)$, $(-2, 3)$, and $(1, 2)$.

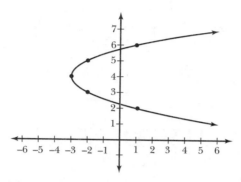

Note: Problems 17.5–17.7 refer to the curve defined by parametric equations $x = t^2 - 3$ and $y = 4 - t$.

17.7 Rewrite the parametrically defined curve in rectangular form.

Solve the parametric equation $y = 4 - t$ for the parameter.

$$y = 4 - t$$
$$y + t = 4$$
$$t = 4 - y$$

Substitute $t = 4 - y$ into the parametric equation $x = t^2 - 3$ and solve for x to create a function in terms of y.

$$x = t^2 - 3$$
$$x = (4 - y)^2 - 3$$
$$x = 16 - 8y + y^2 - 3$$
$$x = y^2 - 8y + 13$$

> You could also solve the equation for y to create functions in terms of x, but that's more complicated. You have to use a technique called completing the square. See Problem 17.10 for more information.

Note: Problems 17.8–17.10 refer to the curve defined by parametric equations $x = t^2 + 2t$ and $y = t^2 - t$.

17.8 Identify points on the parametric curve by substituting the following values into the parameter: $-2, -1, 0, 1$, and 2.

Generate a table of values by substituting $t = -2, -1, 0, 1$, and 2 into the parametric equations to identify coordinates of points on the curve.

> If a problem doesn't give you values of t, these numbers (-2, -1, 0, 1, and 2) are good values to start with. You can always use more values of t if the shape of the graph isn't obvious.

t	$x = t^2 + 2t$	$y = t^2 - t$	(x, y)
-2	$x = (-2)^2 + 2(-2)$ $= 4 - 4$ $= 0$	$y = (-2)^2 - (-2)$ $= 4 + 2$ $= 6$	$(0, 6)$
-1	$x = (-1)^2 + 2(-1)$ $= 1 - 2$ $= -1$	$y = (-1)^2 - (-1)$ $= 1 + 1$ $= 2$	$(-1, 2)$
0	$x = 0^2 + 2(0)$ $= 0$	$y = (0)^2 - (0)$ $= 0$	$(0, 0)$
1	$x = 1^2 + 2(1)$ $= 1 + 2$ $= 3$	$y = (1)^2 - (1)$ $= 1 - 1$ $= 0$	$(3, 0)$
2	$x = 2^2 + 2(2)$ $= 4 + 4$ $= 8$	$y = (2)^2 - 2$ $= 4 - 2$ $= 2$	$(8, 2)$

Note: Problems 17.8–17.10 refer to the curve defined by parametric equations $x = t^2 + 2t$ and $y = t^2 - t$.

17.9 Plot the curve.

In other words, the parabola doesn't open straight up, down, left, or right. Instead, it opens in the upper-right direction.

According to Problem 17.8, the curve passes through points $(0, 6)$, $(-1, 2)$, $(0, 0)$, $(3, 0)$, and $(8, 2)$. Although the parametric equations that define the curve are simple quadratics, the resulting parabolic curve has an axis of symmetry that is neither horizontal nor vertical. This produces a graph that appears to be rotated in the coordinate plane.

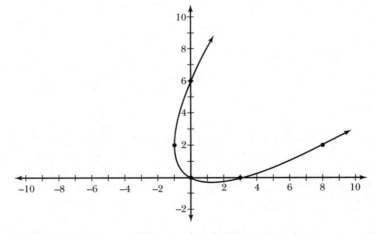

As this example illustrates, parametric equations can produce very complex curves—curves that are far more difficult to describe in rectangular form, as you discover in Problem 17.10.

Note: Problems 17.8–17.10 refer to the curve defined by parametric equations $x = t^2 + 2t$ and $y = t^2 - t$.

17.10 Rewrite the parametrically defined curve in rectangular form.

In order to eliminate the parameter t, you need to solve one of the parametric equations for either x or y. This requires you to complete the square. Begin by adding 1 to both sides of the equation $x = t^2 + 2t$.

$$x + 1 = t^2 + 2t + 1$$

Rewrite the right side of the equation as a perfect square.

$$x + 1 = (t + 1)^2$$

Solve for t.

$$\pm\sqrt{x+1} = \sqrt{(t+1)^2}$$
$$\pm\sqrt{x+1} = t + 1$$
$$-1 \pm \sqrt{x+1} = t$$

> It isn't hard to complete the square, but it's not something you do very often. If you need to review the technique, see Problems 14.12–14.19 in The Humongous Book of Algebra Problems.

The parameter t is equivalent to two different expressions in terms of x: $-1 + \sqrt{x+1}$ and $-1 - \sqrt{x+1}$. Substitute each into the parametric equation $y = t^2 - t$.

$$y = t^2 - t \qquad\qquad y = t^2 - t$$
$$y = \left(-1 + \sqrt{x+1}\right)^2 - \left(-1 + \sqrt{x+1}\right) \qquad y = \left(-1 - \sqrt{x+1}\right)^2 - \left(-1 - \sqrt{x+1}\right)$$
$$y = 1 - 2\sqrt{x+1} + (x+1) + 1 - \sqrt{x+1} \qquad y = 1 + 2\sqrt{x+1} + (x+1) + 1 + \sqrt{x+1}$$
$$y = x - 3\sqrt{x+1} + 3 \qquad\qquad y = x + 3\sqrt{x+1} + 3$$

The rectangular form of the parametric curve is $y = x \pm 3\sqrt{x+1} + 3$.

Note: Problems 17.11–17.12 refer to the curve defined by parametric equations $x = \cos\theta$ and $y = \sin\theta$.

17.11 Graph the parametric curve.

The parametric equations in this problem contain parameter θ instead of t. Furthermore, the parameter is substituted into a pair of trigonometric equations. Therefore, you should substitute angles from the unit circle into the parameter (such as $\theta = \pi/4$ and $\theta = \pi/2$) rather than integers (such as $t = -2$, -1, 0, 1, and 2). In the following table of values, unit circle angles in (and bounding) the first quadrant are substituted into $x = \cos\theta$ and $y = \sin\theta$.

θ	$x = \cos\theta$	$y = \sin\theta$	(x, y)
0	$x = \cos 0$ $= 1$	$y = \sin 0$ $= 0$	$(1, 0)$
$\pi/6$	$x = \cos\dfrac{\pi}{6}$ $= \dfrac{\sqrt{3}}{2}$	$y = \sin\dfrac{\pi}{6}$ $= \dfrac{1}{2}$	$\left(\dfrac{\sqrt{3}}{2}, \dfrac{1}{2}\right)$
$\pi/4$	$x = \cos\dfrac{\pi}{4}$ $= \dfrac{\sqrt{2}}{2}$	$y = \sin\dfrac{\pi}{4}$ $= \dfrac{\sqrt{2}}{2}$	$\left(\dfrac{\sqrt{2}}{2}, \dfrac{\sqrt{2}}{2}\right)$
$\pi/3$	$x = \cos\dfrac{\pi}{3}$ $= \dfrac{1}{2}$	$y = \sin\dfrac{\pi}{3}$ $= \dfrac{\sqrt{3}}{2}$	$\left(\dfrac{1}{2}, \dfrac{\sqrt{3}}{2}\right)$
$\pi/2$	$x = \cos\dfrac{\pi}{2}$ $= 0$	$y = \sin\dfrac{\pi}{2}$ $= 1$	$(0, 1)$

Each angle θ produces the corresponding coordinate $(\cos\theta, \sin\theta)$ on the unit circle. For example, parameter $\theta = \pi/4$ produces point $\left(\sqrt{2}/2, \sqrt{2}/2\right)$. Thus, the graph of the parametric curve is the unit circle.

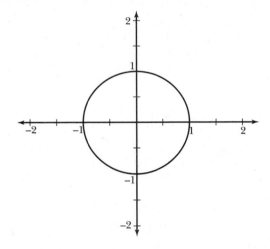

Note: Problems 17.11–17.12 refer to the curve defined by parametric equations x = cos θ and y = sin θ.

17.12 Rewrite the parametrically defined curve in rectangular form.

According to a Pythagorean identity, $\cos^2\theta + \sin^2\theta = 1$. The parametric equations state that $x = \cos\theta$ and $y = \sin\theta$, so replace $\cos\theta$ and $\sin\theta$ in the Pythagorean identity with x and y, respectively.

$$\cos^2\theta + \sin^2\theta = 1$$
$$(\cos\theta)^2 + (\sin\theta)^2 = 1$$
$$(x)^2 + (y)^2 = 1$$
$$x^2 + y^2 = 1$$

The parametric curve has rectangular form $x^2 + y^2 = 1$.

Note: Problems 17.13–17.14 refer to the curve defined by parametric equations x = 2 cos θ and y = 5 sin θ.

17.13 Based on the graph you generated in Problem 17.11, sketch the curve without using a table of values.

Each of the points on this parametric curve, when compared to the points on the unit circle, is 2 times as far from the origin horizontally and 5 times as far from the origin vertically. Whereas the unit circle had a leftmost point of $(-1, 0)$ and a rightmost point of $(1, 0)$, this parametric curve will stretch twice as far, from point $(-2, 0)$ to point $(2, 0)$. Similarly, whereas the unit circle has a maximum height of 1 and a minimum height of –1, this parametric curve will have a maximum height of 5 and a minimum height of –5.

> Because x = 2 cos θ is twice as large as x = cos θ, and y = 5 sin θ is 5 times as large as y = sin θ.

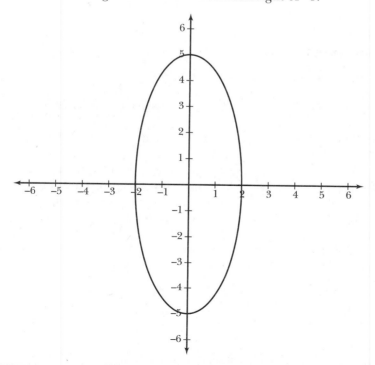

Note: Problems 17.13–17.14 refer to the curve defined by parametric equations $x = 2\cos\theta$ and $y = 5\sin\theta$.

17.14 Rewrite the parametrically defined curve in rectangular form.

Solve $x = 2\cos\theta$ for $\cos\theta$ and solve $y = 5\sin\theta$ for $\sin\theta$.

$$x = 2\cos\theta \qquad y = 5\sin\theta$$

$$\frac{x}{2} = \cos\theta \qquad \frac{y}{5} = \sin\theta$$

Substitute $x/2$ for $\cos\theta$ and $y/5$ for $\sin\theta$ in the Pythagorean identity $\cos^2\theta + \sin^2\theta = 1$.

$$\cos^2\theta + \sin^2\theta = 1$$

$$(\cos\theta)^2 + (\sin\theta)^2 = 1$$

$$\left(\frac{x}{2}\right)^2 + \left(\frac{y}{5}\right)^2 = 1$$

$$\frac{x^2}{4} + \frac{y^2}{25} = 1$$

If you wish to eliminate the fractions from the equation $x^2/4 + y^2/25 = 1$, multiply each term by the least common denominator: 100.

$$100\left(\frac{x^2}{4}\right) + 100\left(\frac{y^2}{25}\right) = 100(1)$$

$$\frac{100}{4}\left(x^2\right) + \frac{100}{25}\left(y^2\right) = 100$$

$$25x^2 + 4y^2 = 100$$

Polar Coordinates
Plot points using distances and angles

Note: In Problems 17.15–17.17, point A has polar coordinates $(r, \theta) = \left(3, \frac{\pi}{6}\right)$.

17.15 Explain how to interpret the polar coordinate 3 of point A.

The first coordinate of a point in polar form is called the radial coordinate or radius, usually written r in the coordinate pair (r, θ). This value represents the distance between the point and a fixed location called the pole. The pole is typically defined as the origin of the coordinate system, so point A is a distance of 3 units from the origin. This is not sufficient information to plot point A, because there are infinitely many points on the plane that are $r = 3$ units away from the origin.

Note: In Problems 17.15–17.17, point A has polar coordinates $(r,\theta) = \left(3,\dfrac{\pi}{6}\right)$.

17.16 Explain how to interpret the polar coordinate $\pi/6$ of point A.

The second coordinate of a point in polar form is called the angular coordinate or polar angle, θ in the coordinate pair (r,θ). This value identifies an angle in the coordinate plane, measured from a ray called the polar axis. Typically, the polar axis is the positive x-axis.

Recall that positive angles in the coordinate plane are measured in the counterclockwise direction and negative angles are measured in the clockwise direction. Thus, point A lies along the ray that forms a $\pi/6 = 30°$ angle with the positive x-axis.

Note: In Problems 17.15–17.17, point A has polar coordinates $(r,\theta) = \left(3,\dfrac{\pi}{6}\right)$.

17.17 Plot A in the coordinate plane.

Consider the following diagram. The point represents the pole, the ray represents the polar axis, and the concentric circles centered at the pole represent fixed distances from the pole. In other words, any point with a radial coordinate of 3 lies on the circle centered at the origin with radius 3—the darkened circle in the diagram.

> Concentric circles have the same center, in this case the pole.

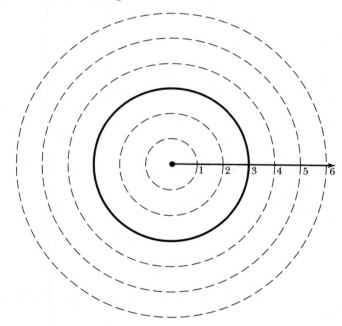

According to Problem 17.16, the angular coordinate identifies the ray upon which point A lies. The angular coordinate is $\theta = \pi/6$, so point A lies at the intersection of the darkened circle with radius 3 and the ray that forms a $\pi/6$, or 30°, angle with the polar axis.

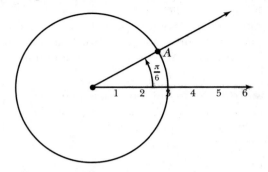

Note: In Problems 17.18–17.21, you plot four similar points to explore how negative signs affect polar coordinates.

17.18 Plot the following point, expressed in polar coordinates: $C_1 = \left(4, \dfrac{\pi}{2}\right)$.

The positive y-axis forms a $\pi/2$ (or 90°) angle with the polar axis, so C_1 is located on that ray. Furthermore, C_1 is located 4 units away from the pole, as illustrated below.

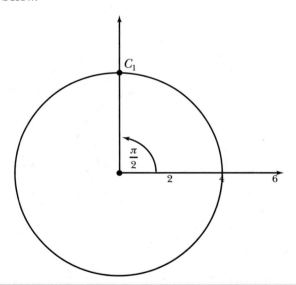

Note: In Problems 17.18–17.21, you plot four similar points to explore how negative signs affect polar coordinates.

17.19 Plot the following point, expressed in polar coordinates: $C_2 = \left(4, -\dfrac{\pi}{2}\right)$.

Negative angular coordinates are measured in the clockwise direction, so C_2 lies on the ray directly below the pole, whereas point C_1 in Problem 17.18 lies on the ray directly above the pole.

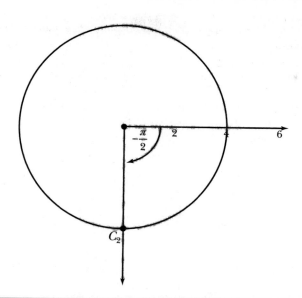

Note: In Problems 17.18–17.21, you plot four similar points to explore how negative signs affect polar coordinates.

17.20 Plot the following point, expressed in polar coordinates: $C_3 = \left(-4, \dfrac{\pi}{2}\right)$.

When the radial coordinate of a point is negative, the point does not lie on the ray described by the angular coordinate. Instead, it lies on the opposite ray. In this problem, C_3 does not lie along the ray measuring $\theta = \pi/2$; it lies along the opposite ray, the dotted ray in the figure below.

> The second coordinate identifies a specific ray. If the first coordinate is positive, then the point lies on that ray. If the first coordinate is negative, the point lies on the opposite ray.

Note: In Problems 17.18–17.21, you plot four similar points to explore how negative signs affect polar coordinates.

17.21 Plot the following point, expressed in polar coordinates: $C_4 = \left(-4, -\dfrac{\pi}{2}\right)$.

Like Problem 17.19, this point has an angular coordinate of $-\pi/2$. However, the radial coordinate is also negative, so C_4 lies on the opposite ray, the dotted ray in the diagram below.

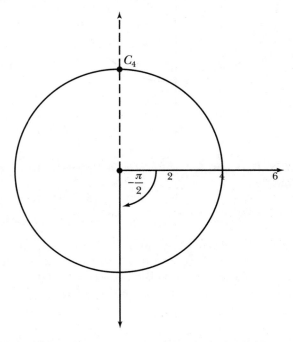

17.22 Plot the following point, expressed in polar coordinates: $P = \left(2, \dfrac{3\pi}{4}\right)$.

Plot the ray in the counterclockwise direction that forms a $3\pi/4$ angle with the polar axis. Because the radial coordinate is positive, P is located 2 units away from the pole along that ray.

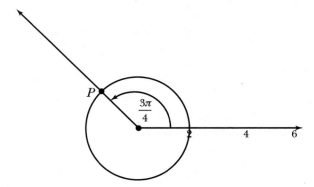

Note: In Problems 17.23–17.24, point Q has polar coordinates (–5, 210°).

17.23 Plot point Q.

In the diagram below, the solid ray is 210° from the polar axis in the counterclockwise direction. Because the radial coordinate is negative, Q lies on the opposite ray, 5 units from the pole.

> The angular coordinate could be measured in degrees or radians.

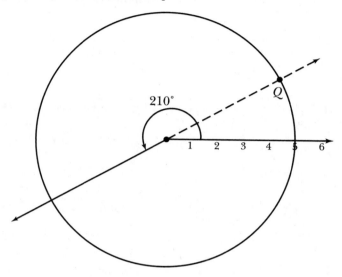

Note: In Problems 17.23–17.24, point Q has polar coordinates (–5, 210°).

17.24 Construct polar coordinates (r, θ) for point Q such that the radial coordinate is positive and the angular coordinate is both positive and acute.

> There is only one way to write a point using rectangular coordinates, but there are infinitely many ways to identify the same point using polar coordinates.

The reference angle for a 210° angle is a 30° angle, as illustrated below. Point Q lies 5 units away from the pole on the dotted ray in the diagram, which is 30° from the polar axis. Therefore, $Q = (5, 30°)$.

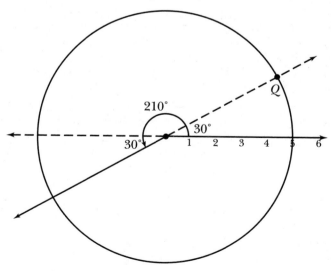

17.25 Given point $P = (1, 0)$ expressed in rectangular coordinates, identify two points in polar form that coincide with P.

Point P lies on the polar axis, exactly $r = 1$ unit away from the pole. The angular coordinates $\theta = 0$ and $\theta = 2\pi$ (or $\theta = 0°$ and $\theta = 360°$) both describe the polar axis, so the points $(1, 0)$ and $(1, 2\pi)$—or $(1, 0°)$, and $(1, 360°)$—both represent P in polar form. All of these points coincide; they describe the same point in the coordinate plane.

These are not the only solutions to this problem. If the radial coordinate is $r = 1$, any angle coterminal to 2π is also a valid angular coordinate, so $(1, 4\pi)$, $(1, 6\pi)$, and $(1, -2\pi)$ are also polar representations of P.

Note: In Problems 17.26–17.29, you identify four different points in polar form that coincide with point A, illustrated below.

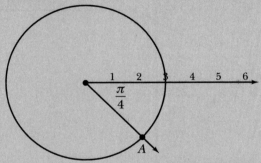

17.26 Identify coordinates of point $A = (r, \theta)$, such that $r > 0$ and $\theta < 0$.

This problem asks you to identify a positive radial coordinate and a negative angular coordinate for A. Note that A is 3 units from the pole, so $r = 3$. Furthermore, A lies on the ray that is $45°$ from the polar axis in the clockwise direction. Recall that angles measured in the clockwise direction have negative measures, so $\theta = -\pi/4$. Thus, A has polar coordinates $(3, -\pi/4)$.

Note: In Problems 17.26–17.29, you identify four different points in polar form that coincide with point A, illustrated in Problem 17.26.

17.27 Identify coordinates of point $A = (r, \theta)$, such that $r > 0$ and $0 < \theta < 2\pi$.

According to Problem 17.26, A could have angular coordinate $\theta = -\pi/4$ when $r = 3$. Add 2π to $-\pi/4$ to identify a positive angle coterminal to $-\pi/4$.

$$-\frac{\pi}{4} + 2\pi = -\frac{\pi}{4} + \frac{8\pi}{4} = \frac{7\pi}{4}$$

The diagram does not tell you that the angle measurement should be negative—it only tells you that the angle between the polar axis and the ray containing A measures $\pi/4$. It's up to you to know that clockwise = negative angle.

Thus, point A has polar coordinates $(3, 7\pi/4)$, as illustrated in the following diagram.

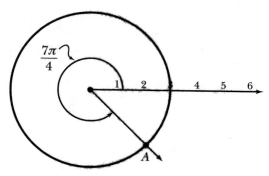

Note: In Problems 17.26–17.29, you identify four different points in polar form that coincide with point A, illustrated in Problem 17.26.

17.28 Identify coordinates of point $A = (r, \theta)$, such that $r > 0$ and $2\pi < \theta < 4\pi$.

According to Problem 17.27, $A = (3, 7\pi/4)$. To identify an angular coordinate for A that is between 2π and 4π, add 2π to the angular coordinate to calculate a positive angle coterminal to $7\pi/4$.

$$\frac{7\pi}{4} + 2\pi = \frac{7\pi}{4} + \frac{8\pi}{4} = \frac{15\pi}{4}$$

The polar coordinate pair $(3, 15\pi/4)$ coincides with point A.

Note: In Problems 17.26–17.29, you identify four different points in polar form that coincide with point A, illustrated in Problem 17.26.

17.29 Identify coordinates of point $A = (r, \theta)$, such that $r < 0$ and $0 < \theta < 2\pi$.

Extend the ray containing point A, and note that it forms a $3\pi/4$ angle with the polar axis. Recall that a negative radial coordinate indicates that you should travel along the opposite ray. Thus, the point $(-3, 3\pi/4)$ is located on the ray opposite $\theta = 3\pi/4$, on the solid ray in the following diagram.

> A ray and its opposite ray form a straight line, which measures 180°, or π radians. Therefore, you can subtract π/4 from π to get 3π/4.

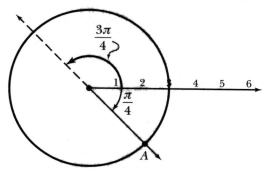

Converting Between Polar and Rectangular Form

Given points or equations

17.30 In the diagram below, point P has rectangular coordinates (x, y) and polar coordinates (r, θ). Apply trigonometric ratios to express x and y in terms of r and θ.

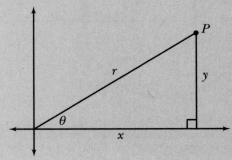

Note that the diagram contains a right triangle, so you can apply trigonometric ratios to describe the relationship between r, θ, x, and y.

$$\cos\theta = \frac{\text{length of leg adjacent to } \theta}{\text{length of hypotenuse}} \qquad \sin\theta = \frac{\text{length of leg opposite } \theta}{\text{length of hypotenuse}}$$

$$\cos\theta = \frac{x}{r} \qquad\qquad\qquad \sin\theta = \frac{y}{r}$$

Cross-multiply to solve the equations for x and y, respectively.

$$\frac{\cos\theta}{1} = \frac{x}{r} \qquad\qquad \frac{\sin\theta}{1} = \frac{y}{r}$$
$$x = r\cos\theta \qquad\qquad y = r\sin\theta$$

You conclude that a point with polar coordinates (r, θ) has rectangular coordinates $(x, y) = (r\cos\theta,\ r\sin\theta)$.

Note: In Problems 17.31–17.32, point B has polar coordinates $(r, \theta) = (6, \pi)$.

17.31 Calculate the x-coordinate of B in rectangular form.

According to Problem 17.30, $x = r\cos\theta$. Substitute $r = 6$ and $\theta = \pi$ into this formula to calculate r.

$$\begin{aligned} x &= r\cos\theta \\ &= 6\cos\pi \\ &= 6(-1) \\ &= -6 \end{aligned}$$

Note: In Problems 17.31–17.32, point B has polar coordinates (r,θ) = (6,π).

17.32 Calculate the *y*-coordinate of *B* in rectangular form and rewrite *B* in rectangular form.

Apply the formula $y = r \sin \theta$, generated in Problem 17.30.

$$\begin{aligned} y &= r \sin \theta \\ &= 6 \sin \pi \\ &= 6(0) \\ &= 0 \end{aligned}$$

According to Problem 17.31, the *x*-coordinate of *B* is −6. You conclude that the rectangular form of *B* is $(x, y) = (-6, 0)$. ←

> You can double-check your answer by plotting both the rectangular and polar coordinates to make sure the points overlap.

17.33 Convert the point from polar to rectangular form: $\left(2, \dfrac{5\pi}{6}\right)$.

Substitute $r = 2$ and $\theta = 5\pi/6$ into the formulas $x = r \cos \theta$ and $y = r \sin \theta$ to calculate the rectangular coordinates of the point.

$$\begin{array}{ll} x = r \cos \theta & y = r \sin \theta \\[6pt] = 2 \cos \dfrac{5\pi}{6} & = 2 \sin \dfrac{5\pi}{6} \\[8pt] = \cancel{2}\left(-\dfrac{\sqrt{3}}{\cancel{2}}\right) & = \cancel{2}\left(\dfrac{1}{\cancel{2}}\right) \\[8pt] = -\sqrt{3} & = 1 \end{array}$$

You conclude that the point with polar coordinates $(2, 5\pi/6)$ coincides with the point $\left(-\sqrt{3}, 1\right)$ in rectangular form.

17.34 Convert the point from polar to rectangular form: $\left(-1, -\dfrac{\pi}{3}\right)$.

Add 2π to $-\pi/3$ to calculate a positive angle coterminal to $-\pi/3$ that lies on the standard unit circle.

$$-\frac{\pi}{3} + 2\pi = -\frac{\pi}{3} + \frac{6\pi}{3} = \frac{5\pi}{3}$$

Substitute $r = -1$ and $\theta = 5\pi/3$ into the formulas $x = r \cos \theta$ and $y = r \sin \theta$ to convert the polar coordinates into rectangular coordinates.

$$\begin{array}{ll} x = r \cos \theta & y = r \sin \theta \\[6pt] = -1\left(\cos \dfrac{5\pi}{3}\right) & = -1\left(\sin \dfrac{5\pi}{3}\right) \\[8pt] = -1\left(\dfrac{1}{2}\right) & = -1\left(-\dfrac{\sqrt{3}}{2}\right) \\[8pt] = -\dfrac{1}{2} & = \dfrac{\sqrt{3}}{2} \end{array}$$

You conclude that the point with polar coordinates $(-1, -\pi/3)$ has rectangular coordinates $\left(-1/2, \sqrt{3}/2\right)$.

17.35 Rewrite the polar equation $r = 2\cos\theta$ in rectangular form.

Whereas a polar equation is written in terms of r and θ, a rectangular equation is written in terms of x and y. According to Problem 17.30, the following statements are true.

$$\frac{x}{r} = \cos\theta \qquad x = r\cos\theta \qquad \frac{y}{r} = \sin\theta \qquad y = r\sin\theta$$

Thus, you can replace the expression $\cos\theta$ in this problem with the equivalent fraction x/r.

$$r = 2\cos\theta$$
$$r = 2\left(\frac{x}{r}\right)$$
$$r = \frac{2x}{r}$$

Cross-multiply to eliminate the fraction.

$$r^2 = 2x$$

Consider the diagram in Problem 17.30. It contains a right triangle with legs of length x and y and a hypotenuse of length r. Apply the Pythagorean theorem: $x^2 + y^2 = r^2$. This allows you to rewrite r^2 in the equation $r^2 = 2x$ in terms of x and y.

$$x^2 + y^2 = 2x$$

You conclude that the rectangular equation $x^2 + y^2 = 2x$ is equivalent to the polar equation $r = 2\cos\theta$.

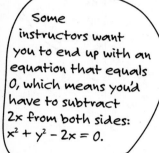

Some instructors want you to end up with an equation that equals 0, which means you'd have to subtract 2x from both sides: $x^2 + y^2 - 2x = 0$.

17.36 Rewrite the polar equation $r^2 = 3\sec\theta\csc\theta$ in rectangular form.

Apply the reciprocal identities to express $\sec\theta$ and $\csc\theta$ in terms of cosine and sine.

$$r^2 = 3\sec\theta\csc\theta$$
$$r^2 = 3\left(\frac{1}{\cos\theta}\right)\left(\frac{1}{\sin\theta}\right)$$
$$r^2 = \frac{3}{\cos\theta\sin\theta}$$

Cross-multiply to eliminate the fraction.

$$r^2(\cos\theta\sin\theta) = 3$$

According to Problem 17.35, $\cos\theta = x/r$ and $\sin\theta = y/r$.

$$r^2\left(\frac{x}{r} \cdot \frac{y}{r}\right) = 3$$

$$r^2\left(\frac{xy}{r^2}\right) = 3$$

$$xy = 3$$

You conclude that $xy = 3$ is the rectangular form of the polar equation $r^2 = 3\sec\theta \csc\theta$.

17.37 In the diagram below, point P has rectangular coordinates (x, y) and polar coordinates (r, θ). Based on the diagram, express r^2 and $\tan\theta$ in terms of x and y.

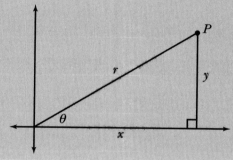

According to the Pythagorean theorem, $x^2 + y^2 = r^2$. The tangent ratio establishes the relationship between θ, x, and y.

$$\tan\theta = \frac{\text{length of leg opposite } \theta}{\text{length of leg adjacent to } \theta}$$

$$\tan\theta = \frac{y}{x}$$

Note: In Problems 17.38–17.39, point A has rectangular coordinates $(x, y) = (3, 0)$.

17.38 Calculate the two possible radial coordinates of A in polar form.

A point in polar form has coordinates (r, θ). You are asked to calculate r, the radial coordinate. Substitute $x = 3$ and $y = 0$ into the formula $x^2 + y^2 = r^2$, presented in Problem 17.37.

$$x^2 + y^2 = r^2$$

$$3^2 + 0^2 = r^2$$

$$9 + 0 = r^2$$

$$9 = r^2$$

$$\sqrt{9} = \sqrt{r^2}$$

$$\pm 3 = r$$

Both $r = 3$ and $r = -3$ are possible radial coordinates of A.

Note: In Problems 17.38–17.39, point A has rectangular coordinates (x, y) = (3, 0).

17.39 Calculate the angular coordinates of *A* that correspond with the radial coordinates you calculated in Problem 17.38.

Substitute $x = 3$ and $y = 0$ into the formula $\tan \theta = y/x$, presented in Problem 17.37.

$$\tan \theta = \frac{0}{3}$$
$$\tan \theta = 0$$

Many angles have a tangent of 0. One such angle is $\theta = 0$ and another is $\theta = \pi$. The points $(3, 0)$ and $(-3, \pi)$ are polar representations of the rectangular coordinate pair $(3, 0)$.

17.40 Convert the point from rectangular to polar form: $(1, -1)$.

Apply the formula $x^2 + y^2 = r^2$ to calculate the radial coordinate of the point.

$$1^2 + (-1)^2 = r^2$$
$$1 + 1 = r^2$$
$$2 = r^2$$
$$\pm \sqrt{2} = r$$

Select one value of *r* to represent the radial coordinate. You now must calculate a corresponding angular coordinate θ using the formula $\tan \theta = y/x$. Note that the point $(1, -1)$ lies in the fourth quadrant of the coordinate plane, so you must select a value of θ that lies in the fourth quadrant.

$$\tan \theta = \frac{y}{x}$$
$$\tan \theta = \frac{-1}{1}$$
$$\tan \theta = -1$$

Both $\theta = -\pi/4$ and $\theta = 7\pi/4$ are angular coordinates that identify the correct ray in the fourth quadrant. Thus, $\left(\sqrt{2}, -\pi/4\right)$ and $\left(\sqrt{2}, 7\pi/4\right)$ are both valid polar representations of the point $(x, y) = (1, -1)$.

The point (3,0) in rectangular form lies on the x-axis, three units right of the origin. In polar form, it lies $r = 3$ units from the pole on the ray $\theta = 0$. If you select the ray $\theta = \pi$ instead, you have to travel in the negative direction, along its opposite ray, to reach the same point. In that case, $r = -3$.

If you selected $r = -\sqrt{2}$, you'd need to pick a value of θ that lies in the second quadrant, which is opposite the fourth quadrant. That's because a negative radial coordinate indicates movement in the opposite direction, along the opposite ray.

17.41 Convert the point from rectangular to polar form: $(-2,-5)$.

Apply the formula $\tan\theta = y/x$ to calculate the angular coordinate.

$$\tan\theta = \frac{y}{x}$$

$$\tan\theta = \frac{-5}{-2}$$

$$\tan\theta = \frac{5}{2}$$

$$\theta = \arctan\frac{5}{2}$$

Apply the formula $x^2 + y^2 = r^2$ to calculate possible radial coordinates.

$$x^2 + y^2 = r^2$$

$$(-2)^2 + (-5)^2 = r^2$$

$$4 + 25 = r^2$$

$$29 = r^2$$

$$\pm\sqrt{29} = r$$

The arctangent function only reports values on the restricted range $-\pi/2 \le \theta \le \pi/2$. Because $5/2$ is positive, $\arctan 5/2$ is positive as well. However, the point $(-2,-5)$ lies in the third quadrant, which is opposite the quadrant containing $\theta = \arctan(5/2)$.

Thus, you must select the negative value of r to ensure that you are traveling along the ray opposite $\theta = \arctan(5/2)$. You conclude that the polar form of point $(-2,-5)$ is $\left(-\sqrt{29}, \arctan(5/2)\right)$. ←

> Or you could add π to θ to get a coterminal angle that lies in the third quadrant. If you do that, the final answer is $\left(\sqrt{29}, \arctan\left(\frac{5}{2} + \pi\right)\right)$.

17.42 Rewrite the rectangular equation $y = 2x + 1$ in polar form.

A rectangular equation is written in terms of x and y, whereas a polar equation is written in terms of r and θ. Recall that $x = r\cos\theta$ and $y = r\sin\theta$. Substitute these values into the rectangular equation.

$$y = 2x + 1$$

$$r\sin\theta = 2(r\cos\theta) + 1$$

$$r\sin\theta = 2r\cos\theta + 1$$

Move the terms containing r to the left side of the equation and factor the shared r out of both expressions.

$$r\sin\theta - 2r\cos\theta = 1$$

$$r(\sin\theta - 2\cos\theta) = 1$$

Solve for r.

$$r = \frac{1}{\sin\theta - 2\cos\theta}$$

Polar Graphs

They can be hard to BEAR (get it?)

17.43 Plot the graph of the polar equation $r = 3$.

The equation $r = 3$ describes the set of points that are $r = 3$ units away from the pole. All points in the plane that are a fixed distance r from a point (in this case the pole), form a circle of radius r around that point (called the center). In other words, $r = 3$ is the polar equation for a circle with radius 3 that is centered at the pole.

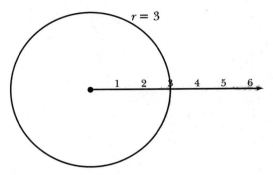

17.44 Based on your answer to Problem 17.43, construct the polar equation of a circle with radius 5 that is centered at the pole.

All points on the circle are a distance of $r = 5$ from the pole, so the polar equation is $r = 5$.

17.45 Plot the graph of polar equation $\theta = \dfrac{\pi}{6}$.

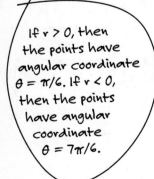

If $r > 0$, then the points have angular coordinate $\theta = \pi/6$. If $r < 0$, then the points have angular coordinate $\theta = 7\pi/6$.

All of the points on the graph of $\theta = \pi/6$ lie either on the ray that is $\pi/6$ radians from the polar axis in the counterclockwise direction or on the opposite ray. Thus, the graph is a line consisting of that ray and its opposite ray.

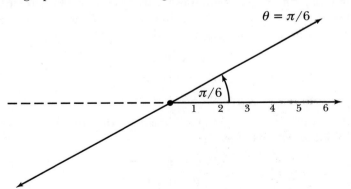

17.46 Based on your answer to Problem 17.45, express the *y*-axis of the rectangular coordinate system in polar form.

The *y*-axis of the rectangular coordinate system forms an angle measuring $\theta = \pi/2$ with the positive *x*-axis (or polar axis). Thus, one polar equation representing the *y*-axis is $\theta = \pi/2$. Other correct solutions include (but are not limited to): $\theta = 5\pi/2$, $\theta = -\pi/2$, and $\theta = -5\pi/2$.

17.47 Rewrite the polar equation $r = 4\csc\theta$ in rectangular form and graph it.

Rewrite the cosecant function in terms of sine.

$$r = 4\left(\frac{1}{\sin\theta}\right)$$

$$r = \frac{4}{\sin\theta}$$

Cross-multiply to eliminate the fraction.

$$r\sin\theta = 4$$

As Problem 17.30 explains, $y = r\sin\theta$.

$$y = 4$$

Thus, $r = 4\csc\theta$ is equivalent to the rectangular equation $y = 4$, the horizontal line 4 units above the *x*-axis.

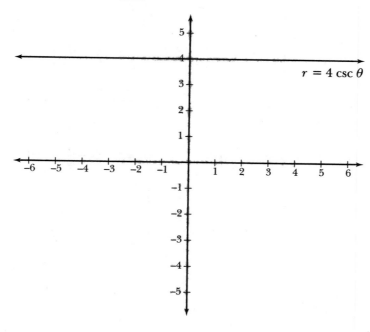

17.48 Graph the polar equation $r = 1 + 2\sin\theta$ by constructing a table of values that includes the following values of θ.

$$0, \frac{\pi}{6}, \frac{\pi}{4}, \frac{\pi}{2}, \frac{3\pi}{4}, \frac{5\pi}{6}, \pi, \frac{7\pi}{6}, \frac{5\pi}{4}, \frac{3\pi}{2}, \frac{7\pi}{4}, \frac{11\pi}{6}, \text{ and } 2\pi$$

> Plug 13 different values into θ to figure out the corresponding value of r. This gives you 11 points on the graph (in polar form).

Substitute each value of θ into the equation $r = 1 + 2\sin\theta$.

θ	$r = 1 + 2\sin\theta$	(r, θ)
0	$r = 1 + 2\sin 0$ $= 1 + 0$ $= 1$	$(1, 0)$
$\dfrac{\pi}{6}$	$r = 1 + 2\sin\dfrac{\pi}{6}$ $= 1 + 2\left(\dfrac{1}{2}\right)$ $= 2$	$\left(2, \dfrac{\pi}{6}\right)$
$\dfrac{\pi}{4}$	$r = 1 + 2\sin\dfrac{\pi}{4}$ $= 1 + 2\left(\dfrac{\sqrt{2}}{2}\right)$ ≈ 2.41	$\left(2.41, \dfrac{\pi}{4}\right)$
$\dfrac{\pi}{2}$	$r = 1 + 2\sin\dfrac{\pi}{2}$ $= 1 + 2(1)$ $= 3$	$\left(3, \dfrac{\pi}{2}\right)$
$\dfrac{3\pi}{4}$	$r = 1 + 2\sin\dfrac{3\pi}{4}$ $= 1 + 2\left(\dfrac{\sqrt{2}}{2}\right)$ ≈ 2.41	$\left(2.41, \dfrac{3\pi}{4}\right)$
$\dfrac{5\pi}{6}$	$r = 1 + 2\sin\dfrac{5\pi}{6}$ $= 1 + 2\left(\dfrac{1}{2}\right)$ $= 2$	$\left(2, \dfrac{5\pi}{6}\right)$
π	$r = 1 + 2\sin\pi$ $= 1 + 0$ $= 1$	$(1, \pi)$

θ	$r = 1 + 2\sin\theta$	(r, θ)
$\dfrac{7\pi}{6}$	$r = 1 + 2\sin\dfrac{7\pi}{6}$ $= 1 + 2\left(-\dfrac{1}{2}\right)$ $= 0$	$\left(0, \dfrac{7\pi}{6}\right)$
$\dfrac{5\pi}{4}$	$r = 1 + 2\sin\dfrac{5\pi}{4}$ $= 1 + 2\left(-\dfrac{\sqrt{2}}{2}\right)$ ≈ -0.41	$\left(-0.41, \dfrac{5\pi}{4}\right)$
$\dfrac{3\pi}{2}$	$r = 1 + 2\sin\dfrac{3\pi}{2}$ $= 1 + 2(-1)$ $= -1$	$\left(-1, \dfrac{3\pi}{2}\right)$
$\dfrac{7\pi}{4}$	$r = 1 + 2\sin\dfrac{7\pi}{4}$ $= 1 + 2\left(-\dfrac{\sqrt{2}}{2}\right)$ ≈ -0.41	$\left(-0.41, \dfrac{7\pi}{4}\right)$
$\dfrac{11\pi}{6}$	$r = 1 + 2\sin\dfrac{11\pi}{6}$ $= 1 + 2\left(-\dfrac{1}{2}\right)$ $= 0$	$\left(0, \dfrac{11\pi}{6}\right)$
2π	$r = 1 + 2\sin(2\pi)$ $= 1 + 2(0)$ $= 1$	$(1, 2\pi)$

Plot each of the points generated by the table of values to create the following graph, called a limaçon.

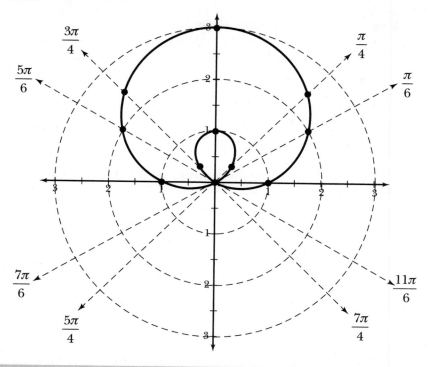

17.49 Plot the graph of the polar equation: $r = 2 \sin 3\theta$.

Apply the technique demonstrated in Problem 17.48, creating the table of values that contains the following points: $(0, 0)$, $(\pi/6, 2)$, $(\pi/3, 0)$, $(\pi/2, -2)$, $(2\pi/3, 0)$, $(5\pi/6, 2)$, and $(\pi, 0)$. This polar graph is called a three-petal rose curve.

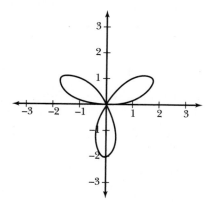

17.50 Plot the graph of the polar equation: $r^2 = 4 \cos 2\theta$.

Substitute unit circle values into θ to generate the points on the graph. For example, substituting $\theta = \pi$ into the equation produces the following values of r.

$$r^2 = 4 \cos 2\theta$$
$$r^2 = 4 \cos (2\pi)$$
$$r^2 = 4(1)$$
$$r = \pm\sqrt{4}$$
$$r = \pm 2$$

Thus, $(2,\pi)$ and $(-2,\pi)$ are points on the polar graph, which is called a lemniscate.

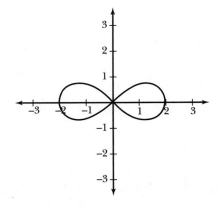

Chapter 18
TRIGONOMETRY OF COMPLEX NUMBERS

You can't spell "trig" without "i"

In the final chapter of this book, you briefly explore complex numbers. Unlike the real numbers in the preceding chapters, complex numbers are two-dimensional. Every complex number is comprised of two distinct parts: a real number and an imaginary number that is multiplied by $i = \sqrt{-1}$.

The first problems of this chapter introduce complex numbers, including how to graph them and how to combine them with other complex numbers via addition, subtraction, multiplication, and division. You then use a familiar technique to express complex numbers in a unique, trigonometric form, which allows you to perform arithmetic operations on complex numbers with ease.

This chapter is all about complex numbers. After a quick intro to remind you how complex numbers work, you'll quickly transition into a new way to write complex numbers. Basically, you use the same strategies you applied to polar coordinates, measuring how far they are from the origin and what angles are formed with the x-axis. Finally, you use a handful of formulas to multiply and divide complex numbers, raise them to powers, and calculate square and cube roots.

Rectangular Form of Complex Numbers
Add, subtract, multiply, divide, and graph

Note: Problems 18.1–18.3 refer to the complex number z = 5 + 3i.

18.1 Identify the real and imaginary parts of z.

> The imaginary part is the real number that is multiplied by i. In other words, 3 is the imaginary part of 5 + 3i, not 3i.

A complex number is the sum of two distinct parts, a real number part and an imaginary number part. In this problem, the complex number $z = 5 + 3i$ equals the sum of its real number part 5 and the imaginary part 3 that is multiplied by i. Note that $i = \sqrt{-1}$; it is considered imaginary because you cannot calculate the square root of a negative number.

Note: Problems 18.1–18.3 refer to the complex number z = 5 + 3i.

18.2 Identify the conjugate of z.

To derive the conjugate of a complex number, change the sign of its imaginary part: $5 - 3i$ is the conjugate of $5 + 3i$.

Note: Problems 18.1–18.3 refer to the complex number z = 5 + 3i.

18.3 Plot z on the complex plane.

The complex plane is similar to the coordinate plane, but the horizontal and vertical axes do not represent x and y. Instead, the horizontal axis represents the real part of the complex number, and the vertical axis represents the imaginary part. The real part of $z = 5 + 3i$ is 5, so z is located 5 units right of the origin. The imaginary part of z is 3, so z is located 3 units above the origin, as illustrated below.

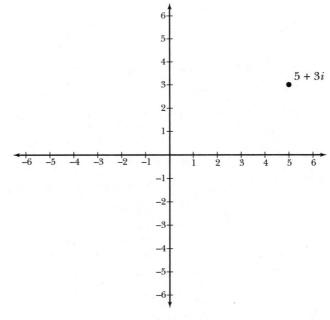

> To plot a complex point, treat the real part as the x-coordinate and the number multiplied by i as the y-coordinate.

Therefore, $z = 5 + 3i$ has rectangular coordinates $(5,3)$. More generally, any complex number $a + bi$ has rectangular coordinates (a, b).

18.4 Given the complex number $z_1 = -1 + 5i$, identify its conjugate z_2 and plot z_2 in the complex plane.

To calculate the conjugate of $z_1 = -1 + 5i$, change the sign of $5i$, the imaginary part: $z_2 = -1 - 5i$. Therefore, z_2 is located 1 unit left of, and 5 units below, the origin. The rectangular coordinates of z_2 are $(-1, -5)$.

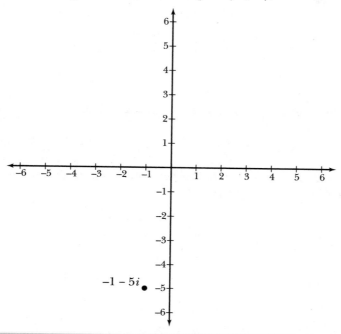

18.5 Simplify the expression: i^6.

Recall that $i = \sqrt{-1}$. Therefore, $i^2 = \left(\sqrt{-1}\right)^2 = -1$. Rewrite i^6 as a series of i^2 factors multiplied together.

$$i^6 = i^2 \cdot i^2 \cdot i^2$$
$$= (-1)(-1)(-1)$$
$$= (1)(-1)$$
$$= -1$$

You conclude that $i^6 = -1$.

Note: Problems 18.6–18.9 refer to complex numbers $c_1 = 2 - 3i$ and $c_2 = 5 + i$.

18.6 Calculate $c_1 + c_2$.

To add complex numbers, combine their real and imaginary parts independently.

$$c_1 + c_2 = (2 - 3i) + (5 + i)$$
$$= (2 + 5) + (-3i + i)$$
$$= 7 - 2i$$

Note: Problems 18.6–18.9 refer to complex numbers $c_1 = 2 - 3i$ and $c_2 = 5 + i$.

18.7 Calculate $c_1 - c_2$.

In order to subtract c_2 from c_1, subtract the real and imaginary parts of c_2 from the corresponding parts of c_1.

$$
\begin{aligned}
c_1 - c_2 &= (2 - 3i) - (5 + i) \\
&= 2 - 3i - 5 - i \\
&= (2 - 5) + (-3i - i) \\
&= -3 - 4i
\end{aligned}
$$

Note: Problems 18.6–18.9 refer to complex numbers $c_1 = 2 - 3i$ and $c_2 = 5 + i$.

18.8 Calculate $c_1 c_2$.

Apply the FOIL method to multiply complex numbers, the same technique you use to multiply two binomials: $(a + b)(c + d) = ac + ad + bc + bd$.

$$
\begin{aligned}
c_1 c_2 &= (2 - 3i)(5 + i) \\
&= 10 + 2i - 15i - 3i^2 \\
&= 10 - 13i - 3i^2
\end{aligned}
$$

Recall that $i^2 = -1$.

$$
\begin{aligned}
&= 10 - 13i - 3(-1) \\
&= 10 - 13i + 3 \\
&= 13 - 13i
\end{aligned}
$$

Note: Problems 18.6–18.9 refer to complex numbers $c_1 = 2 - 3i$ and $c_2 = 5 + i$.

18.9 Calculate $c_1 \div c_2$.

Begin by writing the quotient as a fraction.

$$
\begin{aligned}
c_1 \div c_2 &= \frac{c_1}{c_2} \\
&= \frac{2 - 3i}{5 + i}
\end{aligned}
$$

Multiply the numerator and denominator of the fraction by $5 - i$, the conjugate of the denominator.

$$
= \frac{2 - 3i}{5 + i} \cdot \frac{5 - i}{5 - i}
$$

Apply the FOIL method to calculate the product $(2 - 3i)(5 - i)$ in the numerator and the product $(5 + i)(5 - i)$ in the denominator.

$$= \frac{10 - 2i - 15i + 3i^2}{25 - 5i + 5i - i^2}$$

$$= \frac{10 - 17i + 3i^2}{25 - i^2}$$

$$= \frac{10 - 17i + 3(-1)}{25 - (-1)}$$

$$= \frac{10 - 3 - 17i}{25 + 1}$$

$$= \frac{7 - 17i}{26}$$

Most instructors require you to express the real and imaginary parts of the quotient as individual values, as demonstrated below.

$$c_1 \div c_2 = \frac{7}{26} - \frac{17}{26}i$$

> Divide each term of the numerator $7 - 17i$ by the denominator 26.

Note: Problems 18.10–18.12 refer to complex numbers $z_1 = 9 + 4i$ and $z_2 = 9 - 4i$.

18.10 Calculate $z_1 - z_2$.

> z_1 and z_2 are conjugates.

Subtract the real and imaginary parts of z_2 from the corresponding parts of z_1.

$$z_1 - z_2 = 9 + 4i - (9 - 4i)$$

$$= 9 + 4i - 9 - (-4i)$$

$$= 9 + 4i - 9 + 4i$$

$$= (9 - 9) + (4i + 4i)$$

$$= 8i$$

Note: Problems 18.10–18.12 refer to complex numbers $z_1 = 9 + 4i$ and $z_2 = 9 - 4i$.

18.11 Calculate $z_1 z_2$.

Apply the FOIL method to calculate the product of two complex numbers.

$$z_1 z_2 = (9 + 4i)(9 - 4i)$$

$$= 81 - 36i + 36i - 16i^2$$

$$= 81 - 16i^2$$

$$= 81 - 16(-1)$$

$$= 81 + 16$$

$$= 97$$

Notice that the product of a complex number and its conjugate is a real number—it contains no imaginary part.

> That's why you multiply by the conjugate when you divide complex numbers. You need to get rid of the complex number in the denominator.

> *Note: Problems 18.10–18.12 refer to complex numbers $z_1 = 9 + 4i$ and $z_2 = 9 - 4i$.*
>
> **18.12** Calculate $z_1 \div z_2$.

Express the quotient as a fraction and multiply the numerator and denominator by $9 + 4i$, the conjugate of the denominator.

$$
\begin{aligned}
z_1 \div z_2 &= \frac{9+4i}{9-4i} \\
&= \frac{9+4i}{9-4i} \cdot \frac{9+4i}{9+4i} \\
&= \frac{81+36i+36i+16i^2}{97} \\
&= \frac{81+72i+16(-1)}{97} \\
&= \frac{81-16+72i}{97} \\
&= \frac{65+72i}{97} \\
&= \frac{65}{97} + \frac{72}{97}i
\end{aligned}
$$

According to Problem 18.11, $(9 - 4i)(9 + 4i) = 97$.

Trigonometric Form of Complex Numbers

The reason you learned polar coordinates

> *Note: Problems 18.13–18.16 explain how to convert the complex number $c = \sqrt{3} + i$ into trigonometric form.*
>
> **18.13** Calculate $|c|$, the absolute value of c.

The absolute value of the complex number $c = a + bi$ is equal to $\sqrt{a^2 + b^2}$. In this problem, $a = \sqrt{3}$ and $b = 1$.

$$
\begin{aligned}
|c| &= \sqrt{a^2 + b^2} \\
&= \sqrt{\left(\sqrt{3}\right)^2 + (1)^2} \\
&= \sqrt{3+1} \\
&= \sqrt{4} \\
&= 2
\end{aligned}
$$

Note that the absolute value of c is the distance between c and the origin of the complex plane.

Note: Problems 18.13–18.16 explain how to convert the complex number $c = \sqrt{3} + i$ into trigonometric form.

18.14 Calculate the modulus, r, of c in trigonometric form.

The modulus r of a complex number in trigonometric form is equal to the absolute value of the complex number. According to Problem 18.13, $r = |c| = 2$.

Note: Problems 18.13–18.16 explain how to convert the complex number $c = \sqrt{3} + i$ into trigonometric form.

18.15 Calculate the argument, θ, of c in trigonometric form, and report the answer in radians.

To calculate the argument θ of the complex number $a + bi$, evaluate arctan (b/a). In this problem, $a = \sqrt{3}$ and $b = 1$.

$$\theta = \arctan \frac{b}{a}$$

$$= \arctan \frac{1}{\sqrt{3}}$$

Divide the numerator and denominator by 2 to produce familiar values from the unit circle.

$$= \arctan \frac{1/2}{\sqrt{3}/2}$$

$$= \frac{\pi}{6} \longleftarrow$$

Tan = sine/cosine. This angle has a cosine of √3/2 and a sine of 1/2.

Note: Problems 18.13–18.16 explain how to convert the complex number $c = \sqrt{3} + i$ into trigonometric form.

18.16 Express c in trigonometric form.

A complex number c with modulus r and argument θ has the following trigonometric form.

$$c = r(\cos \theta + i\sin \theta)$$

According to Problems 18.14–18.15, complex number $c = \sqrt{3} + i$ has modulus $r = 2$ and argument $\theta = \pi/6$.

$$c = 2\left(\cos \frac{\pi}{6} + i\sin \frac{\pi}{6}\right)$$

18.17 Given a complex number $z = a + bi$, such that $a > 0$ and $b > 0$, plot z in the complex plane and compare the polar and trigonometric forms of z.

If $a > 0$ and $b > 0$, then complex number $z = a + bi$ appears in the first quadrant, as illustrated below. Let r represent the length of the segment connecting point z to the origin, and let θ represent the angle formed by the x-axis and that segment.

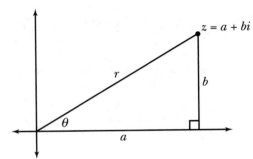

In polar form, z has coordinates (r, θ). Note that the radial coordinate r is equal to the modulus of z, and the angular coordinate θ is equal to the argument of z. Therefore, in trigonometric form, $z = r(\cos\theta + i\sin\theta)$.

Note: Problems 18.18–18.20 refer to the complex number c = –4 + 4i.

18.18 Calculate the modulus of c in trigonometric form.

Apply the modulus formula stated in Problem 18.13, noting that the modulus of a complex number is equal to its absolute value. In this problem, $a = -4$ and $b = 4$.

$$r = \sqrt{a^2 + b^2}$$
$$= \sqrt{(-4)^2 + 4^2}$$
$$= \sqrt{16 + 16}$$
$$= \sqrt{32}$$
$$= 4\sqrt{2}$$

Note: Problems 18.18–18.20 refer to the complex number c = –4 + 4i.

18.19 Calculate the argument of c in trigonometric form.

Substitute $a = -4$ and $b = 4$ into the argument formula stated in Problem 18.15.

$$\theta = \arctan\frac{b}{a}$$
$$= \arctan\left(\frac{4}{-4}\right)$$
$$= \arctan(-1)$$

Technically, arctan $(-1) = -\pi/4$ because of arctangent's restricted range. However, the angle $-\pi/4$ lies in the fourth quadrant, and $c = -4 + 4i$ lies in the second quadrant of the complex plane. Add π (the period of tangent) to $-\pi/4$ to calculate a coterminal angle that lies in the second quadrant.

$$-\frac{\pi}{4} + \pi = -\frac{\pi}{4} + \frac{4\pi}{4}$$
$$= \frac{3\pi}{4}$$

If you use degrees instead of radians, the answer would be 135°.

You conclude that the argument of c is $\theta = 3\pi/4$.

Note: Problems 18.18–18.20 refer to the complex number c = −4 + 4i.

18.20 Express c in trigonometric form.

Substitute $r = 4\sqrt{2}$ and $\theta = 3\pi/4$ into the formula stated in Problem 18.16.

$$c = r(\cos\theta + i\sin\theta)$$
$$= 4\sqrt{2}\left(\cos\frac{3\pi}{4} + i\sin\frac{3\pi}{4}\right)$$

18.21 Express the following complex number in rectangular form and identify its real and imaginary parts.

$$c = 5\left(\cos\frac{5\pi}{3} + i\sin\frac{5\pi}{3}\right)$$

Distribute 5 to the terms in parentheses and evaluate the trigonometric expressions.

$$c = 5\cos\frac{5\pi}{3} + 5i\sin\frac{5\pi}{3}$$
$$= 5\left(\frac{1}{2}\right) + 5i\left(-\frac{\sqrt{3}}{2}\right)$$
$$= \frac{5}{2} - \frac{5i\sqrt{3}}{2}$$
$$= \frac{5}{2} - \frac{5\sqrt{3}}{2}i$$

The complex number c has real part $\dfrac{5}{2}$ and imaginary part $-\dfrac{5\sqrt{3}}{2}$.

18.22 Express the following complex number in rectangular form and identify its real and imaginary parts.

$$z = 3\left(\cos\frac{\pi}{2} + i\sin\frac{\pi}{2}\right)$$

Distribute 3 to each of the terms in parentheses and simplify the expression.

$$z = 3\cos\frac{\pi}{2} + 3i\sin\frac{\pi}{2}$$
$$= 3(0) + 3i(1)$$
$$= 0 + 3i$$
$$= 3i$$

The complex number z has real part 0 and imaginary part 3.

Multiplying and Dividing Trigonometric Form
By plugging into one of two formulas

Note: In Problems 18.23–18.25, $z_1 = 4\left(\cos\dfrac{\pi}{3} + i\sin\dfrac{\pi}{3}\right)$ and $z_2 = 2\left(\cos\dfrac{2\pi}{3} + i\sin\dfrac{2\pi}{3}\right)$.

18.23 Calculate $z_1 z_2$ using the trigonometric form of the complex numbers.

Given complex numbers $z_1 = r_1(\cos\theta_1 + i\sin\theta_1)$ and $z_2 = r_2(\cos\theta_2 + i\sin\theta_2)$, you can apply the formula below to compute the product $z_1 z_2$.

$$z_1 z_2 = r_1 \cdot r_2\left[\cos(\theta_1 + \theta_2) + i\sin(\theta_1 + \theta_2)\right]$$

In this problem, $r_1 = 4$, $\theta_1 = \pi/3$, $r_2 = 2$, and $\theta_2 = 2\pi/3$.

$$z_1 z_2 = 4 \cdot 2\left[\cos\left(\frac{\pi}{3} + \frac{2\pi}{3}\right) + i\sin\left(\frac{\pi}{3} + \frac{2\pi}{3}\right)\right]$$
$$= 8\left[\cos\frac{3\pi}{3} + i\sin\frac{3\pi}{3}\right]$$
$$= 8[\cos\pi + i\sin\pi]$$
$$= 8[-1 + i(0)]$$
$$= 8[-1]$$
$$= -8$$

Note: In Problems 18.23–18.25, $z_1 = 4\left(\cos\dfrac{\pi}{3} + i\sin\dfrac{\pi}{3}\right)$ and $z_2 = 2\left(\cos\dfrac{2\pi}{3} + i\sin\dfrac{2\pi}{3}\right)$.

18.24 Express z_1 and z_2 in rectangular form.

Distribute the modulus of each complex number through the parentheses and simplify the expressions.

$$z_1 = 4\left(\cos\frac{\pi}{3} + i\sin\frac{\pi}{3}\right) = 4\cos\frac{\pi}{3} + 4i\sin\frac{\pi}{3} = 4\left(\frac{1}{2}\right) + 4i\left(\frac{\sqrt{3}}{2}\right) = 2 + 2\sqrt{3}\,i$$

$$z_2 = 2\left(\cos\frac{2\pi}{3} + i\sin\frac{2\pi}{3}\right) = 2\cos\frac{2\pi}{3} + 2i\sin\frac{2\pi}{3} = 2\left(-\frac{1}{2}\right) + 2i\left(\frac{\sqrt{3}}{2}\right) = -1 + \sqrt{3}\,i$$

Note: In Problems 18.23–18.25, $z_1 = 4\left(\cos\dfrac{\pi}{3} + i\sin\dfrac{\pi}{3}\right)$ and $z_2 = 2\left(\cos\dfrac{2\pi}{3} + i\sin\dfrac{2\pi}{3}\right)$.

18.25 Verify your solution to Problem 18.23 by multiplying the rectangular forms of the complex numbers.

According to Problem 18.23, $z_1 z_2 = -8$. In Problem 18.24, you identify the rectangular forms of the complex numbers. Multiply the rectangular forms of z_1 and z_2 to verify that the product is equal to -8.

$$
\begin{aligned}
z_1 z_2 &= \left(2 + 2\sqrt{3}\,i\right)\left(-1 + \sqrt{3}\,i\right) \\
&= -2 + 2\sqrt{3}\,i - 2\sqrt{3}\,i + 2(3)i^2 \\
&= -2 + 6(-1) \\
&= -2 - 6 \\
&= -8
\end{aligned}
$$

> You could write the product as $-8 + 0i$ to make it clear that the product is a complex number with an imaginary part equal to zero.

The product of z_1 and z_2 is -8, whether you express the complex numbers in rectangular or trigonometric form.

18.26 Calculate the product: $[\,3\,(\cos 40° + i\sin 40°)\,]\,[\,7\,(\cos 225° + i\sin 225°)\,]$.

Substitute $r_1 = 3$, $\theta_1 = 40°$, $r_2 = 7$, and $\theta_2 = 225°$ into the product formula, presented in Problem 18.23.

$$
\begin{aligned}
z_1 z_2 &= r_1 \cdot r_2\left[\cos(\theta_1 + \theta_2) + i\sin(\theta_1 + \theta_2)\right] \\
&= 3 \cdot 7\left[\cos(40° + 225°) + i\sin(40° + 225°)\right] \\
&= 21\left(\cos 265° + i\sin 265°\right)
\end{aligned}
$$

> Multiply the moduli $(3 \cdot 7)$ and add the arguments $(40° + 225°)$.

18.27 Given $z_1 = 10\left(\cos\dfrac{\pi}{8} + i\sin\dfrac{\pi}{8}\right)$ and $z_1 z_2 = 5\left(\cos\dfrac{17\pi}{24} + i\sin\dfrac{17\pi}{24}\right)$, calculate z_2.

Let $r(\cos\theta + i\sin\theta)$ represent z_2 in trigonometric form. Therefore, its modulus is r and its argument is θ. To calculate r, note that the modulus of $z_1 z_2$ is equal to the product of the moduli of z_1 and z_2.

$$(\text{modulus of } z_1)(\text{modulus of } z_2) = \text{modulus of } (z_1 z_2)$$
$$10(r) = 5$$
$$r = \frac{5}{10}$$
$$r = \frac{1}{2}$$

The argument of the product $z_1 z_2$ is $17\pi/24$. Note that this argument is equal to the sum of $\pi/8$ and θ, the respective arguments of z_1 and z_2. Solve the following equation for θ to identify the argument of z_2.

$$\text{argument of } z_1 + \text{argument of } z_2 = \text{argument of } (z_1 z_2)$$
$$\frac{\pi}{8} + \theta = \frac{17\pi}{24}$$
$$\theta = \frac{17\pi}{24} - \frac{\pi}{8}$$
$$\theta = \frac{17\pi}{24} - \frac{3\pi}{24}$$
$$\theta = \frac{14\pi}{24}$$
$$\theta = \frac{7\pi}{12}$$

Now that you have calculated r and θ, you can express z_2 in trigonometric form.

$$z_2 = r(\cos\theta + i\sin\theta)$$
$$z_2 = \frac{1}{2}\left(\cos\frac{7\pi}{12} + i\sin\frac{7\pi}{12}\right)$$

> *Note: In Problems 18.28–18.30, $c_1 = 6(\cos\pi + i\sin\pi)$ and $c_2 = 2\left(\cos\frac{3\pi}{4} + i\sin\frac{3\pi}{4}\right)$.*
>
> **18.28** Calculate $c_1 \div c_2$ using the trigonometric form of the complex numbers.

To calculate the quotient of complex numbers $c_1 = r_1(\cos\theta_1 + i\sin\theta_1)$ and $c_2 = r_2(\cos\theta_2 + i\sin\theta_2)$, apply the following formula.

$$\frac{c_1}{c_2} = \frac{r_1}{r_2}\left[\cos(\theta_1 - \theta_2) + i\sin(\theta_1 - \theta_2)\right]$$

In this problem, $r_1 = 6$, $\theta_1 = \pi$, $r_2 = 2$, and $\theta_2 = 3\pi/4$.

$$\frac{c_1}{c_2} = \frac{6}{2}\left[\cos\left(\pi - \frac{3\pi}{4}\right) + i\sin\left(\pi - \frac{3\pi}{4}\right)\right]$$
$$= 3\left[\cos\left(\frac{4\pi}{4} - \frac{3\pi}{4}\right) + i\sin\left(\frac{4\pi}{4} - \frac{3\pi}{4}\right)\right]$$
$$= 3\left(\cos\frac{\pi}{4} + i\sin\frac{\pi}{4}\right)$$

When you multiply complex numbers, you multiply the moduli and add the arguments. When you divide complex numbers, you divide the moduli and subtract the arguments.

Note: In Problems 18.28–18.30, $c_1 = 6\left(\cos\pi + i\,\sin\pi\right)$ *and* $c_2 = 2\left(\cos\dfrac{3\pi}{4} + i\,\sin\dfrac{3\pi}{4}\right)$.

18.29 Express c_1, c_2, and $c_1 \div c_2$ in rectangular form.

Distribute the modulus in each complex number and evaluate the trigonometric expressions.

$$c_1 = 6\cos\pi + 6i\sin\pi = 6(-1) + 6i(0) = -6$$

$$c_2 = 2\cos\frac{3\pi}{4} + 2i\sin\frac{3\pi}{4} = \cancel{2}\left(-\frac{\sqrt{2}}{\cancel{2}}\right) + \cancel{2}i\left(\frac{\sqrt{2}}{\cancel{2}}\right) = -\sqrt{2} + \sqrt{2}\,i$$

$$\frac{c_1}{c_2} = 3\cos\frac{\pi}{4} + 3i\sin\frac{\pi}{4} = 3\left(\frac{\sqrt{2}}{2}\right) + 3i\left(\frac{\sqrt{2}}{2}\right) = \frac{3\sqrt{2}}{2} + \frac{3\sqrt{2}}{2}\,i$$

Note: In Problems 18.28–18.30, $c_1 = 6\left(\cos\pi + i\,\sin\pi\right)$ *and* $c_2 = 2\left(\cos\dfrac{3\pi}{4} + i\,\sin\dfrac{3\pi}{4}\right)$.

18.30 Verify your solution to Problem 18.28 by dividing the rectangular forms of the complex numbers.

According to Problem 18.29, $c_1 = -6$ and $c_2 = -\sqrt{2} + \sqrt{2}\,i$. Calculate $c_1 \div c_2$ to verify that the quotient is equal to $\dfrac{3\sqrt{2}}{2} + \dfrac{3\sqrt{2}}{2}\,i$. ⟵

The answer to Problem 18.28 written in rectangular form

Apply the technique described in Problem 18.9, multiplying the numerator and denominator of c_1 / c_2 by the conjugate of c_2.

$$\frac{c_1}{c_2} = \frac{-6}{-\sqrt{2} + \sqrt{2}\,i}$$

$$= \frac{-6}{-\sqrt{2} + \sqrt{2}i} \cdot \frac{-\sqrt{2} - \sqrt{2}\,i}{-\sqrt{2} - \sqrt{2}\,i}$$

$$= \frac{6\sqrt{2} + 6\sqrt{2}\,i}{2 + 2i - 2i - 2i^2}$$

$$= \frac{6\sqrt{2} + 6\sqrt{2}\,i}{2 - 2(-1)}$$

$$= \frac{6\sqrt{2} + 6\sqrt{2}\,i}{2 + 2}$$

$$= \frac{6\sqrt{2} + 6\sqrt{2}\,i}{4}$$

Divide each of the terms in the numerator by the denominator.

$$= \frac{6\sqrt{2}}{4} + \frac{6\sqrt{2}}{4}\,i$$

$$= \frac{3\sqrt{2}}{2} + \frac{3\sqrt{2}}{2}\,i$$

This verifies the quotient that you calculated in Problem 18.28 and converted into rectangular form in Problem 18.29.

18.31 Calculate the quotient: $12 (\cos 120° + i\sin 120°) \div [3 (\cos 25° + i\sin 25°)]$.

Substitute $r_1 = 12$, $\theta_1 = 120°$, $r_2 = 3$, and $\theta_2 = 25°$ into the formula introduced in Problem 18.28.

$$\frac{c_1}{c_2} = \frac{r_1}{r_2}\left[\cos\left(\theta_1 - \theta_2\right) + i\sin\left(\theta_1 - \theta_2\right)\right]$$

$$= \frac{12}{3}\left[\cos\left(120° - 25°\right) + i\sin\left(120° - 25°\right)\right]$$

$$= 4\left(\cos 95° + i\sin 95°\right)$$

18.32 Given $c_1 = 18\left(\cos\dfrac{11\pi}{5} + i\sin\dfrac{11\pi}{5}\right)$ and $\dfrac{c_1}{c_2} = 2\left(\cos\dfrac{\pi}{3} + i\sin\dfrac{\pi}{3}\right)$, calculate c_2.

Let $c_2 = r(\cos\theta + i\sin\theta)$ represent the unknown complex number. The modulus of the quotient (c_1/c_2) is equal to the quotient of the moduli (r_1/r_2).

$$\frac{\text{modulus of } c_1}{\text{modulus of } c_2} = \text{modulus of } \frac{c_1}{c_2}$$

$$\frac{18}{r} = 2$$

Cross-multiply and solve for r to calculate the modulus of c_2.

$$2r = 18$$

$$r = \frac{18}{2}$$

$$r = 9$$

The argument of the quotient is equal to the difference of the arguments. Solve the following equation to calculate the argument θ of c_2.

$$\text{argument of } c_1 - \text{argument of } c_2 = \text{argument of } \frac{c_1}{c_2}$$

$$\frac{11\pi}{5} - \theta = \frac{\pi}{3}$$

$$\frac{11\pi}{5} - \frac{\pi}{3} = \theta$$

$$\frac{33\pi}{15} - \frac{5\pi}{15} = \theta$$

$$\frac{28\pi}{15} = \theta$$

Thus, c_2 has modulus $r = 9$ and argument $\theta = 28\pi/15$.

$$c_2 = r\left(\cos\theta + i\sin\theta\right)$$

$$c_2 = 9\left(\cos\frac{28\pi}{15} + i\sin\frac{28\pi}{15}\right)$$

De Moivre's Theorem
Raising complex numbers to powers

Note: In Problems 18.33–18.36, z = 5 (cos 30° + i sin 30°).

18.33 Apply De Moivre's theorem to calculate z^2.

Given a complex number written in trigonometric form, you can apply De Moivre's theorem to raise that complex number to exponents. Specifically, given complex number $z = r(\cos \theta + i\sin \theta)$, the following formula represents z^n, such that n is an integer.

$$\left[r(\cos\theta + i\sin\theta) \right]^n = r^n \left[\cos(n\theta) + i\sin(n\theta) \right]$$

In this problem, $r = 5$, $\theta = 30°$, and $n = 2$. ← *You're squaring the complex number, so n = 2. If you were raising z to the ninth power, you would set n = 9.*

$$\left[5(\cos 30° + i\sin 30°) \right]^2 = 5^2 \left[\cos(2 \cdot 30°) + i\sin(2 \cdot 30°) \right]$$
$$= 25(\cos 60° + i\sin 60°)$$

Note: In Problems 18.33–18.36, z = 5 (cos 30° + i sin 30°).

18.34 Verify your answer to Problem 18.33 by expressing z in rectangular form and calculating z^2.

Begin by converting z to rectangular form.

$$z = 5(\cos 30° + i\sin 30°)$$
$$= 5\cos 30° + 5i\sin 30°$$
$$= 5\left(\frac{\sqrt{3}}{2}\right) + 5i\left(\frac{1}{2}\right)$$
$$= \frac{5\sqrt{3}}{2} + \frac{5}{2}i$$

Calculate z^2 in rectangular form using the FOIL method.

$$z^2 = \left(\frac{5\sqrt{3}}{2} + \frac{5}{2}i \right)^2$$
$$= \left(\frac{5\sqrt{3}}{2} + \frac{5}{2}i \right)\left(\frac{5\sqrt{3}}{2} + \frac{5}{2}i \right)$$
$$= \frac{25(3)}{4} + \frac{25\sqrt{3}}{4}i + \frac{25\sqrt{3}}{4}i + \frac{25}{4}i^2$$
$$= \frac{75}{4} + \frac{50\sqrt{3}}{4}i + \frac{25}{4}(-1)$$
$$= \frac{50}{4} + \frac{50\sqrt{3}}{4}i$$
$$= \frac{25}{2} + \frac{25\sqrt{3}}{2}i$$

This complex number is equal to 25 (cos 60° + $i$sin 60°), the solution to Problem 18.33 in rectangular form.

$$z^2 = 25\cos 60° + 25\,i\sin 60°$$

$$= 25\left(\frac{1}{2}\right) + 25\left(\frac{\sqrt{3}}{2}\right)i$$

$$= \frac{25}{2} + \frac{25\sqrt{3}}{2}\,i$$

Note: In Problems 18.33–18.36, z = 5 (cos 30° + i sin 30°).

18.35 Calculate z^3.

Substitute $r = 5$, $\theta = 30°$, and $n = 3$ into De Moivre's theorem.

$$\left[r(\cos\theta + i\sin\theta)\right]^n = r^n\left[\cos(n\theta) + i\sin(n\theta)\right]$$

$$\left[5(\cos 30° + i\sin 30°)\right]^3 = 5^3\left[\cos(3\cdot 30°) + i\sin(3\cdot 30°)\right]$$

$$= 125(\cos 90° + i\sin 90°)$$

Note: In Problems 18.33–18.36, z = 5 (cos 30° + i sin 30°).

18.36 Calculate z^5.

Substitute $r = 5$, $\theta = 30°$, and $n = 5$ into De Moivre's theorem.

$$\left[r(\cos\theta + i\sin\theta)\right]^n = r^n\left[\cos(n\theta) + i\sin(n\theta)\right]$$

$$\left[5(\cos 30° + i\sin 30°)\right]^5 = 5^5\left[\cos(5\cdot 30°) + i\sin(5\cdot 30°)\right]$$

$$= 3{,}125(\cos 150° + i\sin 150°)$$

18.37 Given $c = 2\left(\cos\dfrac{7\pi}{4} + i\sin\dfrac{7\pi}{4}\right)$, calculate c^4.

Substitute $r = 2$, $\theta = 7\pi/4$, and $n = 4$ into De Moivre's theorem.

$$\left[r(\cos\theta + i\sin\theta)\right]^n = r^n\left[\cos(n\theta) + i\sin(n\theta)\right]$$

$$\left[2\left(\cos\frac{7\pi}{4} + i\sin\frac{7\pi}{4}\right)\right]^4 = 2^4\left[\cos\left(\frac{\cancel{4}}{1}\cdot\frac{7\pi}{\cancel{4}}\right) + i\sin\left(\frac{\cancel{4}}{1}\cdot\frac{7\pi}{\cancel{4}}\right)\right]$$

$$= 16(\cos 7\pi + i\sin 7\pi)$$

Although this answer is correct, it is customary to identify an angle coterminal to 7π that lies between 0 and 2π, the bounds of 1 rotation on the unit circle. You conclude that $c^4 = 16(\cos\pi + i\sin\pi)$.

Subtract 2π from 7π until you get a value between 0 and 2π:

$$7\pi - 2\pi = 5\pi$$
$$5\pi - 2\pi = 3\pi$$
$$3\pi - 2\pi = \pi$$

Note: In Problems 18.38–18.39, $z_1 = 2 (\cos 40° + i \sin 40°)$ and $z_2 = 3 (\cos 15° + i \sin 15°)$.

18.38 Calculate $(z_1 z_2)^2$.

The order of operations dictates that you multiply inside parentheses before you raise the product to the second power. To calculate $z_1 z_2$, apply the formula presented in Problem 18.23.

$$z_1 z_2 = r_1 \cdot r_2 \left[\cos(\theta_1 + \theta_2) + i \sin(\theta_1 + \theta_2) \right]$$
$$= 2 \cdot 3 \left[\cos(40° + 15°) + i \sin(40° + 15°) \right]$$
$$= 6 (\cos 55° + i \sin 55°)$$

Apply De Moivre's theorem to raise the product $z_1 z_2 = 6 (\cos 55° + i \sin 55°)$ to the second power.

$$\left[6(\cos 55° + i \sin 55°) \right]^2 = 6^2 \left[\cos(2 \cdot 55°) + i \sin(2 \cdot 55°) \right]$$
$$= 36 (\cos 110° + i \sin 110°)$$

Thus, $(z_1 z_2)^2 = 36 (\cos 110° + i \sin 110°)$.

Note: In Problems 18.38–18.39, $z_1 = 2 (\cos 40° + i \sin 40°)$ and $z_2 = 3 (\cos 15° + i \sin 15°)$.

18.39 Calculate $\left(\dfrac{z_1}{z_2} \right)^3$.

Calculate the quotient inside the parentheses by applying the formula from Problem 18.28.

$$\frac{z_1}{z_2} = \frac{r_1}{r_2} \left[\cos(\theta_1 - \theta_2) + i \sin(\theta_1 - \theta_2) \right]$$
$$= \frac{2}{3} \left[\cos(40° - 15°) + i \sin(40° - 15°) \right]$$
$$= \frac{2}{3} (\cos 25° + i \sin 25°)$$

Apply De Moivre's theorem to raise the quotient to the third power.

$$\left[\frac{2}{3} (\cos 25° + i \sin 25°) \right]^3 = \left(\frac{2}{3} \right)^3 \left[\cos(3 \cdot 25°) + i \sin(3 \cdot 25°) \right]$$
$$= \frac{8}{27} (\cos 75° + i \sin 75°)$$

Roots of Complex Numbers

Square roots, cube roots, etc.

18.40 State the formula used to calculate the nth roots of complex number $z = r(\cos\theta + i\sin\theta)$ and explain how to apply it.

The number of nth roots of a complex number is equal to n. For example, if you are calculating square roots, then there are exactly $n = 2$ square roots of z. Similarly, there are exactly $n = 3$ cube roots of z. These roots are evenly spaced around the unit circle according to the following formula.

$$p_k = \sqrt[n]{r}\left(\cos\frac{\theta + k\cdot 360°}{n} + i\sin\frac{\theta + k\cdot 360°}{n}\right), \text{ such that } k = 0, 1, \cdots, n-1$$

In the formula, p_k represents one root. The subscript merely numbers the roots, from 0 to $n-1$. For example, when you calculate cube roots using this formula, $n = 3$ and the names of the cube roots are p_0, p_1, and p_2. The highest subscript is 2, which is exactly one less than $n = 3$.

Calculating roots is not complicated if you keep these facts in mind:

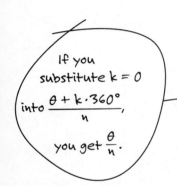

If you substitute $k = 0$ into $\dfrac{\theta + k\cdot 360°}{n}$, you get $\dfrac{\theta}{n}$.

- Each root has the same modulus, $\sqrt[n]{r}$. Only the arguments of the roots vary.
- The argument of the first root, p_0, is equal to θ/n.
- To calculate the remaining arguments, add $360°/n$ to the first argument.

Note: In Problems 18.41–18.42, $c = 16(\cos 100° + i\sin 100°)$.

18.41 Determine how many square roots exist for c and calculate the square root with the smallest argument.

You are asked to calculate the square roots of c, so $n = 2$. The number of roots is equal to n, so there are two square roots of c, named p_0 and p_1. The modulus of the roots is equal to the square root of $r = 16$, the modulus of c. Thus, p_0 and p_1 both have modulus $\sqrt{16} = 4$.

The smaller the subscript of p, the smaller the argument of the root is. Therefore, p_0 always represents the root with the smallest argument. To calculate the argument of p_0, divide $\theta = 100°$ by $n = 2$.

$$\text{argument of } p_0 = \frac{\theta}{n}$$
$$= \frac{100°}{2}$$
$$= 50°$$

You conclude that p_0 is a square root of c; it has modulus 4 and argument 50°.

$$p_0 = 4(\cos 50° + i\sin 50°)$$

Note: In Problems 18.41–18.42, c = 16 (cos 100° + i sin 100°).

18.42 Calculate the remaining square root(s) of c.

As Problem 18.41 explains, c has two square roots: p_0 and p_1. Both have the same modulus, so p_1 also has a modulus of $\sqrt{16} = 4$. To calculate the argument of the remaining root, first divide 360° by $n = 2$.

$$\frac{360°}{n} = \frac{360°}{2} = 180°$$

Add this value to the argument of p_0, which is 50° according to Problem 18.41. That sum is the argument of p_1.

$$\text{argument of } p_1 = \text{argument of } p_0 + \frac{360}{n}$$
$$= 50° + 180°$$
$$= 230°$$

You conclude that $p_1 = 4 (\cos 230° + i\sin 230°)$.

Note: In Problems 18.43–18.44, z = 8 (cos 210° + i sin 210°).

18.43 Determine how many cube roots exist for z and calculate the cube root with the smallest argument.

Exactly $n = 3$ cube roots exist for any complex number: p_0, p_1, and p_2. You are asked to calculate p_0, the cube root with the smallest argument.

Each of the roots has the same modulus: the cube root of the modulus of z.

$$\text{modulus of } p_0, p_1, \text{and } p_2 = \sqrt[n]{\text{modulus of } z}$$
$$= \sqrt[3]{8}$$
$$= 2$$

The argument of p_0 is equal to $\theta = 210°$ divided by n.

$$\text{argument of } p_0 = \frac{\theta}{n}$$
$$= \frac{210°}{3}$$
$$= 70°$$

You conclude that $p_0 = 2 (\cos 70° + i\sin 70°)$.

Not all books name the roots the same way. Your book might use a different letter than p, but the process is the same.

Note: In Problems 18.43–18.44, z = 8 (cos 210° + i sin 210°).

18.44 Calculating the remaining cube roots of z.

In Problem 18.43, you determine that $p_0 = 2(\cos 70° + i\sin 70°)$. The moduli of p_1 and p_2 are equal to the modulus of p_0. To calculate the arguments of p_1 and p_2, you must first divide 360° by $n = 3$. The angles are evenly spaced at an interval of $360/n$ degrees around the unit circle.

$$\frac{360°}{n} = \frac{360°}{3} = 120°$$

Add 120° to the argument of p_0 to calculate the argument of p_1.

$$\text{argument of } p_1 = \text{argument of } p_0 + \frac{360°}{n}$$
$$= 70° + 120°$$
$$= 190°$$

Add 120° to the argument of p_1 to calculate the argument of p_2.

$$\text{argument of } p_2 = \text{argument of } p_1 + \frac{360°}{n}$$
$$= 190° + 120°$$
$$= 310°$$

Recall that both p_1 and p_2 have modulus 2; their arguments are 190° and 310°, respectively.

$$p_1 = 2(\cos 190° + i\sin 190°)$$
$$p_2 = 2(\cos 310° + i\sin 310°)$$

18.45 Calculate the fifth roots of $c = 243\left(\cos\frac{\pi}{3} + i\sin\frac{\pi}{3}\right)$.

You are directed to calculate all of the fifth roots of complex number c. There are $n = 5$ fifth roots, named $p_0, p_1, p_2, p_3,$ and p_4. Each of the roots has the same modulus.

$$\text{modulus of } p_0, p_1, p_2, p_3, \text{ and } p_4 = \sqrt[n]{\text{modulus of } z}$$
$$= \sqrt[5]{243}$$
$$= 3$$

To calculate the argument of p_0, divide $\theta = \pi/3$ by $n = 5$.

$$\text{argument of } p_0 = \frac{\theta}{n}$$
$$= \frac{\pi/3}{5}$$
$$= \frac{\pi}{3} \div 5$$

Dividing by 5 is equivalent to multiplying by 1/5, the reciprocal of 5.

$$= \frac{\pi}{3} \cdot \frac{1}{5}$$

$$= \frac{\pi}{15}$$

You now know the modulus and argument of p_0, so you can express the complex number in trigonometric form.

$$p_0 = 3\left(\cos\frac{\pi}{15} + i\sin\frac{\pi}{15}\right)$$

The remaining roots have arguments that are equally spaced around the unit circle, at an interval of $2\pi/n$. ←

> This angle is measured in radians, so use the formula $2\pi/n$ instead of $360°/n$.

$$\frac{2\pi}{n} = \frac{2\pi}{5}$$

Add $2\pi/5$ to the argument of p_0 to calculate the argument of p_1.

$$\text{argument of } p_0 = \frac{\pi}{15}$$

$$\text{argument of } p_1 = \text{argument of } p_0 + \frac{2\pi}{5}$$

$$= \frac{\pi}{15} + \frac{2\pi}{5}$$

$$= \frac{\pi}{15} + \frac{6\pi}{15}$$

$$= \frac{7\pi}{15}$$

Add $2\pi/5$ to the argument of p_1 to calculate the argument of p_2.

$$\text{argument of } p_2 = \text{argument of } p_1 + \frac{2\pi}{5}$$

$$= \frac{7\pi}{15} + \frac{2\pi}{5}$$

$$= \frac{7\pi}{15} + \frac{6\pi}{15}$$

$$= \frac{13\pi}{15}$$

Add $2\pi/5$ to the argument of p_2 to calculate the argument of p_3.

$$\text{argument of } p_3 = \text{argument of } p_2 + \frac{2\pi}{5}$$

$$= \frac{13\pi}{15} + \frac{2\pi}{5}$$

$$= \frac{13\pi}{15} + \frac{6\pi}{15}$$

$$= \frac{19\pi}{15}$$

Finally, Add $2\pi/5$ to the argument of p_3 to calculate the argument of p_4.

$$\text{argument of } p_4 = \text{argument of } p_3 + \frac{2\pi}{5}$$

$$= \frac{19\pi}{15} + \frac{2\pi}{5}$$

$$= \frac{19\pi}{15} + \frac{6\pi}{15}$$

$$= \frac{25\pi}{15}$$

$$= \frac{5\pi}{3}$$

List all of the fifth roots, recalling that each has modulus $\sqrt[5]{243} = 3$.

$$p_0 = 3\left(\cos\frac{\pi}{15} + i\sin\frac{\pi}{15}\right)$$

$$p_1 = 3\left(\cos\frac{7\pi}{15} + i\sin\frac{7\pi}{15}\right)$$

$$p_2 = 3\left(\cos\frac{13\pi}{15} + i\sin\frac{13\pi}{15}\right)$$

$$p_3 = 3\left(\cos\frac{19\pi}{15} + i\sin\frac{19\pi}{15}\right)$$

$$p_4 = 3\left(\cos\frac{5\pi}{3} + i\sin\frac{5\pi}{3}\right)$$

Appendix A
TABLE OF TRIGONOMETRIC VALUES

Nine pages that answer the question "Why do I need a calculator?"

This table is presented in increments of quarter degrees; each of the other values (including the equivalent radian measure for the angle) is accurate to the thousandths place.

For angles less than 20°, the radian measure of an angle is almost equal to its sine and tangent values. That's why some people use the radian measure of an angle to approximate small sine and tangent values.

Degrees	Radians	Cosine	Sine	Tangent	Secant	Cosecant	Cotangent
0	0	1	0	0	1	—	—
0.25	0.004	0.100	0.004	0.004	1.000	229.184	229.182
0.50	0.009	0.100	0.009	0.009	1.000	114.593	114.589
0.75	0.013	0.100	0.013	0.013	1.000	76.397	76.390
1.00	0.017	0.100	0.017	0.017	1.000	57.299	57.290
1.25	0.022	0.100	0.022	0.022	1.000	45.840	45.829
1.50	0.026	0.100	0.026	0.026	1.000	38.202	38.188
1.75	0.031	0.100	0.031	0.031	1.000	32.746	32.730
2.00	0.035	0.999	0.035	0.035	1.001	28.654	28.636
2.25	0.039	0.999	0.039	0.039	1.001	25.471	25.452
2.50	0.044	0.999	0.044	0.044	1.001	22.926	22.904
2.75	0.048	0.999	0.048	0.048	1.001	20.843	20.819
3.00	0.052	0.999	0.052	0.052	1.001	19.107	19.081
3.25	0.057	0.998	0.057	0.057	1.002	17.639	17.611
3.50	0.061	0.998	0.061	0.061	1.002	16.380	16.350
3.75	0.065	0.998	0.065	0.066	1.002	15.290	15.257
4.00	0.070	0.998	0.070	0.070	1.002	14.336	14.301
4.25	0.074	0.997	0.074	0.074	1.003	13.494	13.457
4.50	0.079	0.997	0.078	0.079	1.003	12.745	12.706
4.75	0.083	0.997	0.083	0.083	1.003	12.076	12.035
5.00	0.087	0.996	0.087	0.087	1.004	11.474	11.430
5.25	0.092	0.996	0.092	0.092	1.004	10.929	10.883
5.50	0.096	0.995	0.096	0.096	1.005	10.433	10.385
5.75	0.100	0.995	0.100	0.101	1.005	9.981	9.931
6.00	0.105	0.995	0.105	0.105	1.006	9.567	9.514

Degrees	Radians	Cosine	Sine	Tangent	Secant	Cosecant	Cotangent
6.25	0.109	0.994	0.109	0.110	1.006	9.186	9.131
6.50	0.113	0.994	0.113	0.114	1.006	8.834	8.777
6.75	0.118	0.993	0.118	0.118	1.007	8.508	8.449
7.00	0.122	0.993	0.122	0.123	1.008	8.206	8.144
7.25	0.127	0.992	0.126	0.127	1.008	7.924	7.861
7.50	0.131	0.991	0.131	0.132	1.009	7.661	7.596
7.75	0.135	0.991	0.135	0.136	1.009	7.416	7.348
8.00	0.140	0.990	0.139	0.141	1.010	7.185	7.115
8.25	0.144	0.990	0.143	0.145	1.010	6.969	6.897
8.50	0.148	0.989	0.148	0.149	1.011	6.765	6.691
8.75	0.153	0.988	0.152	0.154	1.012	6.574	6.497
9.00	0.157	0.988	0.156	0.158	1.012	6.392	6.314
9.25	0.161	0.987	0.161	0.163	1.013	6.221	6.140
9.50	0.166	0.986	0.165	0.167	1.014	6.059	5.976
9.75	0.170	0.986	0.169	0.172	1.015	5.905	5.820
10.00	0.175	0.985	0.174	0.176	1.015	5.759	5.671
10.25	0.179	0.984	0.178	0.181	1.016	5.620	5.530
10.50	0.183	0.983	0.182	0.185	1.017	5.487	5.396
10.75	0.188	0.982	0.187	0.190	1.018	5.361	5.267
11.00	0.192	0.982	0.191	0.194	1.019	5.241	5.145
11.25	0.196	0.981	0.195	0.199	1.020	5.126	5.027
11.50	0.201	0.980	0.199	0.203	1.020	5.016	4.915
11.75	0.205	0.979	0.204	0.208	1.021	4.911	4.808
12.00	0.209	0.978	0.208	0.213	1.022	4.810	4.705
12.25	0.214	0.977	0.212	0.217	1.023	4.713	4.606
12.50	0.218	0.976	0.216	0.222	1.024	4.620	4.511
12.75	0.223	0.975	0.221	0.226	1.025	4.531	4.419
13.00	0.227	0.974	0.225	0.231	1.026	4.445	4.331
13.25	0.231	0.973	0.229	0.235	1.027	4.363	4.247
13.50	0.236	0.972	0.233	0.240	1.028	4.284	4.165
13.75	0.240	0.971	0.238	0.245	1.030	4.207	4.087
14.00	0.244	0.970	0.242	0.249	1.031	4.134	4.011
14.25	0.249	0.969	0.246	0.254	1.032	4.063	3.938
14.50	0.253	0.968	0.250	0.259	1.033	3.994	3.867
14.75	0.257	0.967	0.255	0.263	1.034	3.928	3.798
15.00	0.262	0.966	0.259	0.268	1.035	3.864	3.732
15.25	0.266	0.965	0.263	0.273	1.037	3.802	3.668
15.50	0.271	0.964	0.267	0.277	1.038	3.742	3.606
15.75	0.275	0.962	0.271	0.282	1.039	3.684	3.546
16.00	0.279	0.961	0.276	0.287	1.040	3.628	3.487
16.25	0.284	0.960	0.280	0.291	1.042	3.574	3.431
16.50	0.288	0.959	0.284	0.296	1.043	3.521	3.376

Degrees	Radians	Cosine	Sine	Tangent	Secant	Cosecant	Cotangent
16.75	0.292	0.958	0.288	0.301	1.044	3.470	3.323
17.00	0.297	0.956	0.292	0.306	1.046	3.420	3.271
17.25	0.301	0.955	0.297	0.311	1.047	3.372	3.221
17.50	0.305	0.954	0.301	0.315	1.049	3.326	3.172
17.75	0.310	0.952	0.305	0.320	1.050	3.280	3.124
18.00	0.314	0.951	0.309	0.325	1.051	3.236	3.078
18.25	0.319	0.950	0.313	0.330	1.053	3.193	3.033
18.50	0.323	0.948	0.317	0.335	1.054	3.152	2.989
18.75	0.327	0.947	0.321	0.339	1.056	3.111	2.946
19.00	0.332	0.946	0.326	0.344	1.058	3.072	2.904
19.25	0.336	0.944	0.330	0.349	1.059	3.033	2.864
19.50	0.340	0.943	0.334	0.354	1.061	2.996	2.824
19.75	0.345	0.941	0.338	0.359	1.063	2.959	2.785
20.00	0.349	0.940	0.342	0.364	1.064	2.924	2.747
20.25	0.353	0.938	0.346	0.369	1.066	2.889	2.711
20.50	0.358	0.937	0.350	0.374	1.068	2.855	2.675
20.75	0.362	0.935	0.354	0.379	1.069	2.823	2.639
21.00	0.367	0.934	0.358	0.384	1.071	2.790	2.605
21.25	0.371	0.932	0.362	0.389	1.073	2.759	2.572
21.50	0.375	0.930	0.367	0.394	1.075	2.729	2.539
21.75	0.380	0.929	0.371	0.399	1.077	2.699	2.507
22.00	0.384	0.927	0.375	0.404	1.079	2.669	2.475
22.25	0.388	0.926	0.379	0.409	1.080	2.641	2.444
22.50	0.393	0.924	0.383	0.414	1.082	2.613	2.414
22.75	0.397	0.922	0.387	0.419	1.084	2.586	2.385
23.00	0.401	0.921	0.391	0.424	1.086	2.559	2.356
23.25	0.406	0.919	0.395	0.430	1.088	2.533	2.328
23.50	0.410	0.917	0.399	0.435	1.090	2.508	2.300
23.75	0.415	0.915	0.403	0.440	1.093	2.483	2.273
24.00	0.419	0.914	0.407	0.445	1.095	2.459	2.246
24.25	0.423	0.912	0.411	0.450	1.097	2.435	2.220
24.50	0.428	0.910	0.415	0.456	1.099	2.411	2.194
24.75	0.432	0.908	0.419	0.461	1.101	2.389	2.169
25.00	0.436	0.906	0.423	0.466	1.103	2.366	2.145
25.25	0.441	0.904	0.427	0.472	1.106	2.344	2.120
25.50	0.445	0.903	0.431	0.477	1.108	2.323	2.097
25.75	0.449	0.901	0.434	0.482	1.110	2.302	2.073
26.00	0.454	0.899	0.438	0.488	1.113	2.281	2.050
26.25	0.458	0.897	0.442	0.493	1.115	2.261	2.028
26.50	0.463	0.895	0.446	0.499	1.117	2.241	2.006
26.75	0.467	0.893	0.450	0.504	1.120	2.222	1.984
27.00	0.471	0.891	0.454	0.510	1.122	2.203	1.963

Degrees	Radians	Cosine	Sine	Tangent	Secant	Cosecant	Cotangent
27.25	0.476	0.889	0.458	0.515	1.125	2.184	1.942
27.50	0.480	0.887	0.462	0.521	1.127	2.166	1.921
27.75	0.484	0.885	0.466	0.526	1.130	2.148	1.901
28.00	0.489	0.883	0.469	0.532	1.133	2.130	1.881
28.25	0.493	0.881	0.473	0.537	1.135	2.113	1.861
28.50	0.497	0.879	0.477	0.543	1.138	2.096	1.842
28.75	0.502	0.877	0.481	0.549	1.141	2.079	1.823
29.00	0.506	0.875	0.485	0.554	1.143	2.063	1.804
29.25	0.511	0.873	0.489	0.560	1.146	2.047	1.786
29.50	0.515	0.870	0.492	0.566	1.149	2.031	1.767
29.75	0.519	0.868	0.496	0.572	1.152	2.015	1.750
30.00	0.524	0.866	0.5	0.577	1.155	2	1.732
30.25	0.528	0.864	0.504	0.583	1.158	1.985	1.715
30.50	0.532	0.862	0.508	0.589	1.161	1.970	1.698
30.75	0.537	0.859	0.511	0.595	1.164	1.956	1.681
31.00	0.541	0.857	0.515	0.601	1.167	1.942	1.664
31.25	0.545	0.855	0.519	0.607	1.170	1.928	1.648
31.50	0.550	0.853	0.523	0.613	1.173	1.914	1.632
31.75	0.554	0.850	0.526	0.619	1.176	1.900	1.616
32.00	0.559	0.848	0.530	0.625	1.179	1.887	1.600
32.25	0.563	0.846	0.534	0.631	1.182	1.874	1.585
32.50	0.567	0.843	0.537	0.637	1.186	1.861	1.570
32.75	0.572	0.841	0.541	0.643	1.189	1.849	1.555
33.00	0.576	0.839	0.545	0.649	1.192	1.836	1.540
33.25	0.580	0.836	0.548	0.656	1.196	1.824	1.525
33.50	0.585	0.834	0.552	0.662	1.199	1.812	1.511
33.75	0.589	0.831	0.556	0.668	1.203	1.800	1.497
34.00	0.593	0.829	0.559	0.675	1.206	1.788	1.483
34.25	0.598	0.827	0.563	0.681	1.210	1.777	1.469
34.50	0.602	0.824	0.566	0.687	1.213	1.766	1.455
34.75	0.607	0.822	0.57	0.694	1.217	1.754	1.441
35.00	0.611	0.819	0.574	0.700	1.221	1.743	1.428
35.25	0.615	0.817	0.577	0.707	1.225	1.733	1.415
35.50	0.620	0.814	0.581	0.713	1.228	1.722	1.402
35.75	0.624	0.812	0.584	0.720	1.232	1.712	1.389
36.00	0.628	0.809	0.588	0.727	1.236	1.701	1.376
36.25	0.633	0.806	0.591	0.733	1.240	1.691	1.364
36.50	0.637	0.804	0.595	0.740	1.244	1.681	1.351
36.75	0.641	0.801	0.598	0.747	1.248	1.671	1.339
37.00	0.646	0.799	0.602	0.754	1.252	1.662	1.327
37.25	0.650	0.796	0.605	0.760	1.256	1.652	1.315
37.50	0.655	0.793	0.609	0.767	1.260	1.643	1.303

Degrees	Radians	Cosine	Sine	Tangent	Secant	Cosecant	Cotangent
37.75	0.659	0.791	0.612	0.774	1.265	1.633	1.292
38.00	0.663	0.788	0.616	0.781	1.269	1.624	1.280
38.25	0.668	0.785	0.619	0.788	1.273	1.615	1.268
38.50	0.672	0.783	0.623	0.795	1.278	1.606	1.257
38.75	0.676	0.780	0.626	0.803	1.282	1.598	1.246
39.00	0.681	0.777	0.629	0.810	1.287	1.589	1.235
39.25	0.685	0.774	0.633	0.817	1.291	1.581	1.224
39.50	0.689	0.772	0.636	0.824	1.296	1.572	1.213
39.75	0.694	0.769	0.639	0.832	1.301	1.564	1.202
40.00	0.698	0.766	0.643	0.839	1.305	1.556	1.192
40.25	0.703	0.763	0.646	0.847	1.310	1.548	1.181
40.50	0.707	0.760	0.649	0.854	1.315	1.540	1.171
40.75	0.711	0.758	0.653	0.862	1.320	1.532	1.161
41.00	0.716	0.755	0.656	0.869	1.325	1.524	1.150
41.25	0.720	0.752	0.659	0.877	1.330	1.517	1.140
41.50	0.724	0.749	0.663	0.885	1.335	1.509	1.130
41.75	0.729	0.746	0.666	0.893	1.340	1.502	1.120
42.00	0.733	0.743	0.669	0.900	1.346	1.494	1.111
42.25	0.737	0.740	0.672	0.908	1.351	1.487	1.101
42.50	0.742	0.737	0.676	0.916	1.356	1.480	1.091
42.75	0.746	0.734	0.679	0.924	1.362	1.473	1.082
43.00	0.750	0.731	0.682	0.933	1.367	1.466	1.072
43.25	0.755	0.728	0.685	0.941	1.373	1.459	1.063
43.50	0.759	0.725	0.688	0.949	1.379	1.453	1.054
43.75	0.764	0.722	0.692	0.957	1.384	1.446	1.045
44.00	0.768	0.719	0.695	0.966	1.390	1.440	1.036
44.25	0.772	0.716	0.698	0.974	1.396	1.433	1.027
44.50	0.777	0.713	0.701	0.983	1.402	1.427	1.018
44.75	0.781	0.710	0.704	0.991	1.408	1.420	1.009
45.00	0.785	0.707	0.707	1	1.414	1.414	1
45.25	0.790	0.704	0.710	1.009	1.420	1.408	0.991
45.50	0.794	0.701	0.713	1.018	1.427	1.402	0.983
45.75	0.798	0.698	0.716	1.027	1.433	1.396	0.974
46.00	0.803	0.695	0.719	1.036	1.440	1.390	0.966
46.25	0.807	0.692	0.722	1.045	1.446	1.384	0.957
46.50	0.812	0.688	0.725	1.054	1.453	1.379	0.949
46.75	0.816	0.685	0.728	1.063	1.459	1.373	0.941
47.00	0.820	0.682	0.731	1.072	1.466	1.367	0.933
47.25	0.825	0.679	0.734	1.082	1.473	1.362	0.924
47.50	0.829	0.676	0.737	1.091	1.480	1.356	0.916
47.75	0.833	0.672	0.740	1.101	1.487	1.351	0.908
48.00	0.838	0.669	0.743	1.111	1.494	1.346	0.900

Degrees	Radians	Cosine	Sine	Tangent	Secant	Cosecant	Cotangent
48.25	0.842	0.666	0.746	1.120	1.502	1.340	0.893
48.50	0.846	0.663	0.749	1.130	1.509	1.335	0.885
48.75	0.851	0.659	0.752	1.140	1.517	1.330	0.877
49.00	0.855	0.656	0.755	1.150	1.524	1.325	0.869
49.25	0.860	0.653	0.758	1.161	1.532	1.320	0.862
49.50	0.864	0.649	0.760	1.171	1.540	1.315	0.854
49.75	0.868	0.646	0.763	1.181	1.548	1.310	0.847
50.00	0.873	0.643	0.766	1.192	1.556	1.305	0.839
50.25	0.877	0.639	0.769	1.202	1.564	1.301	0.832
50.50	0.881	0.636	0.772	1.213	1.572	1.296	0.824
50.75	0.886	0.633	0.774	1.224	1.581	1.291	0.817
51.00	0.890	0.629	0.777	1.235	1.589	1.287	0.810
51.25	0.894	0.626	0.780	1.246	1.598	1.282	0.803
51.50	0.899	0.623	0.783	1.257	1.606	1.278	0.795
51.75	0.903	0.619	0.785	1.268	1.615	1.273	0.788
52.00	0.908	0.616	0.788	1.280	1.624	1.269	0.781
52.25	0.912	0.612	0.791	1.292	1.633	1.265	0.774
52.50	0.916	0.609	0.793	1.303	1.643	1.260	0.767
52.75	0.921	0.605	0.796	1.315	1.652	1.256	0.760
53.00	0.925	0.602	0.799	1.327	1.662	1.252	0.754
53.25	0.929	0.598	0.801	1.339	1.671	1.248	0.747
53.50	0.934	0.595	0.804	1.351	1.681	1.244	0.740
53.75	0.938	0.591	0.806	1.364	1.691	1.240	0.733
54.00	0.942	0.588	0.809	1.376	1.701	1.236	0.727
54.25	0.947	0.584	0.812	1.389	1.712	1.232	0.720
54.50	0.951	0.581	0.814	1.402	1.722	1.228	0.713
54.75	0.956	0.577	0.817	1.415	1.733	1.225	0.707
55.00	0.960	0.574	0.819	1.428	1.743	1.221	0.700
55.25	0.964	0.57	0.822	1.441	1.754	1.217	0.694
55.50	0.969	0.566	0.824	1.455	1.766	1.213	0.687
55.75	0.973	0.563	0.827	1.469	1.777	1.210	0.681
56.00	0.977	0.559	0.829	1.483	1.788	1.206	0.675
56.25	0.982	0.556	0.831	1.497	1.800	1.203	0.668
56.50	0.986	0.552	0.834	1.511	1.812	1.199	0.662
56.75	0.990	0.548	0.836	1.525	1.824	1.196	0.656
57.00	0.995	0.545	0.839	1.540	1.836	1.192	0.649
57.25	0.999	0.541	0.841	1.555	1.849	1.189	0.643
57.50	1.004	0.537	0.843	1.570	1.861	1.186	0.637
57.75	1.008	0.534	0.846	1.585	1.874	1.182	0.631
58.00	1.012	0.530	0.848	1.600	1.887	1.179	0.625
58.25	1.017	0.526	0.850	1.616	1.900	1.176	0.619
58.50	1.021	0.523	0.853	1.632	1.914	1.173	0.613

Degrees	Radians	Cosine	Sine	Tangent	Secant	Cosecant	Cotangent
58.75	1.025	0.519	0.855	1.648	1.928	1.170	0.607
59.00	1.030	0.515	0.857	1.664	1.942	1.167	0.601
59.25	1.034	0.511	0.859	1.681	1.956	1.164	0.595
59.50	1.038	0.508	0.862	1.698	1.970	1.161	0.589
59.75	1.043	0.504	0.864	1.715	1.985	1.158	0.583
60.00	1.047	0.5	0.866	1.732	2	1.155	0.577
60.25	1.052	0.496	0.868	1.750	2.015	1.152	0.572
60.50	1.056	0.492	0.870	1.767	2.031	1.149	0.566
60.75	1.060	0.489	0.873	1.786	2.047	1.146	0.560
61.00	1.065	0.485	0.875	1.804	2.063	1.143	0.554
61.25	1.069	0.481	0.877	1.823	2.079	1.141	0.549
61.50	1.073	0.477	0.879	1.842	2.096	1.138	0.543
61.75	1.078	0.473	0.881	1.861	2.113	1.135	0.537
62.00	1.082	0.469	0.883	1.881	2.130	1.133	0.532
62.25	1.086	0.466	0.885	1.901	2.148	1.130	0.526
62.50	1.091	0.462	0.887	1.921	2.166	1.127	0.521
62.75	1.095	0.458	0.889	1.942	2.184	1.125	0.515
63.00	1.100	0.454	0.891	1.963	2.203	1.122	0.510
63.25	1.104	0.450	0.893	1.984	2.222	1.120	0.504
63.50	1.108	0.446	0.895	2.006	2.241	1.117	0.499
63.75	1.113	0.442	0.897	2.028	2.261	1.115	0.493
64.00	1.117	0.438	0.899	2.050	2.281	1.113	0.488
64.25	1.121	0.434	0.901	2.073	2.302	1.110	0.482
64.50	1.126	0.431	0.903	2.097	2.323	1.108	0.477
64.75	1.130	0.427	0.904	2.120	2.344	1.106	0.472
65.00	1.134	0.423	0.906	2.145	2.366	1.103	0.466
65.25	1.139	0.419	0.908	2.169	2.389	1.101	0.461
65.50	1.143	0.415	0.910	2.194	2.411	1.099	0.456
65.75	1.148	0.411	0.912	2.220	2.435	1.097	0.450
66.00	1.152	0.407	0.914	2.246	2.459	1.095	0.445
66.25	1.156	0.403	0.915	2.273	2.483	1.093	0.440
66.50	1.161	0.399	0.917	2.300	2.508	1.090	0.435
66.75	1.165	0.395	0.919	2.328	2.533	1.088	0.430
67.00	1.169	0.391	0.921	2.356	2.559	1.086	0.424
67.25	1.174	0.387	0.922	2.385	2.586	1.084	0.419
67.50	1.178	0.383	0.924	2.414	2.613	1.082	0.414
67.75	1.182	0.379	0.926	2.444	2.641	1.080	0.409
68.00	1.187	0.375	0.927	2.475	2.669	1.079	0.404
68.25	1.191	0.371	0.929	2.507	2.699	1.077	0.399
68.50	1.196	0.367	0.930	2.539	2.729	1.075	0.394
68.75	1.200	0.362	0.932	2.572	2.759	1.073	0.389
69.00	1.204	0.358	0.934	2.605	2.790	1.071	0.384

Degrees	Radians	Cosine	Sine	Tangent	Secant	Cosecant	Cotangent
69.25	1.209	0.354	0.935	2.639	2.823	1.069	0.379
69.50	1.213	0.350	0.937	2.675	2.855	1.068	0.374
69.75	1.217	0.346	0.938	2.711	2.889	1.066	0.369
70.00	1.222	0.342	0.940	2.747	2.924	1.064	0.364
70.25	1.226	0.338	0.941	2.785	2.959	1.063	0.359
70.50	1.230	0.334	0.943	2.824	2.996	1.061	0.354
70.75	1.235	0.330	0.944	2.864	3.033	1.059	0.349
71.00	1.239	0.326	0.946	2.904	3.072	1.058	0.344
71.25	1.244	0.321	0.947	2.946	3.111	1.056	0.339
71.50	1.248	0.317	0.948	2.989	3.152	1.054	0.335
71.75	1.252	0.313	0.950	3.033	3.193	1.053	0.330
72.00	1.257	0.309	0.951	3.078	3.236	1.051	0.325
72.25	1.261	0.305	0.952	3.124	3.280	1.050	0.320
72.50	1.265	0.301	0.954	3.172	3.326	1.049	0.315
72.75	1.270	0.297	0.955	3.221	3.372	1.047	0.311
73.00	1.274	0.292	0.956	3.271	3.420	1.046	0.306
73.25	1.278	0.288	0.958	3.323	3.470	1.044	0.301
73.50	1.283	0.284	0.959	3.376	3.521	1.043	0.296
73.75	1.287	0.280	0.960	3.431	3.574	1.042	0.291
74.00	1.292	0.276	0.961	3.487	3.628	1.040	0.287
74.25	1.296	0.271	0.962	3.546	3.684	1.039	0.282
74.50	1.300	0.267	0.964	3.606	3.742	1.038	0.277
74.75	1.305	0.263	0.965	3.668	3.802	1.037	0.273
75.00	1.309	0.259	0.966	3.732	3.864	1.035	0.268
75.25	1.313	0.255	0.967	3.798	3.928	1.034	0.263
75.50	1.318	0.250	0.968	3.867	3.994	1.033	0.259
75.75	1.322	0.246	0.969	3.938	4.063	1.032	0.254
76.00	1.326	0.242	0.970	4.011	4.134	1.031	0.249
76.25	1.331	0.238	0.971	4.087	4.207	1.030	0.245
76.50	1.335	0.233	0.972	4.165	4.284	1.028	0.240
76.75	1.340	0.229	0.973	4.247	4.363	1.027	0.235
77.00	1.344	0.225	0.974	4.331	4.445	1.026	0.231
77.25	1.348	0.221	0.975	4.419	4.531	1.025	0.226
77.50	1.353	0.216	0.976	4.511	4.620	1.024	0.222
77.75	1.357	0.212	0.977	4.606	4.713	1.023	0.217
78.00	1.361	0.208	0.978	4.705	4.810	1.022	0.213
78.25	1.366	0.204	0.979	4.808	4.911	1.021	0.208
78.50	1.370	0.199	0.980	4.915	5.016	1.020	0.203
78.75	1.374	0.195	0.981	5.027	5.126	1.020	0.199
79.00	1.379	0.191	0.982	5.145	5.241	1.019	0.194
79.25	1.383	0.187	0.982	5.267	5.361	1.018	0.190
79.50	1.388	0.182	0.983	5.396	5.487	1.017	0.185

Degrees	Radians	Cosine	Sine	Tangent	Secant	Cosecant	Cotangent
79.75	1.392	0.178	0.984	5.530	5.620	1.016	0.181
80.00	1.396	0.174	0.985	5.671	5.759	1.015	0.176
80.25	1.401	0.169	0.986	5.820	5.905	1.015	0.172
80.50	1.405	0.165	0.986	5.976	6.059	1.014	0.167
80.75	1.409	0.161	0.987	6.140	6.221	1.013	0.163
81.00	1.414	0.156	0.988	6.314	6.392	1.012	0.158
81.25	1.418	0.152	0.988	6.497	6.574	1.012	0.154
81.50	1.422	0.148	0.989	6.691	6.765	1.011	0.149
81.75	1.427	0.143	0.990	6.897	6.969	1.010	0.145
82.00	1.431	0.139	0.990	7.115	7.185	1.010	0.141
82.25	1.436	0.135	0.991	7.348	7.416	1.009	0.136
82.50	1.440	0.131	0.991	7.596	7.661	1.009	0.132
82.75	1.444	0.126	0.992	7.861	7.924	1.008	0.127
83.00	1.449	0.122	0.993	8.144	8.206	1.008	0.123
83.25	1.453	0.118	0.993	8.449	8.508	1.007	0.118
83.50	1.457	0.113	0.994	8.777	8.834	1.006	0.114
83.75	1.462	0.109	0.994	9.131	9.186	1.006	0.110
84.00	1.466	0.105	0.995	9.514	9.567	1.006	0.105
84.25	1.470	0.100	0.995	9.931	9.981	1.005	0.101
84.50	1.475	0.096	0.995	10.385	10.433	1.005	0.096
84.75	1.479	0.092	0.996	10.883	10.929	1.004	0.092
85.00	1.484	0.087	0.996	11.430	11.474	1.004	0.087
85.25	1.488	0.083	0.997	12.035	12.076	1.003	0.083
85.50	1.492	0.078	0.997	12.706	12.745	1.003	0.079
85.75	1.497	0.074	0.997	13.457	13.494	1.003	0.074
86.00	1.501	0.070	0.998	14.301	14.336	1.002	0.070
86.25	1.505	0.065	0.998	15.257	15.290	1.002	0.066
86.50	1.510	0.061	0.998	16.350	16.380	1.002	0.061
86.75	1.514	0.057	0.998	17.611	17.639	1.002	0.057
87.00	1.518	0.052	0.999	19.081	19.107	1.001	0.052
87.25	1.523	0.048	0.999	20.819	20.843	1.001	0.048
87.50	1.527	0.044	0.999	22.904	22.926	1.001	0.044
87.75	1.532	0.039	0.999	25.452	25.471	1.001	0.039
88.00	1.536	0.035	0.999	28.636	28.654	1.001	0.035
88.25	1.540	0.031	0.100	32.730	32.746	1.000	0.031
88.50	1.545	0.026	0.100	38.188	38.202	1.000	0.026
88.75	1.549	0.022	0.100	45.829	45.840	1.000	0.022
89.00	1.553	0.017	0.100	57.290	57.299	1.000	0.017
89.25	1.558	0.013	0.100	76.390	76.397	1.000	0.013
89.50	1.562	0.009	0.100	114.589	114.593	1.000	0.009
89.75	1.566	0.004	0.100	229.182	229.184	1.000	0.004
90.00	1.571	0	1	—	—	1	0

Appendix B
THE UNIT CIRCLE
Just in case you didn't memorize it

ANGLES MEASURED IN RADIANS

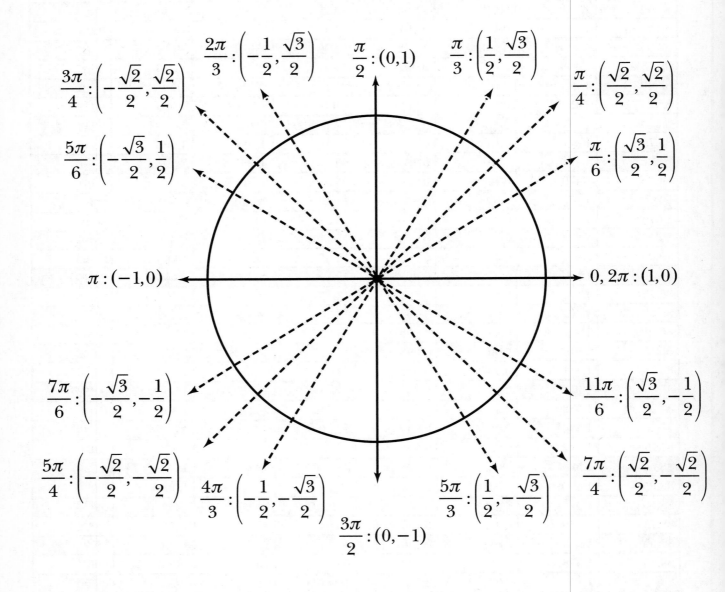

ANGLES MEASURED IN DEGREES

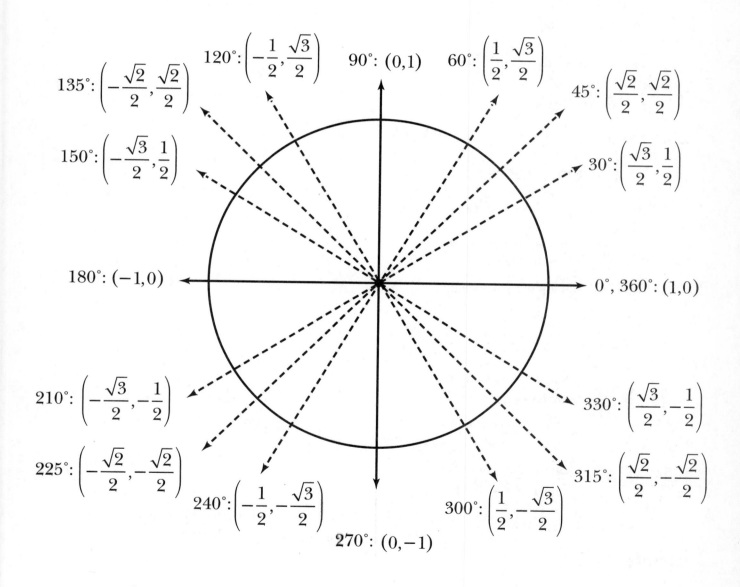

Appendix C
FORMULAS AND IDENTITIES
To memorize or reference

Chapter 1

- Degrees/minutes/seconds to decimal degrees: $d° \ m' \ s'' = d + \dfrac{m}{60} + \dfrac{s}{3,600}$

- Degrees to radians: $\theta \cdot \dfrac{\pi}{180}$

- Radians to degrees: $\theta \cdot \dfrac{180}{\pi}$

- Arc length: $s = r\theta$

Chapter 2

- Pythagorean theorem: $a^2 + b^2 = c^2$

- Distance formula: $D = \sqrt{\left(x_2 - x_1\right)^2 + \left(y_2 - y_1\right)^2}$

- $\sin\theta = \dfrac{\text{length of leg opposite } \theta}{\text{length of hypotenuse}}$

- $\cos\theta = \dfrac{\text{length of leg adjacent to } \theta}{\text{length of hypotenuse}}$

- $\tan\theta = \dfrac{\text{length of leg opposite } \theta}{\text{length of leg adjacent to } \theta}$

- $\csc\theta = \dfrac{\text{length of hypotenuse}}{\text{length of leg opposite } \theta}$

- $\sec\theta = \dfrac{\text{length of hypotenuse}}{\text{length of leg adjacent to } \theta}$

- $\cot\theta = \dfrac{\text{length of leg adjacent to } \theta}{\text{length of leg opposite } \theta}$

- Table interpolation: $\dfrac{\text{small angle difference}}{\text{large angle difference}} = \dfrac{\text{small trigonometric value difference}}{\text{large trigonometric value difference}}$

Chapter 3

- $45° - 45° - 90°$ triangles: $\sqrt{2} \cdot l = h$

- $30° - 60° - 90°$ triangles: $s = \dfrac{1}{2}h, \ l = \sqrt{3} \cdot s$

Chapter 4

- Reference angle α for angle θ terminating in the second quadrant: $m\angle\alpha = \pi - m\angle\theta$

- Reference angle α for angle θ terminating in the third quadrant: $m\angle\alpha = m\angle\theta - \pi$

- Reference angle α for angle θ terminating in the fourth quadrant: $m\angle\alpha = 2\pi - m\angle\theta$

Chapter 5

- Amplitude of a periodic function: $\dfrac{\text{maximum value} - \text{minimum value}}{2}$

- Given periodic function $g(x)$, the period of $g(n \cdot x) = \left|\dfrac{\text{period of } g(x)}{n}\right|$

Chapter 7

- Cofunction identities:

$$\sin\left(\frac{\pi}{2} - x\right) = \cos x \quad \cos\left(\frac{\pi}{2} - x\right) = \sin x \quad \tan\left(\frac{\pi}{2} - x\right) = \cot x$$

$$\csc\left(\frac{\pi}{2} - x\right) = \sec x \quad \sec\left(\frac{\pi}{2} - x\right) = \csc x \quad \cot\left(\frac{\pi}{2} - x\right) = \tan x$$

- Reciprocal identities:

$$\sin x = \frac{1}{\csc x} \quad \cos x = \frac{1}{\sec x} \quad \tan x = \frac{1}{\cot x}$$

$$\csc x = \frac{1}{\sin x} \quad \sec x = \frac{1}{\cos x} \quad \cot x = \frac{1}{\tan x}$$

- Negative identities:

$$\sin(-x) = -\sin x \quad \tan(-x) = -\tan x \quad \cos(-x) = \cos x$$

$$\csc(-x) = -\csc x \quad \cot(-x) = -\cot x \quad \sec(-x) = \sec x$$

- Pythagorean identities:

$$\cos^2 x + \sin^2 x = 1$$

$$1 + \tan^2 x = \sec^2 x$$

$$1 + \cot^2 x = \csc^2 x$$

- Sum and difference formulas:

$$\sin(x \pm y) = \sin x \cos y \pm \cos x \sin y$$

$$\cos(x \pm y) = \cos x \cos y \mp \sin x \sin y$$

Chapter 8

- Double-angle formulas:

$$\sin 2x = 2 \sin x \cos x$$
$$\cos 2x = \cos^2 x - \sin^2 x = 2\cos^2 x - 1 = 1 - 2\sin^2 x$$

- Power-reducing formulas:

$$\cos^2 x = \frac{1 + \cos 2x}{2}$$
$$\sin^2 x = \frac{1 - \cos 2x}{2}$$

- Half-angle formulas:

$$\cos \frac{x}{2} = \pm\sqrt{\frac{1 + \cos x}{2}}$$
$$\sin \frac{x}{2} = \pm\sqrt{\frac{1 - \cos x}{2}}$$

- Product-to-sum formulas:

$$\cos x \cos y = \frac{1}{2}\Big[\cos(x - y) + \cos(x + y)\Big]$$
$$\sin x \sin y = \frac{1}{2}\Big[\cos(x - y) - \cos(x + y)\Big]$$
$$\cos x \sin y = \frac{1}{2}\Big[\sin(x + y) - \sin(x - y)\Big]$$
$$\sin x \cos y = \frac{1}{2}\Big[\sin(x + y) + \sin(x - y)\Big]$$

- Sum-to-product formulas:

$$\cos x + \cos y = 2\cos\left(\frac{x + y}{2}\right)\cos\left(\frac{x - y}{2}\right)$$
$$\cos x - \cos y = -2\sin\left(\frac{x + y}{2}\right)\sin\left(\frac{x - y}{2}\right)$$
$$\sin x + \sin y = 2\sin\left(\frac{x + y}{2}\right)\cos\left(\frac{x - y}{2}\right)$$
$$\sin x - \sin y = 2\cos\left(\frac{x + y}{2}\right)\sin\left(\frac{x - y}{2}\right)$$

- Tangent identities:

$$\tan(x \pm y) = \frac{\tan x \pm \tan y}{1 \mp \tan x \tan y}$$

$$\tan 2x = \frac{2 \tan x}{1 - \tan^2 x}$$

$$\tan^2 x = \frac{1 - \cos 2x}{1 + \cos 2x}$$

$$\tan \frac{x}{2} = \frac{\sin x}{1 + \cos x} = \frac{1 - \cos x}{\sin x}$$

Chapter 10

- Quadratic formula: $x = \dfrac{-b \pm \sqrt{b^2 - 4ac}}{2a}$

Chapter 12

- Triangle area formulas:

$$A = \frac{1}{2}bh$$

$$A = \frac{1}{2}xy \sin Z$$

$$A = \frac{x^2 \sin Y \sin Z}{2 \sin X}$$

- Heron's formula: $A = \sqrt{s(s-x)(s-y)(s-z)}$, where $s = \dfrac{x + y + z}{2}$

- Area of a sector: $\dfrac{\text{measure of central angle}}{\text{measure of one full rotation}} \cdot \pi r^2$

Chapter 13

- Law of sines: $\dfrac{\sin A}{a} = \dfrac{\sin B}{b} = \dfrac{\sin C}{c}$

- Law of cosines:

$$a^2 = b^2 + c^2 - 2bc \cos A$$

$$b^2 = a^2 + c^2 - 2ac \cos B$$

$$c^2 = a^2 + b^2 - 2ab \cos C$$

Chapter 14

- Magnitude of $\mathbf{v} = \langle a, b \rangle$: $\|\mathbf{v}\| = \sqrt{a^2 + b^2}$

- Unit vector of $\mathbf{v}$ in the same direction: $\hat{\mathbf{v}} = \dfrac{\mathbf{v}}{\|\mathbf{v}\|}$

Chapter 15

- Arithmetic vector addition: $\langle a, b \rangle + \langle c, d \rangle = \langle a+c, b+d \rangle$

- Scalar multiplication: $c\langle a, b \rangle = \langle ca, cb \rangle$

- Horizontal and vertical components of a vector:

$$x = \|\mathbf{v}\| \cos\theta$$
$$y = \|\mathbf{v}\| \sin\theta$$
$$\tan\theta = \frac{y}{x}$$

Chapter 16

- Dot product: $\langle a, b \rangle \cdot \langle c, d \rangle = ac + bd$

- Alternate dot product formula: $\mathbf{v} \cdot \mathbf{w} = \|\mathbf{v}\| \|\mathbf{w}\| \cos\theta$

- Modified dot product formula: $\dfrac{\mathbf{v} \cdot \mathbf{w}}{\|\mathbf{v}\| \|\mathbf{w}\|} = \cos\theta$

- Vector projection: $\text{proj}_{\mathbf{w}} \mathbf{v} = \left(\dfrac{\mathbf{v} \cdot \mathbf{w}}{\|\mathbf{w}\|^2} \right) \mathbf{w}$

- Work:

$W = \|\mathbf{F}\| \|\mathbf{d}\|$, when $\mathbf{F}$ and $\mathbf{d}$ have the same direction
$$W = \mathbf{F} \cdot \mathbf{d}$$
$$W = \|\text{proj}_{\mathbf{d}} \mathbf{F}\| \|\mathbf{d}\|$$

Chapter 18

- Absolute value/modulus of complex number $z = a + bi$: $|z| = \sqrt{a^2 + b^2}$

- Argument of complex number $z = a + bi$: $\theta = \arctan\dfrac{b}{a}$

- Product of complex numbers: $z_1 z_2 = r_1 \cdot r_2 \left[\cos(\theta_1 + \theta_2) + i\sin(\theta_1 + \theta_2) \right]$

- Quotient of complex numbers: $\dfrac{z_1}{z_2} = \dfrac{r_1}{r_2} \left[\cos(\theta_1 - \theta_2) + i\sin(\theta_1 - \theta_2) \right]$

- De Moivre's theorem: $\left[r(\cos\theta + i\sin\theta) \right]^n = r^n \left[\cos(n\theta) + i\sin(n\theta) \right]$

- Roots of complex numbers: $p_k = \sqrt[n]{r}\left(\cos\dfrac{\theta + k \cdot 360°}{n} + i\sin\dfrac{\theta + k \cdot 360°}{n} \right)$, such that $k = 0, 1, \cdots, n-1$

Index

ALPHABETICAL LIST OF CONCEPTS WITH PROBLEM NUMBERS

This comprehensive index organizes the concepts and skills discussed within the book alphabetically. Each entry is accompanied by one or more problem numbers, in which the topics are most prominently featured.

All of these numbers refer to problems, not pages, in the book. For example, 4.9 is the ninth problem in Chapter 4.

B–C

D–E–F

I–J–K

i: see *standard unit vector*

identity
  cofunction: 7.1–7.4, 7.6, 7.11–7.12, 7.17–7.18, 7.40–7.41, 8.16
  double-angle: 8.1–8.16, 8.18, 8.39–8.41, 8.45
  half-angle: 8.22–8.27, 8.44
  negative: 7.13–7.22, 7.33, 8.30
  power-reducing: 8.17–8.23, 8.42–8.43
  product-to-sum: 8.28–8.31
  Pythagorean: 7.23–7.33
  reciprocal: 7.5–7.10, 7.12, 7.14–7.15, 7.19–7.20, 7.22, 7.26, 7.30–7.33
  simplifying: 7.1–7.4, 7.6–7.12, 7.14–7.15, 7.18, 7.21, 7.26–7.28, 7.30, 7.43, 8.3, 8.13, 8.16, 8.26–8.27, 8.37–8.38, 8.41
  sum and difference: 7.34–7.45, 8.2, 8.29, 8.36–8.39
  sum-to-product: 8.32–8.35
  verifying: 7.16–7.17, 7.19–7.20, 7.22, 7.24–7.25, 7.29, 7.31–7.33, 7.41, 7.44, 8.2, 8.6–8.7, 8.9, 8.11, 8.14, 8.18, 8.21–8.23, 8.29, 8.31, 8.35, 8.39, 8.42

imaginary number: 18.1, 18.5

initial point: 14.1–14.14, 14.16, 14.18

initial side: 1.6–1.8

interpolation: 2.32–2.36

interval notation: 5.1–5.2, 5.6–5.7

inverse
  functions: 9.1–9.4
  trigonometric functions: 2.40–2.41, 2.43, 2.45, 3.27–3.28, 3.45, 9.5–9.26, 10.4–10.6, 10.11, 10.22–10.23, 10.28–10.30, 10.32, 10.45, 11.12, 11.21, 11.25–11.26, 11.35, 12.43, 13.11, 13.16–13.17, 13.19, 13.21

isosceles triangle: 2.7, 2.13, 3.4, 12.7, 12.15–12.16, 12.28, 12.43

j: see *standard unit vector*

L–M

law of cosines
  formula: 13.13–13.14
  SAS: 13.18–13.21, 13.23
  SSS: 13.14–13.17, 13.21

law of sines
  AAS: 13.2–13.4, 13.8
  ambiguous case: 13.9–13.12
  ASA: 13.5–13.7
  formula: 13.1
  SSA: 13.9–13.12, 13.17, 13.19

leg: 2.1–2.10, 2.13–2.28, 2.42–2.45, 3.6–3.7, 3.11–3.14, 3.22

lemniscate: 17.50

length (of an arc): 1.41–1.45

length, signed: 4.3–4.5, 4.30–4.35, 4.37–4.39, 4.41–4.45

limaçon: 17.48

linear trigonometric equation: see *simple equation*

magnitude: 14.25–14.36, 14.38–14.40, 14.48–14.51, 15.20, 15.24, 15.26–15.28, 15.31, 15.34, 15.38–15.45, 16.7–16.8, 16.49–16.50, 16.52

major arc: 1.45

maximum: 5.3–5.4, 5.8

measure
  angle: 1.9–1.17, 1.20–1.30
  arc: 1.40

minimum: 5.3–5.4, 5.8

minor arc: 1.45

minutes: 1.18–1.19

modulus: 18.13–18.14, 18.18

multiple-angle equation: 10.34–10.45, 11.8

multiplying
  complex numbers: 18.8, 18.11, 18.23, 18.25–18.27, 18.38
  vectors: see scalar multiplication or dot product

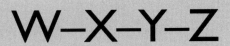